Number Theory and the Periodicity of Matter

Jan C.A. Boeyens · Demetrius C. Levendis

Number Theory and the Periodicity of Matter

Springer

Jan C.A. Boeyens
University of Pretoria
South Africa

Demetrius C. Levendis
University of the Witwatersrand
South Africa

ISBN 978-1-4020-6659-7 e-ISBN 978-1-4020-6660-3

Library of Congress Control Number: 2007937405

Printed on acid-free paper.

9 8 7 6 5 4 3 2 1

springer.com

Preface

Of all the great innovations and intellectual achievements of mankind there is nothing that rivals the invention of counting and discovery of the number system. The way in which this discovery led to the development of abstract higher mathematics is the least of its merits, compared to the universal fascination that the natural numbers hold for all people. Numbers are at the roots of magic, superstition, religion and science. Numerologists can interpret great historical and cosmic events, predict the future and explain human nature. Better informed, sophisticated people may frown upon and ridicule such claims, but the number of incidents that link numbers to physical effects is simply too large to ignore as mere coincidence. It is in cases like these that the more respectable *number theory* is substituted for numerology.

Although it is recognized as the most fundamental branch of mathematics, the vocabulary of number theory includes concepts such as prime number, perfect number, amicable number, square number, triangular number, pyramidal number, and even magic number, none of which sounds too scientific and may suggest a different status for the subject. Not surprisingly, number theory remains the pastime of amateurs and professionals alike – all the way from the great Gauss down. It may be claimed that abstract number theory is more lofty than mundane science, never to be degraded into a servant of physical theory. Even so, a constant stream of books rolls from the printing presses of the world, extolling the wonderful synergy that exists between Fibonacci numbers, the golden ratio and self-similar symmetry on the one hand, with works of art (e.g. Da Vinci), architecture (Parthenon), biological growth, classical music and cosmic structure, on the other.

Despite claims to the contrary some of the profound insights into the understanding of the world were directly inspired by numbers. The best known example is the realization by Pythagoras that harmonious music, produced by a stringed instrument, is dictated by a sequence of rational fractions, defined in terms of natural numbers. Armed with this insight he noted a parallel numerical regularity in the motion of heavenly bodies, could hear the *music of the spheres*, and concluded: *all is number*. Quite remarkably, modern astronomy has confirmed, but cannot explain, the numerical sequence that regulates the orbital motion of the planets in the solar system. The well-known, but often ridiculed, Bode–Titius law correctly predicted all planetary orbits, leaving a single gap, where the asteroid belt was subsequently discovered, and also predicted the correct orbit for the unknown planet Pluto. The rings of Saturn are found to obey the same law, which, however, is still treated by the academic world as no more than entertaining coincidence.

Another entertaining coincidence was discovered by the high-school teacher Johann Balmer, in the form of a simple numerical formula, which accounted for the spectrum of light emitted by incandescent atomic matter, that continued to baffle the physicists of the world. Thirty years later Niels Bohr, on the basis of Balmer's formula, managed to construct the first convincing model of the atom and introduced the quantum theory into atomic physics. The quantum mechanics that developed from Bohr's model managed to extend the Balmer formula into a complete description of atomic structure on the basis of five sets of (integer and half-integer) quantum numbers.

The quantum numbers held out the immediate promise of accounting for the most fundamental concept of chemistry, known as the *Periodic Table of the Elements*, described by one of its discoverers, Alexandre-Émile de Chancourtois, in the statement: *the properties of the elements are the properties of numbers*. Although the quantum-mechanical explanation of elemental periodicity was only partially successful, the scientific world stopped looking (1926) for the numbers of de Chancourtois, until the chance discovery of these numbers by the present authors (2001).

In the interim, the development of atomic theory had been prodigious and impressive. It saw the identification of atomic species, called isotopes or nuclides, not included in the periodic classification, and of antimatter, the mirror image of ordinary matter. Once the proper numerical basis of the periodic classification had been spotted, all the new forms of atomic matter now find their proper place in the extended periodic classification. *Number theory and the Periodicity of Matter* deals with this discovery, its background, significance and predictions. The consequences are enormous. It shows why periodicity cannot be fully described by the quantum theory of electrons. The role of protons and neutrons, the other stable sub-atomic particles, are of equal importance. Only by taking the number of all these particles (called nucleons) into account is it possible to rationalize many aspects of atomic and nuclear physics. These aspects include nuclear synthesis, cosmic abundance of nuclides, nuclear stability, radioactive decay, nuclear spin and parity, nuclear size and shape, details of neutron scattering and superconductivity. At this stage the discovery is at the same level as that of Balmer, with all the science that it promises to produce still in the future.

New ideas on the theme have been communicated at several international conferences and as postgraduate lecture series at the University of Pretoria, South Africa, and of Heidelberg, Germany. The single idea received most enthusiastically was the decisive role of algebraic number theory, the Fibonacci series, the Farey sequence and the golden ratio in shaping the grand periodic system in a closed topological space, best described by the geometry of a Möbius strip. That explains why the packing of nucleons in an atomic nucleus follows the same pattern as the arrangement of florets in a flower head and why only special elements turn superconducting on cooling.

The haunting question, which is constantly insinuated but never answered conclusively, concerns the nature of the natural numbers. Have they been invented or discovered by humans? In other words, do numbers have an independent existence outside of the human mind? The answer to this question is non-trivial and probably of decisive importance for the future development of both science and mathematics. At this stage it is not even clear where to look for understanding – the abstract or the mundane, or where the twain shall never meet.

The theme of this book is to explore the consequences of the serendipitous discovery that stable nuclides obey the same periodic law as the chemical elements; both laws are rooted in elementary number theory. The nature of the discovery is such that, from the related periodic structures that occur in the natural numbers, as well as atomic matter, fundamental details about the electronic configuration of atoms and the baryonic arrangement in atomic nuclei can be derived, without the use of higher mathematics. A high degree of self-consistency substantiates the basic thesis from internal evidence, without assumption. This self-consistency includes convergence of nucleon distribution to the golden ratio and a natural limit on the number of elements and nuclides which are stable against radioactive decay. On this basis all observed properties of atoms and atomic nuclei can be understood as characteristic of a number system defined on a closed interval.

The key to understanding of atomic matter through number theory exists therein that atoms consist of whole numbers of protons, neutrons and electrons. The ratio of protons to neutrons in any nuclide therefore is a simple rational fraction, and this quantity, in relation to mass number, is the important factor that determines the stability of nuclides against radioactive decay. For light nuclei the ratio is $Z/N = 1$ and, for heavy nuclides, it converges to the golden mean ($\tau = 1/\Phi$), as shown by geometrical construction.

The point of convergence is established by plotting a modular(4) set of rational fractions, ordered in Farey sequence. This infinite set contains a modular subset of points that represent stable nuclides of allowed nuclear composition along a set of 11 regularly spaced festoons. A second Farey sequence, defined by Fibonacci fractions on the interval $(1,\tau)$, defines the set of 11 curves. This procedure shows that the limiting golden ratio is approached as $Z \rightarrow 102$, $N \rightarrow 102\Phi$ (165) and $A \rightarrow 102\Phi^2$ (267), which matches experimental observation. The analysis is valid for all mod(4) sets of nuclides, totalling 264, which decomposes into 11 periods of 24.

Periodic laws in terms of atomic and neutron numbers are readily projected out from the general law. In the case of atomic number, four periodic laws that reflect different cosmic environments are obtained, and these are interpreted to define a mechanism of nuclear synthesis by α-particle addition. The neutron-based periodic law, for the first time, rationalizes the empirically derived magic numbers of nuclear physics and provides a rational basis for the analysis of nuclear properties, including spin and parity.

To understand the full impact of the discovery the reader should have a working knowledge of elementary number theory, the periodic table of the elements, introductory atomic physics and elementary cosmography. The layout of the book has been planned in accordance with these needs. The first chapter is an introductory summary of the main thesis, followed by a primer on number theory, and similar chapters on the periodic table and the distribution of matter in the universe. With all background material in place, subsequent chapters re-examine the main arguments in more detail and with more emphasis on the wider implications of the results.

The work is presented without any pretence to expose inadequacies in existing science or provide an alternative, more fundamental model description of any aspect of chemistry, physics or cosmography. It only seeks to highlight an amazing facility of number theory to throw new light, particularly on old chemistry and nuclear physics. This descrial is all the more remarkable when read with the following quotation from Michio Kaku [1]:

> [...]some mathematical structures, such as number theory (which some mathematicians claim to be the purest branch of mathematics), have never been incorporated into any physical theory. Some argue that this situation may always exist: Perhaps the human mind will always be able to conceive of logically consistent structures that cannot be expressed through any physical principle.

Our conclusions indicate a definite link between natural numbers and atomic structure, supported by irrefutable internal evidence. The parallel with conclusions reached by W.D. Harkins, almost a century ago, is of interest.

He displayed the stable nuclides, known at the time, before discovery of the neutron, as a function of proton/neutron ratio, converging to $Z/N \rightarrow 0.62$, apparently without recognizing this limit as the golden ratio. This golden ratio turns out to be pivotal for the understanding of emerging self-similar relationships between different forms of matter, ranging from the sub-atomic to the cosmic.

The basis of atomic periodicity in number theory was explored by extended discussions with Demetrius Levendis during a period of sabbatical leave that he spent with me at the University of Pretoria. This interaction led to the interpretation in terms of Farey sequences and demonstration of the equivalent roles of mass number and nuclear binding energy. Without his insight this work would not have been possible.

Many of the arguments reached maturity at the Ruprecht-Karls-University of Heidelberg where, as a visiting professor, I had the opportunity to explore these ideas with members of the Inorganic Chemistry Institute of Professor Peter Comba in a seminar series. Critical comments and technical assistance from many others, over several conferences, seminars and informal discussions are gratefully acknowledged. In this respect I must single out Sonke Adlung, Pari Antalis, Aloysio Janner, Tibor Koritsanszky, Gert Krynauw, Richard Lemmer, John Ogilvie, Zorka Papadopolos and Casper Schutte. All inaccuracies remain my sole responsibility.

Jan C.A. Boeyens
Pretoria
July 2007

Contents

Chapter 1

Introduction

1.1 Number Magic

There is an ancient belief that the physical world experienced by mortals is a pale reflection of a perfect world carved in numbers. The Pythagorean cult and the name of Plato are commonly mentioned in this context. In modern times special numbers such as 3, 6, 7, 11, 24, 60, 666, etc., are no longer credited with magical or secret powers, but numerology in other guises is still widely practised. As in Pythagorean times the fascination with prime numbers and irrational numbers continues unabated. New records in the calculated number of digits defining π, e, and τ are set and recorded on websites all the time. There is little doubt that these three irrational numbers are indeed fundamental constants of Nature, although their precise relevance may still be debated for a long time to come.

Best known of the three irrationals is π, the ratio between circumference (s) and diameter (m) of a circle in the Euclidean plane. Although this relationship does not hold in curved space the value of π is assumed constant, such that, either $s > \pi m$ or $s < \pi m$ for positive and negative curvature, respectively. The simplest and most familiar example of a curved two-dimensional space is the surface of a sphere, which is approximated by the average surface of planet earth. The shortest distance between two points in this curved space is no longer a straight line, but a *geodesic*, or segment of the great circle that connects the two points. A triangle in this surface has the appearance shown on the left in Figure 1.1. A triangle in negatively curved two-dimensional space is also shown for comparison.

For millenia the inhabitants of planet earth failed to notice the curvature of the surface along which they moved. Today, they still fail to notice any curvature of their three-dimensional living space. It raises the intriguing

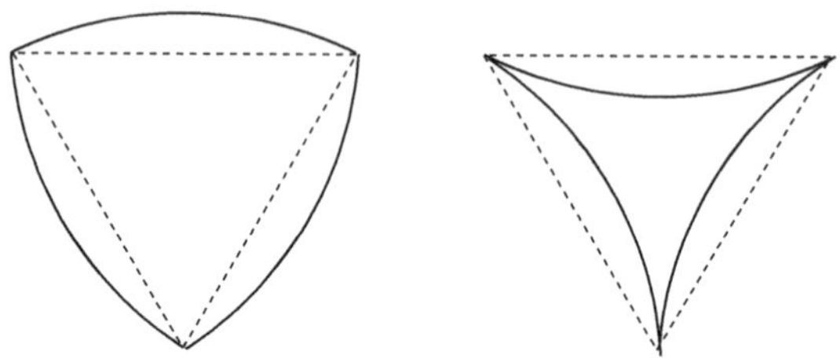

Figure 1.1: *Triangles in spherical and hyperbolic two-dimensional space, respectively.*

possibility that the carefully observed value of π might be characteristic of local non-euclidean space, naïvely interpreted as being flat. Evidence that such is the case will be presented in the course of this work.

The basis of natural logarithms, e is the irrational number that defines exponential growth or decay, which means growth, positive or negative, at a rate proportional to the mass of the growing entity. The constant e relates to π by the remarkably simple equation,

$$e^{i\pi} + 1 = 0 \tag{1.1}$$

in which $i = \sqrt{-1}$. It is almost self-evident that e, like π assumes a special irrational value dictated by the local curvature of space–time.

There is some confusion around the terminology used to specify the third fundamental irrational number, known as the golden mean, golden ratio or golden section. In this monograph this golden constant is defined by the symbols Φ and τ such that

$$\Phi = 1 + \tau = 1.6180... \tag{1.2}$$

which obeys the following set of relationships:

$$\frac{1}{\Phi} = \tau \tag{1.3}$$

$$\Phi^2 = \frac{1+\tau}{\tau} = \Phi + 1 \tag{1.4}$$

$$\tau^2 = 1 - \tau$$

There is a current upsurge of interest in the golden mean [2], which is known to characterize a diversity of natural phenomena such as the leaf and seed arrangements in flower heads and cones, the curvature of elephant tusks, the growth patterns of seashells, the flight path of a perigrine falcon pursuing its prey in free-fall and the structure of spiral galaxies. Central to the theme of this book is the way in which the periodic laws pertaining to atomic matter are dictated by space–time curvature and the golden mean. It is therefore not too surprising to find a simple relationship between Φ and the other irrational constants of Nature, e.g.

$$\Phi = 2\cos\left(\frac{\pi}{5}\right) \tag{1.5}$$

Ramanathan [3] lists continued fractions attributed to Ramanujan relating Φ, π and e in one formula, but these are complicated expressions that defy comprehension.

An irrational number γ is always flanked by two rational fractions in a Farey sequence of order N [4]

$$\frac{a}{b} < \gamma < \frac{c}{d} \tag{1.6}$$

The approximation of γ by rationals improves for large N, although its true value is never reached. This convergence to irrationality is like the approach of some discrete self-similar sequence of entities towards a grainless continuum. For example, the ratio of protons to neutrons is rational by definition. It will be shown how this ratio approaches an ideal irrational value demanded by the curvature of continuous space–time. Only those nuclides with proton/neutron ratios within the converging Farey sequence occur in Nature. Any regularities in the nuclear set must have a counterpart in the Farey sequence as specified by the natural numbers, and vice versa.

Only the properties of numbers are known in advance and the more fruitful approach is therefore to look for structure in the set of numbers and use this structure to derive regularities in the composition of atomic matter. At first sight this one-to-one relationship between numbers and matter strikes the observer, including one of us [5] as almost magical. It is not.

1.2 Periodic Structures

The proton number of an atom is an integer. It may be argued that the known periodicity of the chemical elements should therefore match some periodic structure in the number system. Any modular decomposition defines a periodic distribution. The most striking of these is the distribution of prime numbers at values of $6n \pm 1$.

In order to relate the prime number distribution to elemental periodicity it is instructive to note that the common periodic groups of elements, as defined in terms of an Aufbau procedure based on electronic energy sub-shells in atoms, consist of $2(s)$, $6(p)$, $10(d)$ and $14(f)$ elements respectively.

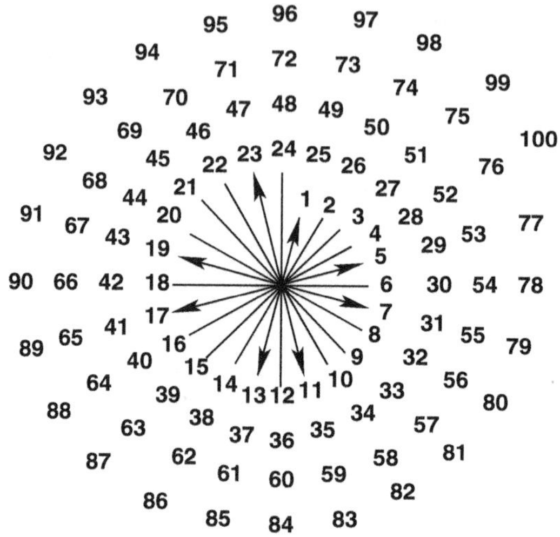

Figure 1.2: *Natural numbers arranged on a spiral of period 24. Arrows identify the eight radial directions where all prime numbers, apart from 2 and 3, are located.*

Combinations such as $2+6$, $6+10$, $2+14$ and $10+14$ are all multiples of 8, which number therefore defines a convenient periodic basis for classification. Such an eight-group arrangement matches the distribution of prime numbers in a number spiral of period 24, shown in Figure 1.2. It will be demonstrated that the 264 stable (non-radioactive) nuclides are thereby divided into 24 groups of 11, and that the familiar periodic table of the elements is a subset of the more general periodic table of the nuclides. However, the periodic table of the elements derived from the general law is not unique and it varies as a function of the proton/neutron ratio at which the subset is defined.

The eight-group arrangement that agrees with the familiar form of the periodic table occurs at $Z/N \rightarrow \tau$. An idealized form of this table, consistent with the electronic energy spectrum of atomic hydrogen as defined by Schrödinger's equation, occurs at $Z/N = \tau - \delta$, in which $\delta \simeq 1.2 - \tau$. Periodic arrangements, derived from these two tables, by inversion of electronic energy levels, such that $E_f < E_d < E_p < E_s$, occur at ratios of $Z/N = 1$ and $Z/N = 1 + \delta$.

To explore the distribution of nuclear levels it is necessary to examine periodic relationships in the nuclear region, defined by the limits $Z/N = \pm 0.22$. The predicted arrangements correspond separately to the energy levels of both neutrons and of protons in the nucleus, as defined by the semi-empirical shell model of nuclear structure. The observation that nuclear particles obey the same periodic law as extranuclear electrons indicates a self-similar relationship between the two regimes.

The number spiral, interpreted as a basis for the periodicity of atomic matter, imposes upper limits of 300 and 100 on the number of stable nuclides and elements respectively. These limits are only reached in periodic systems defined at $Z/N \simeq 1$. At $Z/N \approx \tau$ these numbers are reduced to 264 and 81, respectively. This reduction in number has a clear implication for the process of nuclear synthesis and transformation.

The first important general conclusion to be drawn from the relationship between atomic periodicity and the number spiral is the surprising result that a wealth of detail about nuclear structure and electronic configuration of atoms emerges without the use of higher mathematics. Even an ill-defined concept such as electronegativity acquires new logical meaning on the basis of an improved periodic law.

1.3 Nuclear Synthesis

Known trends, related to the effect of applied pressure on the electronic energy levels of atoms [6], provide convincing evidence that the inversion of electronic energies inferred from periodicities is linked to different thermodynamic conditions associated with Z/N ratios at τ and 1, respectively. Under conditions that favour nuclear synthesis, at high pressure, the optimal ratio is unity. This ratio is the same that characterizes an α-particle, ${}^4_2\mathrm{He}^{2+}$. This observation supports the conjecture that, at sufficiently high pressure, α-particles can be fused in a chain reaction to produce nuclides of steadily increasing mass number. With increasing mass number the ratio $Z/N \to 1$, irrespective of initial structure.

The mechanism of nuclear synthesis by α-addition becomes especially attractive under the observation that progressive α-addition to four elementary units of mass number $N(\mathrm{mod}4) \equiv 0, \pm1, \pm2$, allowing for radioactive decay where empirically indicated, only produce all of the known 264 stable nuclides. Significantly, nuclides of elements 43 and 61 are circumvented.

A consistent picture emerges from the assumption that nuclear synthesis by α-addition occurs in massive stars that favour a Z/N ratio of unity. A total of 300 stable nuclides of 100 elements, are produced in four series of mass number $A = 4n$, $4n \pm 1$, and $4n + 2$. Should the star disintegrate, the 300 types of nuclide, released into regions of lower pressure, suffer a phase transition that inverts electronic configuration and renders a number of previously stable nuclides unstable against radioactive decay. In the solar system only 264 different nuclides of 81 elements survive the phase change.

The proposed scheme provides a simple rationalization of the six periodic systems identified at different Z/N ratios. Four of these systems refer to extranuclear electronic configurations of atoms, apparently associated with different thermodynamic conditions. Alternatively, these different states could be interpreted as different states of space–time curvature. Unit ratio has already been linked to massive stars, known to cause severe curvature of space–time. By analogy, the curvature of planetary space can be correlated with the golden mean τ. It is therefore no accident that biological and crystal growth, like nuclear stability on the planet, is closely linked to the ratio τ, which by implication characterizes near-empty, slightly curved space–time.

Empty euclidean space, devoid of matter, and of zero curvature, probably has no real existence. The hypothetical space, invoked by Schrödinger's model of the hydrogen atom that recognizes no interaction other than the coulombic attraction between proton and electron, is truly empty and flat. Electronic configurations of atoms in such an environment obey Schrödinger's law, but deviate from actual, observed configurations. The inverse of flat space has infinite curvature. It follows that the fourth state of atoms with totally inverted electronic structure, is confined to the singularity supposed to exist at the centre of a black hole.

When the same analysis is repeated in terms of neutron number $N = A - Z$, rather than atomic number, three equivalent periodic systems are projected out at $Z/N = 1, \tau$ and 0, all of them consistent with the empirically derived nuclear energy spectrum defined by the so-called magic numbers. Compared to the flexible variable system defined by atomic number, nuclear periodicity remains invariant under all cosmic thermodynamic conditions.

1.4 Nuclidic Periodicity and Stability

Periodicity of the elements is well known to reflect variation of electronic configuration as a function of atomic number. In the more general case of nuclidic periodicity this cannot be the only factor. A more important factor is the variation of nuclear stability as a function of neutron excess, defined in terms of atomic number Z and mass number A, as $N_e = A - 2Z$. As a general trend N_e increases with increasing mass number. A simple explanation of this trend is that more neutrons are required to screen the coulombic repulsion between the protons in heavy nuclei.

Neutron excess increases in discreet steps. It increases from 0 to 43 in 44 unit steps over the entire range of stable nuclides. Within a (mod 4) family the step size is four units and the linear relationship between Z and A for each set of N_e, always has a gradient of 2, as shown in Figure 1.3. Each nuclidic

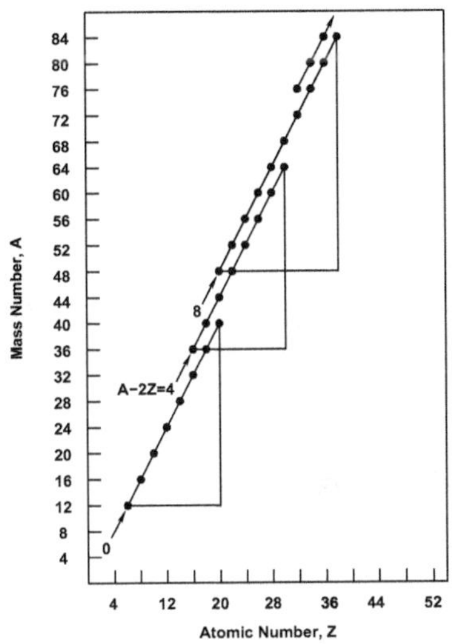

Figure 1.3: *Relationship between atomic and mass numbers for nuclides of constant neutron excess in the* $A(mod\ 4) \equiv 0$ *series.*

group of constant neutron excess terminates at both ends in a radioactive nuclide and since, invariably, $\Delta A = 2\Delta Z$, each region of stability is defined in terms of the golden mean,

$$\sqrt{(\Delta A/2)^2 + (\Delta A)^2} = \frac{\sqrt{5}}{2}\Delta A = \left(\phi - \frac{1}{2}\right)\Delta A \tag{1.7}$$

Furthermore, the maximum stability in each group occurs in the middle of the range and by averaging over neighbouring modular-group segments the 11 periods of 24 nuclides are reproduced as a function periodic in nuclear stability. To confirm this conclusion it is noted that an appropriately scaled plot of experimentally measured nuclear binding energy vs Z is virtually indistinguishable from Figure 1.3-type plots over the same nuclides.

The assumed 11×24 periodicity in mass number can also be demonstrated by plotting nuclide distributions on axes of Z/N vs A. The well-defined periodic function that emerges here is independent of Z/N and confirms the 11-period assumption. The exact inverse of the nuclide distribution, but still in line with the same periodicity, is obtained in a plot of (N_e/Z) vs A. Both of these plots describe neutron imbalance and converge to the same value, which is readily calculated from the condition $Z/N = (N - Z)/Z$, as $Z/N \rightarrow \tau$ as $Z \rightarrow 102$, $N \rightarrow 165$ and $A \rightarrow 267 = Z/\tau^2$. All of the assumptions based on prime-number distribution therefore emerge naturally from internal evidence.

The distribution of stable nuclides is conveniently indicated on a two-dimensional plot of their positions, with respect to axes of $Z/(A-Z)$ vs Z. In this representation a straight line that limits the stability of nuclides against β^+ decay is constructed by connecting points, which define a set of rational Fibonacci fractions that converge to τ, and appear on the 44 curves through the positions of nuclides of constant neutron excess. This line extends between coordinates of (1,0) and $(\tau, 102)$. Since there are no nuclides with atomic numbers 43 and 61, exactly 100 elements are included in the set defined by $0 < Z \leq 102$. A maximal subset of the same Fibonacci fractions, between coordinates of $(\frac{14}{19}, 0)$ and $(\frac{2}{3}, 87)$ defines a straight line that limits nuclear stability against β^- decay, in the same way. The resulting triangle of stability agrees with observation and automatically limits the maximum allowed atomic number to 83.

1.5 Hidden Symmetry

Discovery of the periodic law of the elements by Mendeléeff and others sparked a lot of activity aimed at finding the ultimate symmetric formulation of that law. By 1920 the most successful formulation [7] was modelled on pseudo-periodic vibrations on a string, consisting of 11 unequal periods, approaching eight atomic-number units and excluding hydrogen. After that date attention shifted to redefinition of the law in terms of quantum-mechanical atomic models, in preference to graphic representations as advocated by Stewart [7].

Recognition of the periodic law as a special case of a more general law [5] now reveals that a fully symmetrical formulation is possible after all. Figure 1.4 shows some detail of the analysis that leads to the recognition of the general periodic law. An important feature is the symmetry observed at both the highest and the lowest Z/N ratios. At the high ratio that corresponds to infinite curvature of space–time, the periodic limits indicated as black dots are symmetrically disposed around atomic number 51. At $Z/N = 0$ that represents the nuclear energy spectrum of neutrons an equivalent symmetry, but different in detail, occurs around 51. The connecting lines between the two symmetrical sets generate four additional periodic arrangements of lower symmetry. One of these, at $Z/N = \tau$, refers to the system considered by Stewart [7]. In this instance the symmetry observed at the extreme ratios are broken, or hidden; to explain Stewart's limited success.

The four sets of almost parallel lines in Figure 1.4 create the impression that together they describe a closed function. When the high-ratio points related by reflection across 51, are identified so as to share common positions,

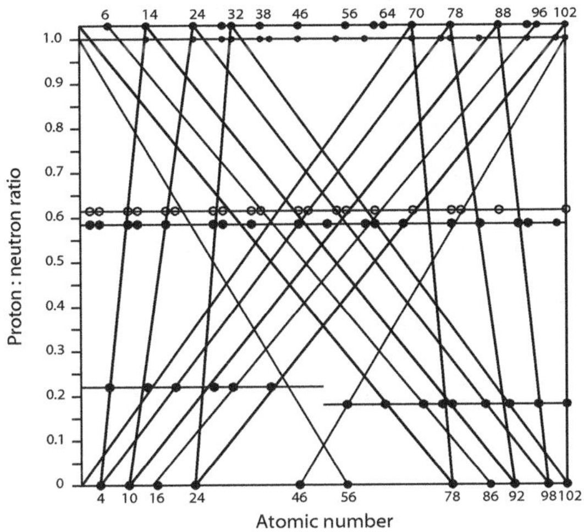

Figure 1.4: *Some of the straight lines that generate periodic relationships at appropriate values of nuclidic proton/neutron ratios and reveal the hidden symmetry.*

the function is seen to close on the double cover of a Möbius surface. Rather than being two sets of lines that define the same periodic relationships twice over, these lines now spontaneously separate to lie on opposite sides in the Möbius double cover. The two sets are interpreted to represent atoms of matter and antimatter respectively.

A sine curve is readily fitted through the set of periodic limiting points of the fully symmetric arrangements. The resulting curves represent closed functions in the interval $(0, 102)$ on identification of the extreme values. Identification of these points is again seen to require a twist and inversion between matter and antimatter. Two elements remain excluded, thereby limiting the number of natural elements to 100. In familiar space $(Z/N = \tau)$ the symmetry is hidden due to the disappearance of 18 elements through instability. Three gaps, each one six elements long, are required to keep electronic structures in register with periodic requirements. A graphic of the resulting arrangement is shown in Figure 1.5. To reveal the hidden symmetry in this representation it is necessary to distinguish between symmetry number (0–101) and atomic number[1] (0–83). Although Stewart, unaware of the hidden symmetry, had no reason to include three blank regions, he came

[1]The neutron is identified as element 0.

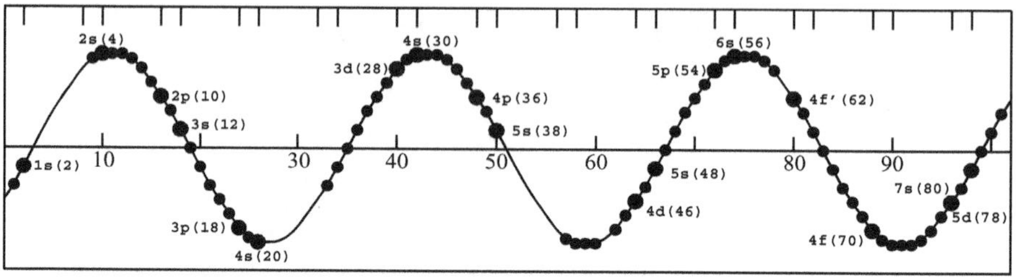

Figure 1.5: *Atomic positions superimposed on the sine wave that describes the hidden symmetry of periodic classification. Short vertical lines define allowed positions for closed electronic sub-shells; the horizontal axis represents symmetry numbers. Atomic numbers are shown in parentheses.*

remarkably close to recognizing the full symmetry that excludes the end members from the periodic region, as in Figure 1.5. This shift of two atomic-number units arises from scaling of the experimental nuclidic stability curve to coincide with predictions based on the parabolic function

$$x^2 - x - n = 0 \tag{1.8}$$

that generates the golden ratio. All aspects of the general periodic function are therefore fully accounted for in terms of aspects of number theory that relates to the golden ratio.

1.6 Number Patterns

An observed periodicity of 24 in the countdown of natural numbers, when used as a model for nuclidic periodicity, has produced the scheme that maps atoms of matter and antimatter on opposite sides of a Möbius interface. It is instructive to note that an identical mapping may be used to represent natural numbers and their conjugates. The crucial assumption required here is an identity of infinities as illustrated in Figure 1.6. Following the arrows, real numbers convert at $R = +\infty$ into their conjugates. At $I = -\infty$ imaginary numbers convert into their real conjugates.[2]

If the numbers are arranged on a spiral as before, each domain, like matter space, will be chiral. Numbers on opposite sides of the Möbius interface will

[2]A familiar example of such an inversion is provided by the trigonometric tan function that approaches $+\infty$ at $\pi/2$ and returns from $-\infty$ immediately beyond that value.

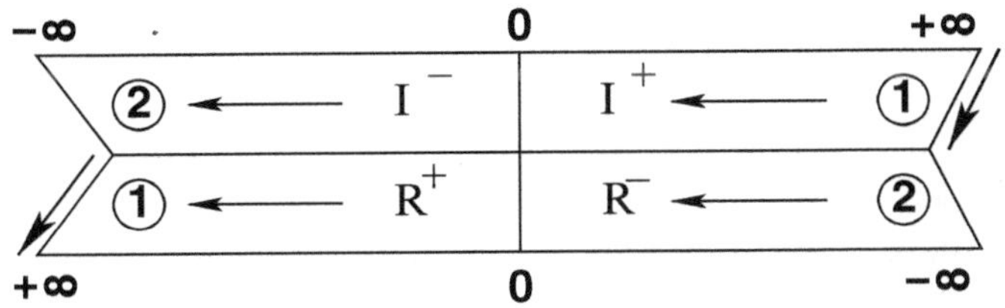

Figure 1.6: *Unfolded double cover of a Möbius strip to model the relationship between real and imaginary numbers. Continuation of R^+ beyond $+\infty$ at 1 proceeds through 1 at $I = +\infty$, with a twist as indicated by the arrows. The conversion from R to I happens gradually in* terra incognito *near infinity.*

be of opposite chirality, but passing along the surface, chirality is gradually inverted from one chiral form to the next.

Real and imaginary numbers in orthogonal relationship, co-exist in the achiral Möbius interface, to be interpreted as the complex plane. It follows that the Möbius representation is an over-simplification. A more realistic representation is provided by the projective plane, or two-dimensional generalization of the one-dimensional Möbius strip. This construct cannot be embedded in three-dimensional space and requires at least four dimensions.

Natural periodic systems and numbers unfold in the same geometric space. Periodic systems that occur in local space–time depend on parameters related to the golden mean. The same parameters, not only describe biological and crystal growth in a local environment, but also the shape of spiral galaxies. It follows, by implication, that a characteristic, locally observed non-euclidean manifold pervades all of space–time as a general global curvature, and that extreme conditions of curvature are only associated with the accumulation of high-density mass. The golden mean and the values of π and e are characteristic of the global curvature.

The two most important features associated with the golden mean is the property of self-similarity and the appearance of rational Fibonacci fractions. Any pattern that depends on either or both of these principles, automatically involves the golden mean. Nuclides are built up in rational proportions from protons and neutrons, consistent with the most efficient packing in space; a property dictated by τ and the Fibonacci fractions. The same principle regulates extranuclear electronic configurations, except for a change of scale. Atomic and nuclear structures are self-similar and obey the same periodic law as natural numbers. The considerations underlying Fibonacci phyllotaxis in biology are exactly the same.

The DNA code is digital and correlation with any number system, especially of base 4, is therefore hardly surprising. More surprising is the high correlation between prime-number values of DNA codons and coded amino acids [8]. This correlation, as a function of evolution may not be equally well defined at all times and, at present, might be hidden. By analogy however, it appears that a number-based periodicity applies to amino acids and codons, on the same basis as to elements and isotopes. The ratio between prime-cross numbers and natural numbers, 8:24 per cycle, is the same as between 21 amino acids and 63 codons, or 100 elements and 300 nuclides. The mysterious link between atomic matter, natural numbers and biological structures can only be the geometry of space–time and its critical parameters π, e and τ.

1.7 Cosmic Structure

The relationship between numbers, atoms and space–time is consistent with the cosmic geometry proposed before [9] on the basis of quantum theory. This proposal portrays the world on the surface of a Möbius strip, or hyper projective plane in more dimensions, such that matter and antimatter are juxtaposed on opposite sides of the interface.

The macroscopic world is known, for instance from the electromagnetic right-hand rule, to be chiral. Because of the Möbius twist, displacement along the surface gradually inverts chirality and converts matter into antimatter. Using state of aggregation as a criterion to differentiate between classical and non-classical entities, the interface between two sides of the Möbius surface, can be viewed as the quantum domain. All massless bosons, such as photons are transmitted in this interface. It is gratifying to note that in the number system this interface contains the complex plane. The fundamental difference between classical and non-classical theories is the complex phase [10] associated with quantum systems.

An important consequence of generally curved space is that it obviates the Doppler interpretation to red shifts in galactic light. Three-dimensional space is proposed to curve into a fourth, time coordinate. Remote objects hence find themselves at different time coordinates. When a photon passes between two sites this time difference has the same effect as a Doppler recession. If therefore, the expanding universe is no longer a logical necessity, big-bang cosmology becomes less attractive.

Without the big-bang time constraint, the issue of nuclear synthesis is open for reconsideration. The mechanism of α-particle fusion may warrant serious consideration. It is well known that ^{4}He gas forms a Bose-Einstein condensate at about 4 K. Like ^{4}He an α-particle is also a boson and a condensate thereof under high pressure may well be the source of heavy nuclei.

The realisation that the electronic configuration of atoms can vary in different thermodynamic environments enables simple new interpretations of several cosmological riddles. It resolves the paradox of high red-shift quasi stellar objects that are physically associated with low red-shift galaxies by providing an alternative explanation of the postulated intrinsic red shifts. The observation of anomalous Fraunhofer absorption effects in the corona of quasars is probably due to the same effect.

1.8 Nuclear Structure and Properties

Nuclide distribution, plotted as a function of neutron imbalance and mass number has been shown to converge to the golden ratio and to fix the limits to the formation and transformation of atomic matter. To confirm that the observed and postulated periodicities are more than a curiosity of number theory, an almost identical plot is obtained by substituting and scaling experimental values of nuclear binding energy, for neutron imbalance on one axis. In the actual plot, the quantity $(BE/N - BE/Z)$ is used as second axis. Convergence is now towards an energy value of 8.5 MeV, instead of τ. Remarkably, when plotting the two parts separately, separate periodic distributions, exactly in line with the postulated variation of stability as a function of neutron excess, is obtained.

1.8.1 Nuclide Abundance

The demonstration that binding energy per neutron or proton follows the trend predicted by number theory, suggests that the disputed mechanism of nucleogenesis could be resolved by testing the data on cosmic abundance against the same periodic relationship. It has been accepted for a long time by astrophysicists that only an equilibrium mechanism can result in abundances that are directly related to the binding energies of nuclides. The alternative, $\alpha - \beta - \gamma$, or big-bang mechanism maps out individual synthetic pathways for all nuclides, independent of binding energy. The most reliable published data on solar abundances were used to investigate their relationship to mass number within modular families, $A(\text{mod}\,4)$. It turns out that nuclide abundance follows the same periodic trend within the 11×24 scheme as binding energy per nucleon. The equilibrium mechanism as assumed in the scheme of α-particle addition is clearly favoured.

1.8.2 Nuclear Spin and Parity

The periodic law based on neutron number provides an unequivocal definition of an energy spectrum, clearly consistent with the accepted shell structure of the nucleus, which is based on magic numbers generated by an assumed strong spin-orbit coupling. Reworking the assignment of nuclear spin and parity in terms of the new magicity failed to improve previous assignments significantly. The agreement for odd-mass nuclei of atomic and/or neutron number less than 50 is reasonable, but it fails completely at higher mass number.

The magic spectrum is clearly not of central-field type. The spin-orbit scheme that assumes a central field probably fails because of this inconsistency. As an alternative the 11×24 periodic law, which is readily interpreted in terms of a simple central-field assignment, may be considered instead. As a first approximation, Hund's rule which is based on maximal quenching of orbital angular momentum of spherically symmetrical atoms [11] succeeds for about two thirds of all spins. In all other cases an unexplained excess over Hund's rule spin is observed. Strong correlation with nuclei known to be distorted, confirmed that the angular momentum of unsymmetrical nuclei contributed the excess spin. It has now been demonstrated that nuclear distortion and excess spin both occur for nuclei where the mass-levels and magic levels are slightly out of register.

Mass number periodicity arises from the stacking of nucleons whereas magic-number periodicity reflects pairwise interaction energies involving neutrons and protons. In many cases the overall shape of a nucleus may depend critically on the symmetry requirements of different interactions, analogous to the Jahn-Teller effect in the ligand fields of coordination complexes. In the latter case, a symmetrical ligand field is distorted by an unsymmetrical electron distribution on the central atom, in a process that lowers the overall energy. Interaction between the two nuclear fields produces distortion by the same mechanism.

Parity is not affected by these distortions and the assignment follows directly in terms of two quantum numbers associated in a logical way with the energy levels defined by the new magic numbers.

1.8.3 Neutron Scattering

The size and shape of atomic nuclei is inferred from experiments such as the way in which they scatter neutron beams. Apparent nuclear size, as measured by low-energy neutrons, is commonly tabulated as thermal-neutron scattering cross section. Nuclides with mass number within a few broad regions have been known to have anomalously large scattering cross sections, ascribed to cooperative properties such as collective vibrational and rotational motion. A direct correlation with the appearance of high-spin nuclei now provides a reasonable explanation of both effects. Cooperative effects are seen to cluster periodically within modular A (mod 4) groups where cross section, polarizability, nuclear size and high spin, all approach local maxima. It is important to note that the concept of high-spin nuclei is undefined within the standard spin–orbit scheme.

An even clearer picture of size periodicity is presented by data on coherent neutron scattering lengths. Within a family of nuclides with common neutron excess, scattering length varies periodically with mass number, to reach maxima close to or at the completion of packing energy levels.

1.8.4 Radioactivity

Having defined stable nuclei as non-radioactive it follows that, like binding energy radioactivity must be a periodic property of nuclides. In fact, radioactivity has been the first property of nuclides found to recur as a function of neutron excess within modular families. Whenever the ratio of Z/N within a modular set exceeds the critical limit, set by a convergent triangle of stability, the next nuclide is an unstable potential positron emitter or it transforms by electron capture. Beyond the low-ratio limit a β^--emitter occurs.

A number of unstable α-emitters occur within the triangle of stability. The most important characteristic that such nuclei have in common, is their proximity to both proton and neutron magic-number levels, within a region of large polarizable nuclei. In this case the opposing effects of proton and neutron level symmetries counteract the tendency to lower nuclear energy by distortion of the nucleus. The combined effect is that the nuclei concerned are rendered unstable with respect to α-emission. The puzzling instability of ^{8}Be is due to the same cause.

Bound-state β-decay provides direct experimental confirmation of continuity between the nucleus and extra-nuclear electron clouds, consistent with the notion that a common periodic law operates in both domains.

1.8.5 Nuclear Structure

The unique properties of atomic nuclei arise from the operation of the strong nuclear force that originates within the quark structure of nucleons. The interaction between proton and neutron, symbolically represented by an interconversion of *up* and *down* quarks

$$\psi(uud) \leftrightarrow \psi(ddu)$$

may also be described as the exchange of a charged virtual pion, $\psi(\bar{u}d)$, consisting of a quark–antiquark pair. The binding energy of 8.5 MeV per nucleon, associated with this process, is shown to represent the energy equivalent of the golden ratio.

Although this simple picture suggests a structureless distribution of nucleons, documented regularities associated with the appearance of excess spin and nuclear polarization, point at a regular pattern of nucleon arrangement. One vivid pattern is the periodic recurrence of excess spin along two parallel mass-number series that stay out of phase by ten mass units. In two-dimensional analogy, two nucleon layers that spiral together from the centre outwards, fall into step after an initial mismatch and maintain this phase difference as successive nucleons occupy progressively larger volumes from the inside out, as shown in Figure 1.7. The analogy to botanical Fibonacci phyllotaxis is unmistakable. Like the florets in a sunflower head nucleons may be viewed as spread along a surface that spirals out in three dimensions. An interesting property of such a spiral is its visibility, which is masked by secondary optical effects. The two drawings shown in Figure 1.8 are developed around the same set of primary points. As on the surface of a pineapple, the fundamental, space-filling spiral goes undetected while the macroscopic pair of Fibonacci spirals presents a more stable image.

It is tempting to correlate respective proton and neutron positions with such Fibonacci spirals. This phyllotaxis predicts the ratio of protons to neutrons to

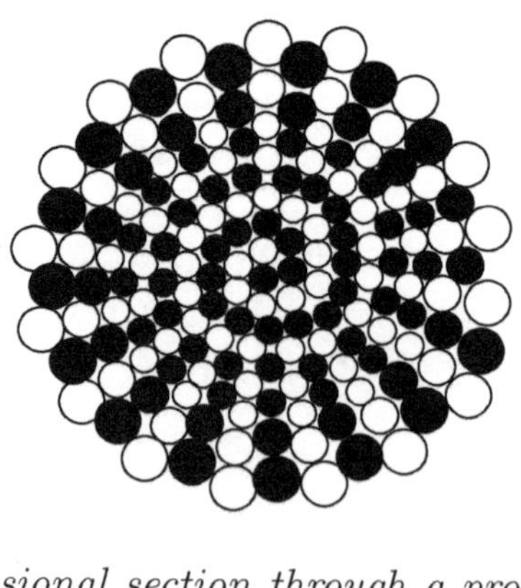

Figure 1.7: *Two dimensional section through a proposed packing model that may account for the observed two-sequence periodicity observed on the nuclear spin function.*

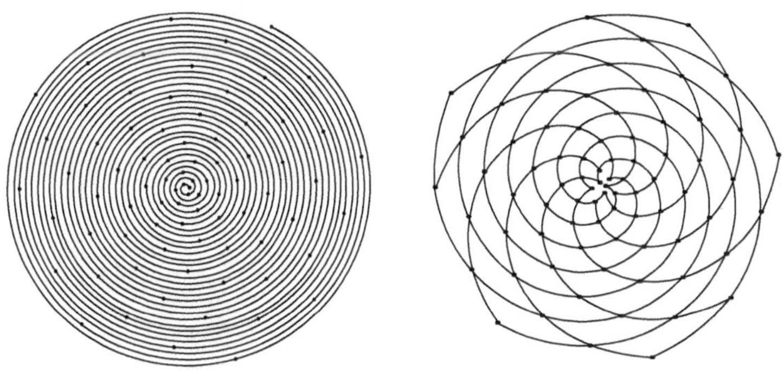

Figure 1.8: *The space-filling spiral (left) defined by a set of points may be invisible to the untrained eye, whereas a pair of Fibonacci spirals (right) that connect the same points, are readily discerned.*

be the ratio of successive Fibonacci numbers, which is known to converge to the golden ratio, as also observed for stable nuclides of high mass number.

Of more fundamental importance is the relationship between nuclear chirality and parity, elucidated by this picture. Chirality depends on the direction of a three-dimensional space-filling spiral, while parity is determined by the sequence of secondary, alternating even and odd Fibonacci spirals. Whereas chirality is fixed for all matter, the parity of nuclei, considered in periodic sequence, inverts at the completion of neutron energy levels.

1.9 Holistic Symmetry

The periodicity of atomic matter arises from the mapping of nuclear compositions on atomic number, in a closed interval. As the closure requires an involution, number periodicity appears in parallel with periodic states of matter in a mapping on to a projective plane in four dimensions, interpreted as a cosmic phase diagram. It implies self-similarity between sub-atomic entities and cosmic domains, regulated by the topology of space–time and dictates a geometrical structure of the universe, based on the golden ratio.

The same symmetry is imposed on the arrangement of nucleons as well as extra-nuclear electrons in an atom. Should this holistic balance be disturbed the atom decays radioactively in order to restore the balance. An important consequence of this holistic balance in an atom is that interaction between the nuclei on regular lattice sites in a solid matrix can lead to cooperative effects such as superconductivity. No other single theory can rationalize all forms of superconduction and, the recognition of new criteria that determine transition temperatures, may lead to the development of improved superconductors.

Chapter 2

Number Theory Primer

2.1 Introduction

Number theory is arguably the purest and most fundamental branch of mathematics, but certainly not the most useful. The claim, often repeated, that mathematics provides the most powerful tool to address social, commercial and engineering needs, may well be correct. To infer that mathematics has been developed for this purpose is an exaggeration. Mathematics found its own direction during classical times[1] when it was converted into an abstract, deductive and axiomatic system of thought. Number theory, above all, constitutes reasoned response to purely intellectual challenges.

The insistence on deductive reasoning as the sole method of proof in mathematics established this branch of learning as a system of thought in the human mind. Reason, not observation, had to decide what was correct. This exclusive use of deduction has differentiated mathematics from all other fields of knowledge. In particular, it defines the sharp distinction between science and mathematics. Science, that also uses conclusions obtained by experimentation and induction often has to revise or discard such conclusions, whereas the conclusions of mathematics have stood for thousands of years.

The Pythagoreans were the first to use the process of deduction exclusively and systematically and to treat mathematical concepts as abstractions. The advantage of abstraction is a gain in generality. A theorem, once proven about an abstract triangle, applies to a triangle formed by matchsticks as well as by the positions of three heavenly bodies. The Greeks distinguished

[1] An historical account of the philosophy and practice of mathematics is given by Kline [12].

mathematical theory from practices such as geodesy and calculation, and developed *arithmetica*, the theory of numbers.

The fascination of number theory is such that after 2,000 years it still is the only branch of mathematics pursued with equal fervour by both professionals and amateurs. The level of interest in the present instance, is somewhere between the two extremes and less pure than either. It is inspired by the realisation that elements of number theory, applied to atomic systems, provide a more fundamental clarification of the periodic law than the advanced theories of physics and chemistry.

Only those aspects of number theory, relevant to the central theme of atomic periodicity will be discussed here, without aiming at mathematical rigour, but with all arguments based on well-established theorems [4, 13, 15].

2.2 Numbers and Arithmetic

The numbers of number theory are familiar to most as the units of counting. The counting numbers belong to a larger class of numbers, called integers.

The Integers

The positive integers or *natural numbers* are the counting numbers

$$1, 2, 3, 4, 5, \ldots\ldots.$$

The *integers* are the counting numbers, their negatives and zero:

$$\{I\} = 0, \pm 1, \pm 2, \pm 3, \ldots.. \tag{2.1}$$

The non-negative integers are the positive integers and zero.

Counting is the simplest example of a mathematical abstraction. The same system is used to count all enumerable entities, be they dogs, diamonds or delinquents. The counting process experienced with small numbers, can be shown to remain valid also for large, untried numbers, by the method of *mathematical induction.* Recall that [12] *deductive reasoning consists of those ways of deriving new statements from accepted facts that compel the acceptance of derived statements*, and that deductive proof is the only proof accepted in mathematics. However, there exists a mathematical induction principle which, unlike induction in science, is independent of experimental observation. It is used when a relation between two functions of an integral variable, n is suspected as a result of trials with small values of n. The suspected result is assumed to hold for a particular value of n, m say, and used to demonstrate that it leads to the same valid result for $n = m + 1$. To complete the proof it is shown that the assumption holds for some special value, such as $n = 1$, and by induction for all values of $n = 2, 3, \ldots$

To validate the counting process, the set S of positive integers is assumed. For each number s in S let S also contain the next integer $s + 1$. Since S contains 1, it contains all positive integers. This is not a proof but a demonstration that counting operates consistently in terms of the mathematical induction principle.[2]

In the simplest mode of counting the members of one set is matched on a one-to-one basis with those of another, like the set of ten fingers, or a pile of markers collected together without internal order. The count generated for an object of the first set is known as its *cardinal* number. However, the reference set could be an ordered aggregate in which every element has a rank or place. Counting against such a set yields an *ordinal* number for each member of the first set. This distinction becomes important when assigning numerical positions to each chemical element in sequence of the periodic classification.

2.2.1 Arithmetic

Addition and multiplication are defined on the counting numbers: addition in terms of counting and multiplication in terms of addition. The positive integers are used to count the members in a set. On counting the members of two separate sets in a single count, without starting over at 1, the result is the *sum* of the two separate counts. Addition can be given a formal definition that shows how to generate sums involving $s + 1$, $n + s + 1$, etc. from sums involving s, by mathematical induction. Multiplication can also be given formal inductive definition based on addition, generating the product $n(s+1)$ from $n \cdot s$.

The order in which two sets are combined during addition or multiplication makes no difference to the outcome. This property is expressed in terms of the following laws:

$$a + b = b + a \qquad a \cdot b = b \cdot a \qquad \text{Commutative law}$$

$$(a + b) + c = a + (b + c) \qquad (ab)c = a(bc) \qquad \text{Associative law}$$

$$a(b + c) = ab + ac \qquad (a + b)c = ac + bc \qquad \text{Distributive law}$$

[2]The mathematical induction principle is not proved to be a property of the positive integers, but is postulated as a defining property. It gives a formal rule for generating the sequence of counting numbers.

Among the integers, 0 is the identity with respect to addition. It is the integer for which

$$n + 0 = 0 + n = n, \quad \text{for every integer } n.$$

For any positive integer m the corresponding negative $-m$ is the additive inverse. It is the integer for which

$$m + (-m) = (-m) + m = 0$$

By definition

$$0 \cdot m = m \cdot 0 = 0, \text{ for each integer } m.$$

It is often necessary to use the fact that positive integers are ordered with a relation $>$, which can be defined in terms of counting or in terms of addition. In terms of counting $n > m$ (read, n is greater than m) if n comes after m in the sequence of counting. In terms of addition:
Let m and n be positive integers. Then $n > m$ if there is a positive integral difference d, such that $n = m + d$.

Binary Numbers The specification of common numbers in decimal notation is almost certainly a remnant of counting practice using a ten-finger base. To base 10 the number 1234 is interpreted as

$$1234 = 10^3 + 2 \times 10^2 + 3 \times 10^1 + 4 \times 10^0.$$

This representation is not unique and the same number can be expressed to any base. Binary, or base 2, notation has become especially important because of computer usage, where it has the advantage of involving only two digits, 0 and 1, like the on and off positions of a switch. The binary number 1000011, for instance, is transcribed into decimal notation, by writing the extended form[3]

$$\begin{aligned} 1000011 &= 1 \times 2^6 + 0 \times 2^5 + 0 \times 2^4 + 0 \times 2^3 + 0 \times 2^2 + 1 \times 2^1 + 1 \times 2^0 \\ &= 64 + 0 + 0 + 0 + 0 + 2 + 1 \\ &= 67 \end{aligned}$$

It may be shown by mathematical induction that any decimal number can be transcribed into binary notation.

[3]The symbol 2 does not exist in binary notation, 2 = 10 in binary.

2.2.2 Divisibility

The positive integers have the Archimedean property: *given positive integers a and b, there is a positive integer n for which $bn > a$.*
This theorem is proved by induction and provides the basis for the *division algorithm*:
Let a be a positive integer and let c be a non-negative integer. Then there is a non-negative integer b for which $c = ab + r$, with $0 \leq r \leq b$.
As a corollary, the integer a divides positive integer c, written $a|c$, if there exists a natural number b, such that $c = ab$. The integer a is called a *divisor* of c.

Let a and b be integers, not both zero. An integer d is a common divisor of a and b if $d|a$ and $d|b$. The integer g is a *greatest common divisor* (GCD) of a and b, written (a, b), if g is a common divisor of a and b, and g is divisible by every common divisor of a and b.

The ideal GCD, (a, b)

Consider all integers of the form

$$n = ax + by \tag{2.2}$$

Suppose that d is a common divisor of a and b such that $a = da'$ and $b = db'$ for some a' and b'. Then

$$n = da'x + db'y = d(a'x + b'y) \quad ; \quad \text{i.e.} \quad d|n.$$

Thus any integer of form (2.2) is a multiple of (a, b) and so $(a, b)|n$. All integers n in (2.2) are multiples of (a, b) and hence (a, b) is the smallest positive value that n can take.

All the integers in (2.2) form what is called an ideal of integers: a set of integers selected so that if two of them are added or subtracted the answer is again one of the set; if one of them is multiplied by an integer, the answer once more belongs to (2.2). From (2.2)

$$y = -\frac{a}{b}x + \frac{n}{b}$$

is an equation of a straight line of slope $-a/b$ and intercept n/b. As n ranges over the multiples of (a, b) the graph ranges over a family of parallel lines. The perpendicular distance d between any of the lines and the origin is $n/\sqrt{a^2 + b^2}$. The graph for $n = 6x + 9y$ is shown in Figure 2.1. The line corresponding to (a, b) is the one with smallest n and passing closest to the

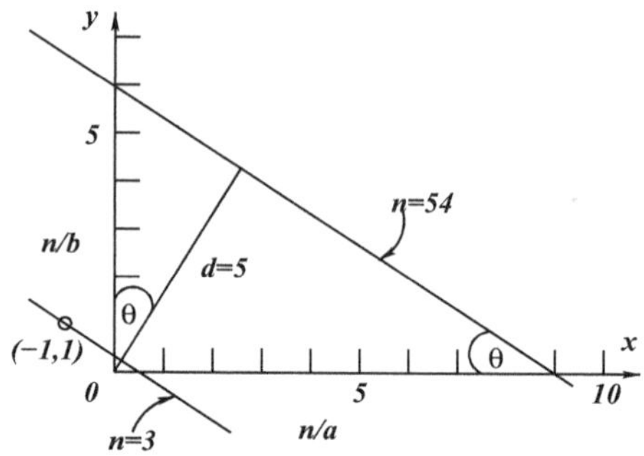

Figure 2.1: *The ideal of integers* $n = 6x + 9y$ *represented as a family of straight lines.*

origin. In this case $n = 3$ and the line goes through the lattice point with integral coordinates $(-1, 1)$,

$$(6, 9) = 3 = -1(6) + 1(9)$$

The GCD of two integers a and b is the smallest positive integer among the linear expressions, $ax + by$, where x and y are integers.

2.2.3 Prime Numbers

Every natural number a is divisible by 1 and a, referred to as *improper divisors*. Thus, 6 has the proper divisors 2 and 3 and improper ones 1 and 6. If the number $n > 1$ has no divisors other than the improper ones n is said to be a *prime number*. Otherwise n is *composite*.

If an integer a divides an integral multiple of another number, $a|bn$, only when $a|n$, the integers a and b are said to be *relatively prime* or *coprime*. This statement is equivalent to the following theorem, given without proof: An integer a is relatively prime to b if and only if $(a, b) = 1$.

An integer p is *irreducible* if $p > 1$ and if p has no divisors strictly between 1 and p; that is no $d|p$ with $1 \leq d < p \rightarrow d = 1$. (The arrow $\rightarrow$ reads, *implies.*)

The Sieve of Erathosthenes

One way to find primes is to wash away the composite integers, leaving the primes caught in the sieve. The procedure is illustrated in Figure 2.2 that lists all numbers from 2 to 100. The first prime 2 is circled, followed by striking out all of its multiples in the table. Next 3 is circled and all of its larger multiples are struck out. Alternate multiples of 3 have already been

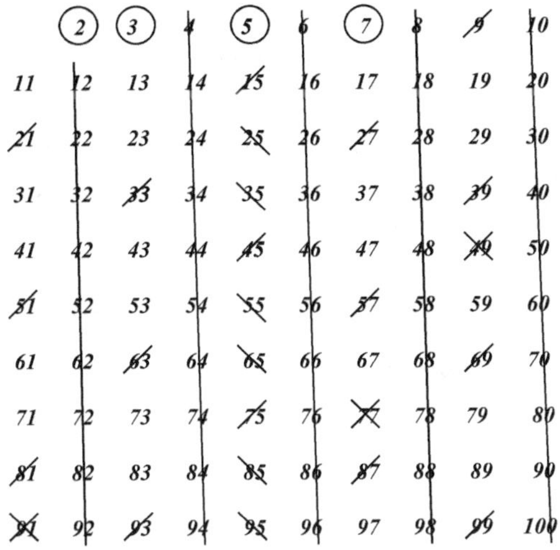

Figure 2.2: *The sieve of Erathostenes to find prime numbers* <100.

washed out as multiples of 2. When the prime number 11 is reached all of its multiples, less than 100 have already been removed. Only 25 prime numbers remain in the sieve. If a number n is composite, at least one of its factors must be $\leq\sqrt{n}$. Thus, the primes between $\sqrt{n}$ and n are obtained by deleting the multiples of all primes up to, but not beyond $\sqrt{n}$.

Euclid's Theorem

The numbers 1 and -1 are neither prime nor composite. They are *units*. Any non-primitive integer >1 is divisible by a prime. The proof of this theorem depends on the definition that a composite integer has a factorization, $n = n_1n_2$. Choose the factorization that minimizes n_1, i.e. such that n_1 is prime. If it is not prime, then $n_1 = n_1'n_2'$ with $1 < n_1' < n_1$, so that $n = n_1n_2 = (n_1'n_2')n_2 = n_1'(n_2'n_2)$. This result contradicts the assumption that n_1 is the smallest divisor of n. The theorem has been proved.

The theorem is now used to prove Euclid's famous theorem that there are infinitely many primes.

Proof Suppose that there is only a finite set of prime numbers, e.g. those between 1 and the largest prime p_n. Form the product of these primes and consider the integer

$$N = (2 \cdot 3 \cdot 5 \cdots p_n) + 1$$

By the previous theorem this number N is either a prime or it is divisible by a prime. However, from $N - (2 \cdot 3 \cdot 5 \cdots p_n) = 1$, it cannot be divisible by any of the primes between 1 and p_n: if some prime from the list, $p_i|N$ then p_i would also have to divide 1. Then either N is prime or some other prime $(p_o \neq p_i)|N$, contrary to the assumption that the finite list contains all the primes.

2.2.4 Magic Numbers

The ancient belief that numbers had magical powers prompted investigation into the internal structure of numbers as reflected by their divisors or their shapes.

One possible measure of the qualities of a number n is provided by the sum $\varsigma(n)$ of its divisors. On this basis numbers are considered *deficient, perfect* or *abundant* for $\varsigma(n) < n$, $\varsigma(n) = n$, or $\varsigma(n) > n$, respectively.

Perfect Numbers

The first perfect number is $6 = 1 + 2 + 3$. In another notation perfect numbers are identified by the sum of all positive integers d such that $d|n$, i.e.

$$\sigma(n) = \sum_{d|n} d \quad \text{and} \quad \tau(n) = \sum_{d|n} 1$$

where $\tau(n)$ is the number of divisors of n. A perfect number is a positive integer for which $\sigma(n) = 2n$. For the perfect number 28,

$$\sigma(28) = \sum d = 1 + 2 + 4 + 7 + 14 + 28 = 56$$
$$\tau(28) = \sum 1 = 6$$

Amicable Numbers

Ancient Greeks called two positive integers m and n amicable (friendly) if the sum of the proper divisors of m equals $n - 1$ and the sum of the proper divisors of n equals $m - 1$.
Suppose that $\sigma(m) = \sigma(n) = m + n$. Then

$$\sigma(m) - m = n$$
$$\sigma(n) - n = m$$

which defines an amicable pair.

Pythagoras found the smallest amicable pair:

$$220 : \sum_{d|n} d = 1 + 2 + 4 + 5 + 10 + 11 + 20 + 22 + 44 + 55 + 110 = 284$$

$$284 : \sum_{d|n} d = 1 + 2 + 4 + 71 + 142 = 220$$

The pair, $m = 3 \times 5 \times 7 \times 9 \times 13 = 12{,}285$ and $n = 3 \times 5 \times 7 \times 139 = 14{,}595$ represent a pair of odd amicable numbers.

m-Gonal Numbers

m-Gonal numbers are defined by generalization of triangular numbers, known to the ancients as 1, 3, 6, 10, 15, etc. Each triangular number counts the dots that can be arranged evenly in equilateral triangles of increasing size (Figure 2.3). The number of points in each row of the large triangle is one more than the number on the previous row. Let t_n represent the nth triangular number. Then $t_n = t_{n-1} + n$, $n > 0$, e.g. $t_0 = 0$, $t_1 = 0 + 1 = 1$, $t_2 = 1 + 2 = 3$, $t_3 = 3 + 3 = 6$, etc.

To find a general formula for t_n, form the sum

$$\sum_{n=1}^{k} t_n = \sum_{n=1}^{k} (t_{n-1} + n) = \sum_{n=1}^{k} t_{n-1} + \sum_{n=1}^{k} n$$

$$= \sum_{i=0}^{k-1} t_i + k(k+1)/2$$

$$\text{so that} \quad t_k = \sum_{n=1}^{k} t_n - \sum_{n=0}^{k} t_n - \sum_{n=0}^{k-1} t_n = k(k+1)/2$$

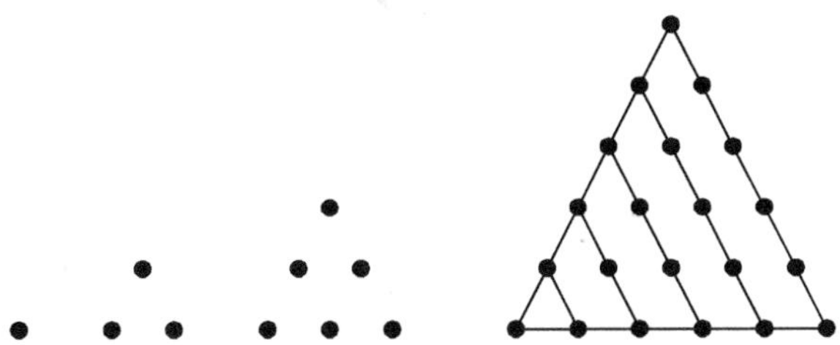

Figure 2.3: *Generation of triangular numbers.*

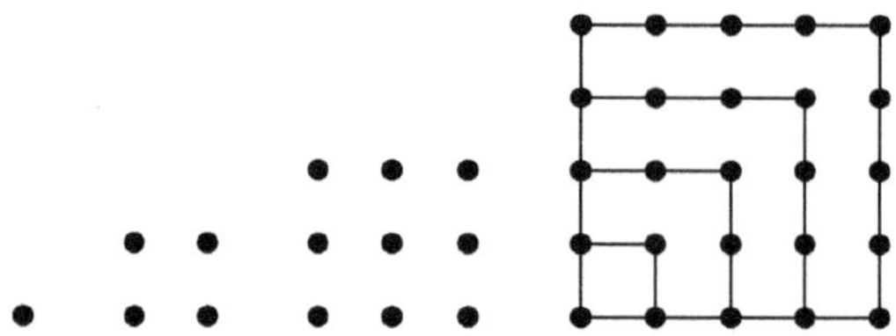

Figure 2.4: *Definition of square numbers.*

Square numbers are defined in strict analogy with triangular numbers (Figure 2.4). Hence, for $s_0 = 0$ and $n > 0$, $s_n = s_{n-1} + 2n - 1$.
Let all $s_i = 0$, for $i < 0$. Then, for $l \geq 0$, $s_l = s_{l-1} + 2l + 1$.
The formula in this form will be shown to describe the orbital angular momentum degeneracy of hydrogen wave functions and the number of electronic energy sublevels for principal quantum number, n,

$$\sum_{l=0}^{n-1}(2l+1) = n^2 \tag{2.3}$$

m-Gonal numbers, $m > 4$, follow the same pattern.

Pyramidal Numbers Pyramidal numbers are generated by generalization of m-gonal numbers from two- to three-dimensional representation. The simplest pyramidal numbers are the tetrahedral numbers, 1, $1+3 = 4$, $4+6 = 10$, $10+10 = 20$, $20+15 = 35$, etc., obtained by adding the next triangular number to the previous tetrahedral number, starting from $p_3^1 = 1$. In geometrical representation the tetrahedral numbers count the number of equal spheres that fill a tetrahedron to a given level (Figure 2.5). Each layer contains a triangular number of points (spheres), i.e. 1, 3, 6, 10, 15, etc. These numbers will be encountered again as half the integral solutions (n) to the radical $\sqrt{4n+1}$ which generate the parabola that describes nuclide stability and periodicity in terms of the golden ratio.

Pascal's Triangle

This ancient array of numbers that provides formulae to the coefficients that occur in binomial expansions, e.g.

$$(1+x)^4 = 1 + 4x + 6x^2 + 4x^3 + x^4$$

was known to Arab mathematicians long before the time of Pascal. Such an array of coefficients is shown in Table 2.1. The table is constructed by

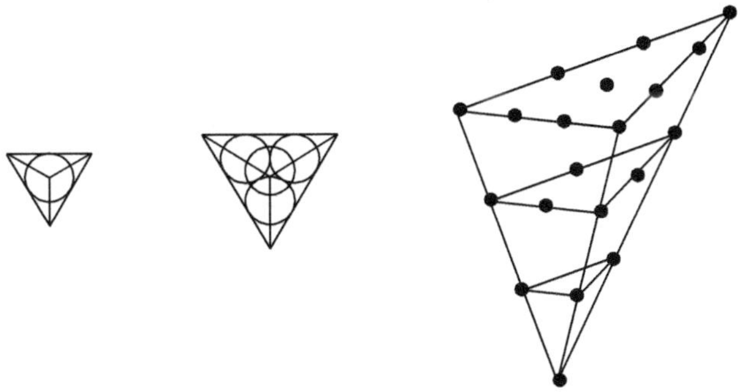

Figure 2.5: *Geometric representation of tetrahedral numbers. The frame at right shows the centres of touching spheres only.*

Table 2.1: *Pascal's triangle.*

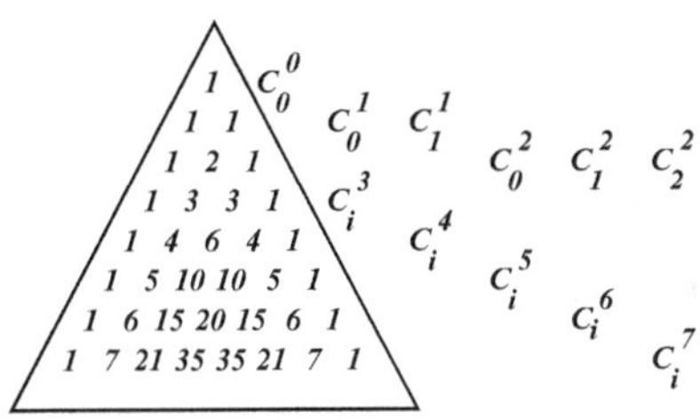

forming the sums $C_i^{n+1} = C_i^n + C_{i-1}^n$ of the two numbers immediately above any given number. Note that when n is prime, the binomial coefficients C_i^p for $i = 1, 2, \ldots, p-1$ are all divisible by p.

Since binomial coefficients can be either even or odd an interesting pattern is generated by colouring all squares containing odd numbers black and leaving all the even ones blank.

The pattern of Figure 2.6 belongs to a class of geometrical objects known as *fractals*. A fractal is something that displays the same structure at any magnification. This symmetry property is known as *self-similarity*.

A surprising property of Pascal's triangle is the preponderance of blank space. As the total area under consideration is increased the total black area becomes insignificant and tends to zero. This trend means that almost all numbers in Pascal's triangle are even.

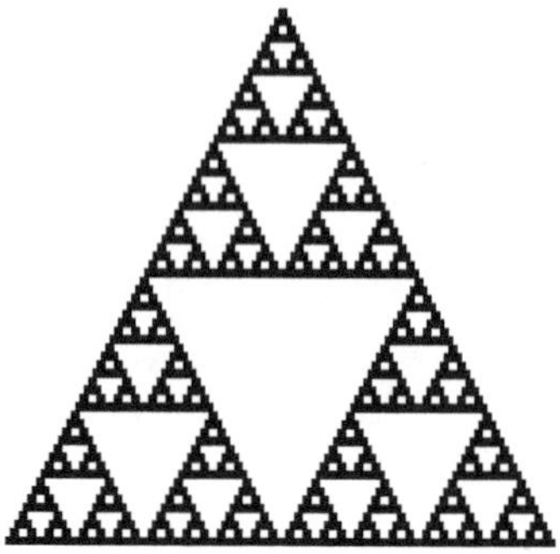

Figure 2.6: *Fractal patterns in Pascal's triangle.*

2.2.5 Fundamental Theorem of Arithmetic

From the definition that $p|bc \to p|b$ or $p|c$ it follows by repeated application that any integer >1 can be factored into a product of primes. It will be shown, not only that each integer has a prime factorization, but also that the factorization is unique. Unit factors will not be shown in the factorization unless the given integer is negative, when the unit factor -1 is included. Positive prime factors will be listed in increasing order of size. The following theorem follows within these conventions.

The fundamental theorem of arithmetic Every integer >1 (or <−1) has exactly one factorization into primes.

Proof Suppose n is an integer >1. Then either n is prime, or it is divisible by a prime, p, say $n = pq$, for some quotient q. Again, either q is prime or it is divisible by a prime, say $q = p_1q_1$. Either q_1 is prime or $q_1 = p_2q_2$. On repeated application of the same rule it is eventually shown that any integer $n > 1$ or (< -1) has some factorization into a product of primes. If not, suppose some integer $n > 1$ has two different factorizations,

$$n = p_1p_2 \dots p_r \quad \text{and} \quad n = q_1q_2 \dots q_s$$

with primes listed in increasing order of size, $p_i \le p_j$ for $i < j$ and similarly for the q's.
Since $p_1|n$ in the p-factorization, p_1 must also divide n in its q-factorization:

$$p_1|n, \quad \text{where} \quad n = q_1q_2 \dots q_s$$

Now compare p_1 and q_1. If $p_1 \neq q_1$, then $p_1|q_2q_3 \dots q_s$. Similarly, either $p_1 = q_2$ or $p_1|q_3 \dots q_s$, and so on. Eventually, p_1 must equal one of the q's. Then,

$p_1 \geq q_1$, the smallest of the q's. Repeat the process for q_1, showing that q_1 equals one of the p's and $q_1 \geq p_1$. Then $p_1 = q_1$. Divide n by $p_1 = q_1$, getting

$$n/p_1 = p_2 p_3 \ldots q_s$$

As before it can now be shown that $p_2 \geq q_2$ and $q_2 \geq p_2$, so that $p_2 = q_2$, and so on.
Assume that $s > r$ such that after r steps

$$n/p_1 p_2 \ldots p_r = 1 = q_{r+1} q_{r+2} \ldots q_s$$

However, each q is prime and >1. Therefore $r = s$. The factorization is unique.

GCD and LCM of factored Integers

It is now possible to prove the

Theorem If $a = p_1^{a_1} p_2^{a_2} \ldots p_r^{a_r}$ and $b = p_1^{b_1} p_2^{b_2} \ldots p_r^{b_r}$ where the p's are prime and the exponents a_i and b_i are non-negative integers, then

$$(a, b) = p_1^{c_1} p_2^{c_2} \ldots p_r^{c_r}$$

where each $c_i = \min\{a_i, b_i\}$.

Proof Note that any exponent equal to zero implies that the corresponding p_i does not occur in the relevant factorization, ($p^0 = 1$).
The divisors of a are the integers of the form $d = p_1^{d_1} p_2^{d_2} \ldots p_r^{d_r}$, where $0 \leq d_i \leq a_i$, for each i. Each $d|a$, therefore

$$a = d \left(p_1^{a_1 - d_1} p_2^{a_2 - d_2} \ldots p_r^{a_r - d_r} \right)$$

No other integers divide a, since[4] $p_i^{g_i} \nmid a_i$ if $g_i > a_i$ or if p_i is not one of the primes in the factorization of a. From this observation, $p_1^{c_1} p_2^{c_2} \ldots p_r^{c_r}$ divides both a and b and is in turn divided by every common divisor.

[4] $\nmid$, does not divide

Corollary Two positive integers are relatively prime iff[5] their factorizations have no primes in common.
$[a, b]$, the least common multiple (LCM) of a and b is a multiple of a and a multiple of b and it divides every common multiple of a and b.

Using arguments like those in the derivation of (a, b) it is shown that $[a, b] = p_1^{m_1} p_2^{m_2} \dots p_r^{m_r}$, where each $m_i = \max\{a_i, b_b\}$.

The GCD and LCM of two positive integers can be found from their factorizations by using the minimum and maximum exponent for each prime power, respectively.

2.2.6 Gaussian Integers

Introduction of the unit -1 is necessary to ensure that the fundamental theorem applies to both positive and negative integers. The question arises whether the fundamental theorem would still hold if the class of integers is defined even wider than in (2.1). The Gaussian integers are members of the imaginary quadratic field $\mathbb{Q}(\sqrt{-1})$ and constitute such an extended class. They form a ring often denoted $\mathbb{Z}[i]$ and have the form $a + bi$, with a and b ordinary integers and $i = \sqrt{-1}$. To avoid possible confusion the ordinary integers may be referred to as *rational* integers, $\mathbb{Z}[1]$, or simply $\mathbb{Z}$. Note that in both sets $\mathbb{Z}$ and $\mathbb{Z}[i]$ the sum, difference and and products of all integers are integers, but $(a + bi)|(c + di)$ only if there is an $e + fi$ such that

$$(a + bi)(e + fi) = (ae - bf) + (af + be)i = s + di$$

An element of $\mathbb{Z}[i]$ is a unit if it divides 1, and hence also every element of $\mathbb{Z}[i]$. The units of $\mathbb{Z}[i]$ are ± 1 and $\pm i$. A number π in $\mathbb{Z}[i]$ is prime if it is not a unit and if in every factorization $\pi = \alpha\beta$ one of α or β is a unit.

A number of properties and theorems which can be proved [13] for Gaussian integers will be given here without proof, noting that in general the proofs have much in parallel with the corresponding proofs for rational integers.

1. If $\alpha = a + bi$ is an element of $\mathbb{Z}[i]$ its norm $N(\alpha)$ or $N\alpha$, is defined to be $\alpha\alpha^* = |\alpha|^2 = a^2 + b^2$, ($\alpha^* = a - bi$ is the *complex conjugate* of α). The norm has the following properties:

    (a) If α is in $\mathbb{Z}$ as well as $\mathbb{Z}[i]$, then $N\alpha = \alpha^2$.

    (b) $N(\alpha\beta) = N\alpha N\beta$

[5] iff, if and only if

(c) $N\alpha = 1$ iff α is a unit.

(d)

$$N\alpha \begin{cases} = 0 & \text{if } \alpha = 0 \\ = 1 & \text{if } \alpha = \pm 1 \text{ or } \pm i \\ > 1 & \text{otherwise} \end{cases}$$

(e) If $N\alpha$ is prime in $\mathbb{Z}$, then α is prime in $\mathbb{Z}[i]$.

The converse of (e) is false: 3 is prime in $\mathbb{Z}$. $N3 = 3^2 = 9$. Suppose $3 = \alpha\beta$. Then $9 = N\alpha N\beta$. If neither α nor β is a unit $N\alpha \neq 1$, $N\beta \neq 1$, so $N\alpha = N\beta = 3$. But this means that if $\alpha = a + bi$, then $a^2 + b^2 = 3$, which is impossible for any pair of integers a, b in $\mathbb{Z}$.

2. If α and β are Gaussian integers, $\beta \neq 0$, then there exist two integers π and ρ such that

$$\alpha = \pi\beta + \rho \qquad\qquad N\rho < N\beta$$

3. If π is prime and $\pi|\alpha\beta$ then $\pi|\alpha$ or $\pi|\beta$.

4. The fundamental theorem of arithmetic holds in $\mathbb{Z}[i]$.

5. Euclid's theorem holds for primes in $\mathbb{Z}[i]$.

2.2.7 The Binomial Equation

Finding the square root of a complex number A is equivalent to finding the solution X of the quadratic equation $X^2 = A$. Let $A = a+ib$ and $X = x+iy$. Then the real numbers x and y must be such that

$$(x + iy)^2 = a + ib = x^2 + y^2 + 2ixy$$

Therefore $x^2 + y^2 = a$, $2xy = b$. Combined with the identity

$$(x^2 + y^2)^2 = (x^2 - y^2)^2 + 4x^2y^2$$

this gives

$$x^2 + y^2 = \sqrt{a^2 + b^2}$$

and hence

$$x^2 = \frac{\sqrt{a^2 + b^2} + a}{2}, \qquad y^2 = \frac{\sqrt{a^2 + b^2} - a}{2}$$

With the correct sign convention [14] the required solutions are

$$X = \pm\left(\sqrt{\frac{\sqrt{a^2+b^2}+a}{2}} + (\text{sign})i\sqrt{\frac{\sqrt{a^2+b^2}+a}{2}}\right) \tag{2.4}$$

where sign= + or − as $b >$ or < 0. For $b = 0$, $X = \pm\sqrt{(\text{sign})\,a}$, sign= + or − as $a >$ or < 0. When $a = b = 0$ there is only one trivial solution, $X = 0$.

The binomial equation $x^n = a$ has a simple trigonometric solution based on De Moivre's formula [14]

$$(\cos\theta + i\sin\theta)^n = \cos n\theta + i\sin n\theta \tag{2.5}$$

$$\text{Write}\quad a = r(\cos\theta + i\sin\theta)$$

$$x = \rho(\cos n\varphi + i\sin n\varphi) = a$$

Hence $\rho^n = r$ and $n\varphi = \theta + 2k\pi$ for any integer k, since the equality holds for all $n\varphi$ that differs from θ by the full angle. The nth roots of the complex angle follows as

$$x = \sqrt[n]{r}\left(\cos\frac{\theta+2k\pi}{n} + i\sin\frac{\theta+2k\pi}{n}\right)$$

Geometrically these roots are the vertices of a regular polygon with n sides.

The equation $x^n = 1$ defines the roots of unity of degree n. In this case $r = 1$, $\theta = 0$ and the roots are obtained from the formula

$$\cos\frac{2k\pi}{n} + i\sin\frac{2k\pi}{n} = e^{2\pi ik/n} \tag{2.6}$$

2.2.8 Algebraic Number Theory

Extension of the number system to include Gaussian integers leads to the much wider notion of algebraic number theory. Any algebraic integer of the form $a + b\sqrt{d}$ where d is square free, forms a quadratic field $\mathbb{Q}(\sqrt{d})$. If $d > 0$, the field is a real quadratic field, and if $d < 0$, it is an imaginary quadratic field. In one example, the integers in $\mathbb{Q}(\sqrt{-3})$ are called Eisenstein integers.

The algebraic integers in an arbitrary quadratic field do not necessarily have unique factorizations. For example, the fields $\mathbb{Q}(\sqrt{-5})$ and $\mathbb{Q}(\sqrt{-6})$ are not uniquely factorable, since

$$21 = 3\cdot 7 = (1+2\sqrt{-5})(1-2\sqrt{-5})$$

$$6 = \sqrt{-6}(\sqrt{-6}) = 2\cdot 3$$

although the above factors are all prime within their fields.

The totality of all algebraic integers is said to form a *ring*, which is a set of numbers that contains $\alpha + \beta$, $\alpha - \beta$, $\alpha\beta$ when it contains α and β. For a ring in general, but not necessarily always:

(a) The commutative law of multiplication $ab = ba$, does not hold

(b) There is no identity element of multiplication

(c) The cancellation law of multiplication does not hold

An *integral domain* is a specialized ring and a *field* is a specialized integral domain.

There are no primes in the ring of all algebraic numbers. For let α be an integer different from zero or a unit. One can always write $\alpha = \sqrt{\alpha}\sqrt{\alpha}$. If α satisfies $p(x) = 0$, then $\sqrt{\alpha}$ satisfies $p(x^2) = 0$, so $\sqrt{\alpha}$ is an integer. This condition does not prevail in the ring of integers in a fixed algebraic number field $\mathbb{K}$. Then $\alpha|\beta$ if β/α is an integer of $\mathbb{K}$; ϵ is a unit if $\epsilon|1$; α is prime if it is not zero or a unit, and if any factorization $\alpha = \beta\gamma$ into integers of $\mathbb{K}$ implies that either β or γ is a unit.

A set A of integers in $\mathbb{K}$ is an *ideal* if, together with any pair of integers α and β in A, the set also contains $\lambda\alpha + \mu\beta$ for any integers λ and μ in $\mathbb{K}$. Two ideals in $\mathbb{K}$ are equivalent, $A \sim B$, if there are two non-zero integers α and β in $\mathbb{K}$ such that $\alpha(A) = \beta(B)$. The totality of ideals in $\mathbb{K}$ equivalent to a fixed ideal $A \neq (0)$ is said to constitute a *class*. The number of classes is called the *class number* h of $\mathbb{K}$. It turns out that [13] a field has unique factorization of integers into primes iff its class number is 1.

Functions In defining a function $f : X \to Y$, it means that a unique element of Y is assigned to each element of X. If no element of Y is assigned to more than one element of X, the function f is called an *injection* or one-to-one [16]. In symbols it is implied ($\Rightarrow$) that

$$\forall x_1, x_2 \in X, [x_1 \neq x_2 \Rightarrow f(x_1) \neq f(x_2)]$$

If each element of Y is assigned to some ($\exists$) element of X, the function f is a *surjection*, $\forall y \in Y, \exists x \in X, y = f(x)$. If f is both an injection and a surjection, it is called a *bijection*.

Suppose that $f : X \to Y$ is a function and A is a subset of X, i.e. $A \subseteq X$. The function $g : A \to Y$ can be defined by $g(a) = f(a)$ for all $a \in A$. This function is called a *restriction* of f to A and is denoted by $f|A$.

2.3 Distribution of Prime Numbers

The distribution of prime numbers is a constant theme of number theory. Euclid's theorem that proves the existence of an infinite number of primes is well known. Equally famous is Dirichlet's theorem:
The arithmetic progression $a + kn$, $n = 1, 2, 3, \ldots$ *contains infinitely many primes if* a *and* k *are coprime.*
This theorem can be proved in several ways [4]. Neither of the theorems however, explain the rather irregular distribution of primes among the natural numbers. As conjectured by Gauss the number of primes $\pi(x)$ less than or equal to a given number x is asymptotic to $x/\log x$. Stated differently

$$\lim_{x \to \infty} \frac{\pi(x) \log x}{x} = 1$$

this conjecture, after satisfactory proof, is now known as the prime number theorem. The sieve of Erastosthenes identifies all primes less than a given maximum, but it also fails to solve the problem of prime-number distribution.

A much celebrated computer-generated graphic that appears to reveal a mysterious regularity in the distribution of prime numbers was published some time ago [17], made it to the cover of Scientific American [19] and features on many websites [18]. It is known as a prime-number spiral and the positions of primes show up visibly along lines, parallel to the main diagonals of the dense spiral. The first 700 integers, (mod 100), arranged on a spiral are shown in Figure 2.7.

The method whereby the spiral is constructed is shown on the right in Figure 2.7. The anticlockwise progression that defines the spiral is shown to be made up by the alternating addition of square numbers (2.2.4) from the right and the left, respectively. All the even squares are thereby constructed to appear along the NW diagonal and the odd squares along the SE diagonal. This construction explains the diagonal distribution of primes, since in any E-W or N-S column even and odd numbers alternate, whereas they are segregated in diagonal directions. By filling all blocks that represent odd numbers a chequerboard pattern is generated. All primes, with the exception of 2, are thus constructed to occur along these diagonal lines. The pattern of Figure 2.7 (left) arises from distinction between composite (white) and prime numbers (black), rather than between even and odd integers. Deviations from the chequerboard pattern that create the irregular appearance of Figure 2.7 are readily explained:

(a) The parallel lines adjacent to the square diagonals contain the odd numbers $n^2 - 1 = (n + 1)(n - 1)$ and must be prime free. Like the diagonal of square numbers these lines therefore have no black marks.

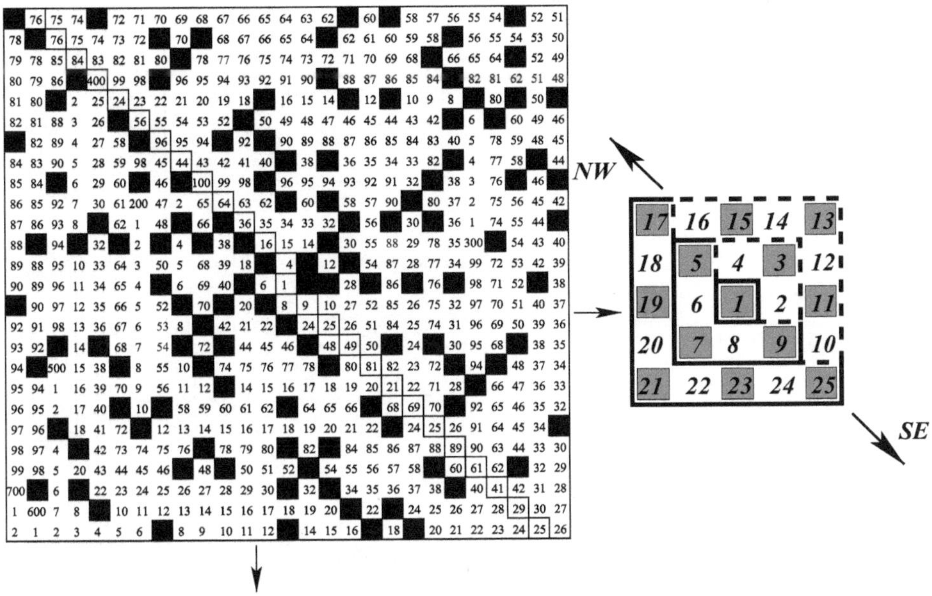

Figure 2.7: *Reconstruction of the number spiral claimed to exhibit a strongly nonrandom appearance [17]. The positions of all prime numbers are indicated in black. Squares of the natural numbers are shown blocked in along one main diagonal. The small diagram on the right shows how the spiral is made up of square numbers. The shading shows that odd and even numbers define a chequerboard pattern.*

(b) The lines $(n^2 \pm 1)$ adjacent to the diagonal of odd squares contain only even integers and apart from 2, no other primes.

(c) All numbers on the line $n-2$ parallel to the even-square diagonal, are even.

(d) Another prominent prime-free diagonal array of odd numbers, i.e. 21, 45, 77, etc. is of the form $n^2 - 4 = (n-2)(n+2)$.

(e) In general any set of numbers $n^2 - m^2 = (n-m)(n+m)$, for n odd and m even, is prime free and occurs along a line parallel to the diagonal and situated between two lines of even integers.

(f) Two of the most conspicuous features clearly visible on extended computer maps of the spiral are two clear streaks at right angles marked on Figure 2.7 by two small arrows. The vertical streak consists of two empty columns. The odd numbers in the column on the left are composite numbers related to the squares n^2 by $n^2 - \frac{1}{2}(n+3)$, factoring into $(2n-3)(\frac{n-1}{2})$. Those in the second column are

$$n^2 - \frac{1}{2}(n+1) = (2n+1)\left(\frac{n-1}{2}\right)$$

The contiguous prime-free horizontal rows are

$$n^2 + \frac{1}{2}(n-1) = (2n-1)\left(\frac{n+1}{2}\right)$$

$$\text{and} \qquad n^2 + \frac{1}{2}(n-3) = (2n+3)\left(\frac{n-1}{2}\right)$$

The appearance of diagonal blocks of three adjacent prime-free lines is explained in full. These blocks are striking features of the computer graphic. As shown for the vertical and horizontal empty blocks, all other features of the prime-number spiral can be explained by similar simple arguments. The spiral display does not reveal any new information or pattern related to the distribution of prime numbers and the entire construct is no more than a special arrangement of the sieve of Erasthosthenes.

2.3.1 Twin Primes

The number spiral of Figure 1.2 is more informative than that of Figure 2.7, since it clearly demonstrates the occurrence of twin primes $(p, p+2)$, cousin primes $(p, p+4)$ and sexy primes $(p, p+6)$ [20]. It confirms that all twin primes, excepting 2 and 3 are of the form $6n \pm 1$. As a matter of fact, it further shows that all primes, except 2 and 3 are of the form $6n \pm 1$. The periodicity of the spiral is 24 and hence all numbers on the radial lines through the small even numbers $n = 2$ to 24 are of the form $n + 24m$, $m = 1, 2, 3 \ldots$, and therefore include all even positive integers. The radial lines through odd multiples of $3 < 24$ cover all of the remaining multiples of 3 as in the sieve of Erasthostenes. The eight remaining radial lines, of the form $6n \pm 1$, contain all remaining numbers i.e. the prime numbers ≥ 5 and their common multiples. The prime number distribution according to the 24-number spiral is shown in Figure 2.8. A remarkable feature demonstrated in Figure 2.8 is that the squares of all prime numbers ≥ 5 occur along the radial line through 1, together with prime and composite numbers of the form $24m + 1$. The squares of familiar primes occur at $m = 1, 2, 5 \ldots$, etc. Squares of squared primes occur at $m = 25, 100, \ldots$. For those values of m that do not label

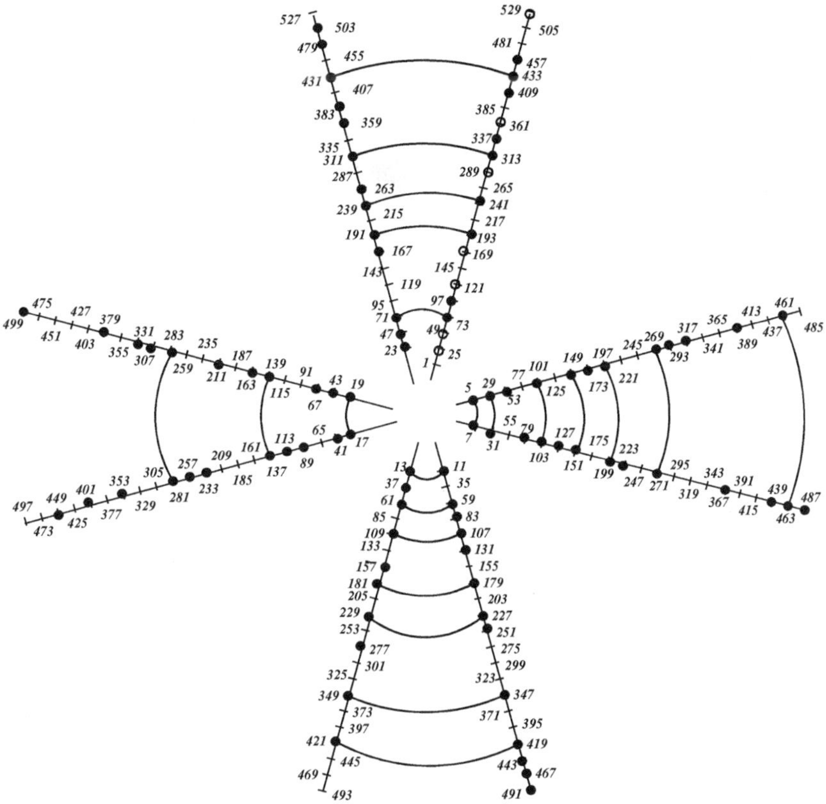

Figure 2.8: *The* $6n \pm 1$ *prime number cross, showing prime numbers in black and squares of primes as open circles. Twin primes are connected by arcs.*

the positions of prime numbers, $24m + 1$ is not the square of an integer, i.e. $m = 3, 4, 6, \ldots$. This observation suggests that all integers of the form $j = \sqrt{24m+1}$ should be prime. However, for $m = 51, 126, \ldots$ etc., $j = 5 \times 7$, 5×11, etc. There is no obvious way to distinguish between these prime and composite square roots.

It is remarkable that the set of integers generated by the sequence $j = \sqrt{24m+1}$ is identical to the set $i = 6n \pm 1$, distributed over the eight arms of the cross of Figure 2.8. This observation is summarized by the formula

$$m = \frac{n}{2}(3n \pm 1), \quad n = 1, 2, 3 \ldots$$

that applies to all primes ≥ 5 and their common multiples. The m-values of possible twin primes differ by n.

2.3.2 The Sieve Revisited

The $6n \pm 1$ prime cross of Figure 2.8 corresponds to the first stage of sifting out composite numbers by the Erastosthenes procedure, i.e. removal of all multiples of the first two prime numbers 2 and 3. The same procedure can be continued to screen out the multiples of successive higher prime numbers, $p = 5, 7, 11, 13, \ldots$ etc., from the set of Figure 2.8, by rearrangement on spirals of period $2p$. A spiral of period 14 is shown in Figure 2.9. The multiples of 7 are found grouped together on two radial lines $c_m^+ = p(6m + 1)$ and $c_m^- = p(6m - 1)$, $p = 7$. The same result is found for spirals of period 10, 22, $\ldots 2p \ldots$, with multiples of p located at positive values of

$$c_m^{\pm} = p(6m \pm 1),\ m = 0, 1, 2, \ldots \tag{2.7}$$

For a given prime its multiples in common with those of lower primes occur as c_m with $6m < p$. It is shown in Table 2.2 how this procedure screens out composite numbers less than 300. The procedure outlined here may be used as an algorithm to identify prime numbers, but not as a formula to generate them. The problem is that each formula such as (2.7) generates, not

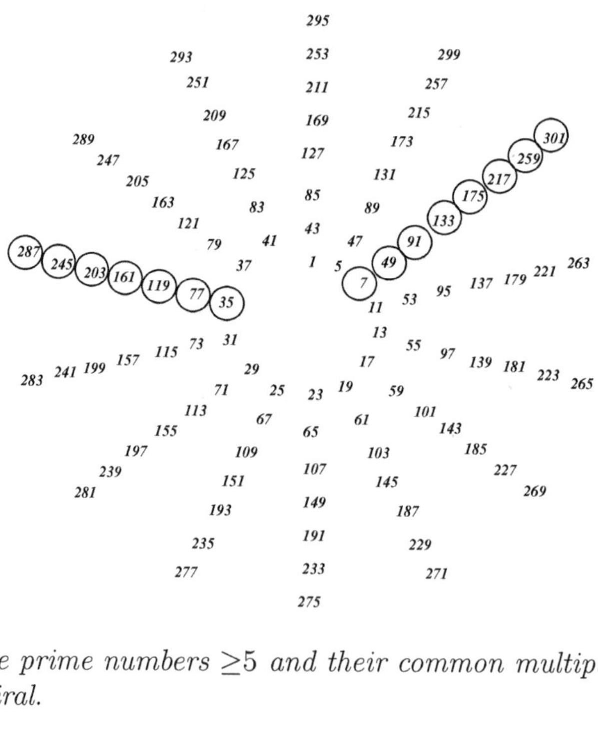

Figure 2.9: *The prime numbers ≥5 and their common multiples arranged on a 14-period spiral.*

Table 2.2: *Composite numbers of the form* $p(6m \pm 1) < 300$.

p \ m		1	2	3	4	5	6	7	8	9	10
5	−	25	55	85	115	145	175	205	235	265	295
	+	35	65	95	125	155	185	215	245	275	
7	−		77	119	161	203	245	287			
	+	49	91	133	175	217	259				
11	−		121	187	253						
	+		143	209	275						
13	−			221	299						
	+		169	247							
17	−										
	+			289							

only multiples of p, but also the common multiples of all other primes. Any formula that generates only prime numbers must therefore be complicated.

2.3.3 Prime-generating Polynomials

It is generally agreed that no rational algebraic function can consistently yield prime numbers. However, several low-order polynomials are known to generate primes between well-defined limits. The best known example is the family of polynomials

$$P = n^2 + n + p,\ n = 0, 1, \ldots, p - 2$$

in which P is prime iff the field $\mathbb{Q}(\sqrt{1-4p})$ has class number $h = 1$, restricting the allowed values of p to $p = 2,\ 3,\ 5,\ 11,\ 17$ and 41 [13, 21]. The case $p = 41$ defines Euler's formulae, $n^2 \pm n + 41$, that give distinct primes for the 40 consecutive integers $n = 0(1)$ to $39(40)$.

The variable part of the Euler formula

$$n^2 \pm n = k \tag{2.8}$$

is related to tetrahedral numbers, the golden ratio, twin primes and the periodic table.

Prime Spirals

Approximation [23] of the zth prime number by

$$p_z = z \ln z, \qquad \text{i.e.} \quad z = e^{p_z/z},$$

suggests that the position of a prime number may be represented by its phase angle on a logarithmic spiral of the form [22] $z = ke^{c\theta}$ ($c = \cot\varphi$). φ denotes the constant angle of the spiral with radii through the pole, and k is a scale factor. In a simplified demonstration, shown in Figure 2.10, the phase angle is chosen as z radians, such that the angular difference between successive primes is 1 radian and prime positions appear at points $p_z = kz$ on an Archimedean spiral. The positions of the first 100 primes are shown on smaller scale in Figure 2.11. The spiral that connects the points is not shown and a set of spiral arms shows up more conspicuously. This pattern is related to Fibonacci phyllotaxis that occurs in botanical growth and other logarithmic spirals in Nature with a divergence angle of $\chi = 137.5°$, rather than 1 radian as used in the demonstration. In radians, $\chi = 2\pi/\Phi^2$, where $\Phi = 1.6180\ldots$ is the golden number. Without further proof this relationship is interpreted to indicate a connection of the prime number distribution with the golden section and Fibonacci numbers.

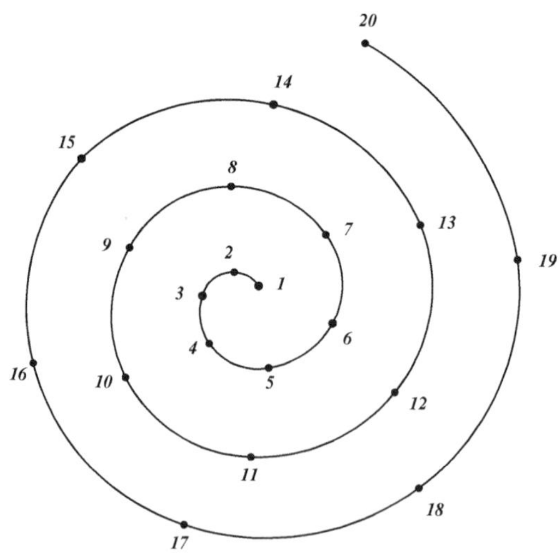

Figure 2.10: *Distribution of prime numbers on a spiral.*

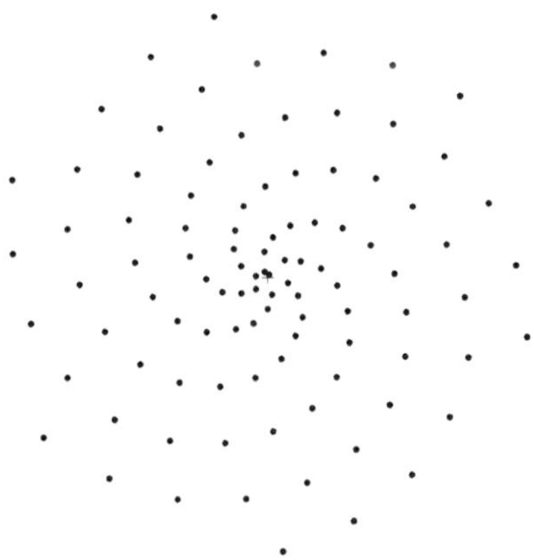

Figure 2.11: *The first 100 prime numbers distributed on a spiral.*

2.4 Fibonacci Numbers

Leonardo Fibonacci of Pisa in his book *Liber Abaci* (Book of the Abacus) published in 1202, posed the following question:

> How many pairs of rabbits will be produced in a year, beginning with a single pair, if in every month each pair bears a new pair which becomes productive from the second month on?

The problem is analyzed in Table 2.3: Each month some baby pairs grow up and must be added to the number of adult pairs of the previous month. Each adult pair of the previous month produces one baby pair so that the current number of baby pairs is the same as the number of adult pairs of the previous month. Except for the starting point, the three rows of Table 2.3 are alike, and each represents a *Fibonacci sequence*, consisting of the Fibonacci numbers,

$$F_0 = 0,\ F_1 = F_2 = 1,\ F_3 = 2,\ F_4 = 3,\ F_5 = 5,\ \ldots$$

in general

$$F_{n+1} = F_n + F_{n-1}\,;\ n > 0,\ F_1 = 1,\ F_2 = 1$$

This kind of *recursion* formula is cast into the *generalized Fibonacci sequence*

$$u_{n+1} = u_n + u_{n-1}, \quad u_1,\ u_2 \text{ to be chosen.}$$

The case $u_1 = 1$, $u_2 = 3$ is called the *Lucas sequence*, with Lucas numbers 1, 3, 4, 7, 11, 18, etc.

Table 2.3: *Rabbit breeding in Fibonacci sequence.*

	Months elapsed												
	Start	1	2	3	4	5	6	7	8	9	10	11	12
Adult pairs	1	1	2	3	5	8	13	21	34	55	89	144	233
Baby pairs	0	1	1	2	3	5	8	13	21	34	55	89	144
No. of pairs	1	2	3	5	8	13	21	34	55	89	144	233	377

2.4.1 The Golden Ratio

The most important feature of the Fibonacci sequence is that the ratio between successive Fibonacci numbers converges to an irrational number, known as the golden ratio, Φ, i.e.

$$\lim_{n\to\infty} \frac{F_{n+1}}{F_n} = \Phi = 1.61803\ldots$$

This irrational number has several unique properties, established by simple calculation, e.g.

$$\Phi = \frac{1}{\Phi} + 1 \quad (2.9)$$

$$\Phi^2 = \Phi + 1 \quad (2.10)$$

The golden ratio is an irrational number, but not a transcendental one (such as π) since it can be obtained as a solution to the polynomial Equation (2.10), i.e.

$$\begin{aligned} \Phi &= \frac{1 \pm \sqrt{1+4}}{2} \\ &= 1.61803\ldots \quad \text{or} \quad -0.61803\ldots \end{aligned}$$

The first of these is generally considered to be the golden ratio, whereas the second solution, $\tau = -\frac{1}{\Phi}$.

Any power of the golden ratio can be broken down into the sum of smaller powers, such as

$$\Phi^5 = \Phi^4 + \Phi^3$$

and ultimately reduced to the sum of an integer and an integer multiple of the golden ratio. Take

$$\begin{aligned} \Phi^6 &= \Phi^5 + \Phi^4 = 2\Phi^4 + \Phi^3 \\ &= 2(\Phi^3 + \Phi^2) + (\Phi^2 + \Phi) \\ &= 2(2\Phi^2 + \Phi) + (2\Phi + 1) \\ &= 8\Phi + 5 \end{aligned}$$

The two coefficients are successive Fibonacci numbers. In general

$$\Phi^n = F_n\Phi + F_{n-1}$$

The golden ratio can also be derived from trigonometric functions

$$\Phi = 2\sin\left(\frac{3\pi}{10}\right) = 2\cos\left(\frac{\pi}{5}\right)$$
$$\tau = 2\sin\left(\frac{\pi}{10}\right) = 2\cos\left(\frac{2\pi}{5}\right)$$

These angles are all fractions of $3\pi/5 = 108°$, characteristic of a regular pentagon. It is therefore not unexpected to find that the golden ratio turns up in examples of fivefold symmetry. An interesting example is provided by a regular decagon, radius R, unit sides, $\theta = \pi/5$, shown in Figure 2.12. Draw BD to bisect ABC. From similar triangles BCD and ABC it follows that

$$\frac{BD}{DC} = \frac{1}{R-1} = \frac{AB}{BC} = R$$

Hence $R^2 - R = 1$, i.e. $R = \Phi$.
Joining together pairs of congruent triangles gives rise to two geometrical shapes $ABCE$ and $ABDE$, known as *kite* and *dart*, the respective objects of Penrose tiling [25]. With these tiles a surface can be covered completely in an asymmetric and non-periodic way with small areas of five-fold symmetry occurring, as shown on the left in Figure 2.12.

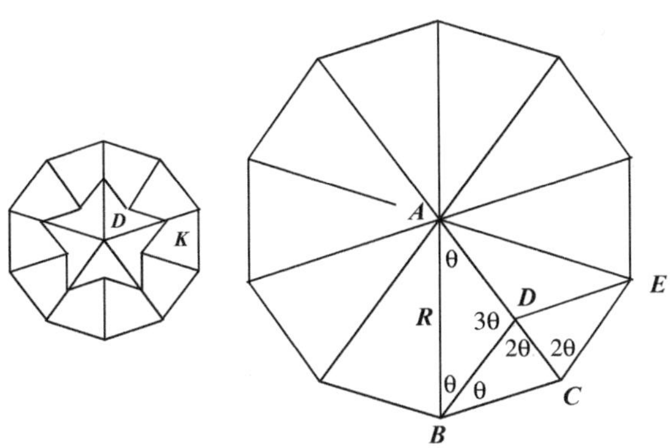

Figure 2.12: *Diagram to demonstrate the relationship between Φ, fivefold symmetry and the objects of Penrose tiling.*

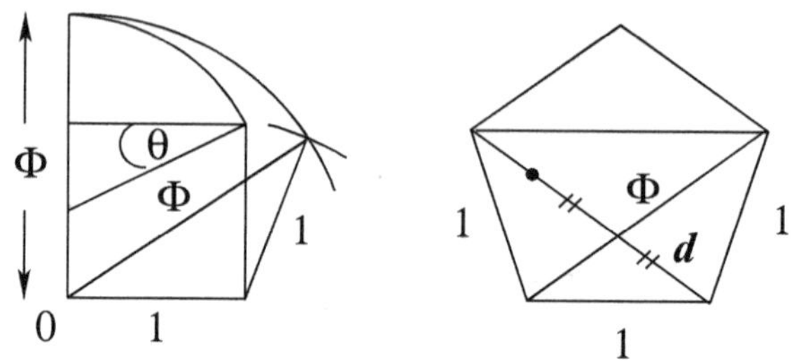

Figure 2.13: *Euclid's construction of the golden ratio.*

Euclid's Construction

In order to construct a regular pentagon, Euclid derived the golden ratio geometrically. The circle from the midpoint on one side of a unit square through an opposite corner, as in Figure 2.13, intersects the extension of that side at a distance ϕ. The unit pentagon is obtained by construction around a diagonal of length ϕ. The angle $\theta = \tan^{-1}(1/2) = 26.6°$.

Self-similarity

Kepler has been quoted [24] to write

> [T]he golden ratio served the creator as an idea when he introduced the continuous generation of similar objects from similar objects.

Adding the same length again to the short segment (d) of intersecting diagonals of the pentagon, its ratio to the remaining section

$$\frac{\phi - 2d}{2d} = \frac{1}{\phi}$$

again equals the golden ratio. This process may be continued indefinitely to generate a sequence of ever smaller copies of the original diagonal divided into two parts related by ϕ. The same process may be unfolded in the opposite direction to create larger copies. This property is called self-similarity. Setting $2d = 1$, the equation $\phi^2 - \phi = 1$ yields the value of the golden ratio, $\phi = (1 + \sqrt{5})/2$.

The property of self similarity of the golden section becomes evident from its definition as a continued fraction

$$x = 1 + \frac{1}{1 + \frac{1}{1 + \frac{1}{1 + \frac{1}{1 + \dots}}}} = 1 + \frac{1}{x}, \quad \text{i.e. } x = \Phi$$

Table 2.4: *Pascal triangle to show relationship with Fibonacci sequence.*

1										
	1	1								
		1	2	1						
			1	3	3	1				
				1	4	6	4	1		
					1	5	10	10	5	...
						1	6	15	20	...
							1	7	21	...
								1	8	...
									1	...
1	1	2	3	5	8	13	21	34	55	.

Self similarity as a symmetry is best known from fractal structures [29]. The chaotic nature of turbulent flow, for instance, is fractal and the self-similar symmetry that develops spontaneously can also be considered as related to the golden section.

To demonstrate without proof that the golden section and the Fibonacci sequence also relate to fractal structures the Pascal triangle of Table 2.1 is rearranged by shifting each row two positions to the right with respect to the previous row, to give the arrangement shown in Table 2.4. Summation over the columns generates the Fibonacci numbers.

2.4.2 Phyllotaxis and Growth

The spiral patterns in composite flower heads and cones, which almost invariably involve Fibonacci numbers, have been a topic of discussion for centuries. The mechanism leading to the appearance of Fibonacci numbers is well known [26] and accepted: successive leaves appear on the stem at an angle of about 137.5° ($360/\Phi^2$), the golden section of the unit circle, which is between 3/8 and 2/5 of a revolution. Thus five leaves make up just under two revolutions and eight leaves just over three revolutions. In a composite flower head, the individual florets are usually formed in leaf axils, resulting in the same rule of placement, but in a far more concentrated fashion. The number of opposing spirals, or helical whorls (parastichies) in many flower heads correspond [22, 27] to adjacent Fibonacci numbers, e.g. 21 and 34, 34 and 55, etc.

These growth spirals are often stated to be logarithmic spirals, like those that occur in shells and horns, as noted by D'Arcy Thomson [28]. However,

Figure 2.14: *Archimedean spiral through a set of points with divergence angle of 137.5°.*

the cell areas between flower heads do not increase exponentially and the angle between spiral and radius vector is not constant, but increases towards 90°.

Patterns of Fibonacci phyllotaxis are shown superimposed on the Archimedean spiral of Figure 2.14 in Figure 2.15.

The same set of points, generated by the divergence angle of 137.5° form the basis of all frames in Figure 2.15. The spiral arms are constructed by connecting points, separated in numerical sequence by the Fibonacci numbers 2, 3, 5 and 8. The fifth frame shows the five- and eightfold arms in superposition. The final frame represents $F_7 = 13$. It is clear that all Fibonacci spirals occur in each flower head. Their visibility is related to and may be masked by the number of florets.

2.5 Rational Fractions

Fair division of an integer number (a) of objects among a smaller number of recipients (b) is not always possible, i.e.

$$\frac{a}{b} = q + r\,,\ r \geq 0 \tag{2.11}$$

The remainder $r \neq 0$ whenever $b \nmid a$. However, in the field of *rational* numbers that contains an inverse element a^{-1} for each integer $a \neq 0$, such that $aa^{-1} = 1$, division, except by 0 is always possible. Some rational numbers ab^{-1} are integers $(r = 0)$. If not,

$$ab^{-1} = \frac{a}{b}$$

is called a *rational fraction*; a and b integer.

Figure 2.15: *Fibonacci spirals: From the top (left to right) the number of spiral arms corresponds to* F_3, F_4, F_5, F_6, $F_5 + F_6$ *and* F_7.

The ordering of rational fractions is more complicated than for integers. Firstly, *equality* is defined by the equivalence relation:

$$\frac{a}{b} = \frac{c}{d} \quad \text{is equivalent to} \quad ad = bc$$

It has the required properties of:

Reflexivity: $\frac{a}{b} = \frac{a}{b}$ since $ab = ba$.

Symmetry: If $\frac{a}{b} = \frac{c}{d}$ then $\frac{c}{d} = \frac{a}{b}$ ($ad = bc$ implies $cb = da$)

Transitivity: If $\frac{a}{b} = \frac{c}{d}$ and $\frac{c}{d} = \frac{e}{f}$, the $\frac{a}{b} = \frac{e}{f}$ because $ad = bc$, $cf = de$, and thus $adf = bcf = bde$. Therefore $af = be$, since $d \neq 0$.

Secondly, inequality:

$$\frac{a}{b} < \frac{c}{d} \quad \text{is equivalent to} \quad ad < dc.$$

This relation is

Irreflexive: $\frac{a}{b} \not< \frac{a}{b}$

Asymmetric If $\frac{a}{b} < \frac{c}{d}$, then $\frac{c}{d} \not< \frac{a}{b}$

Transitive: If $\frac{a}{b} < \frac{c}{d}$ and $\frac{c}{d} < \frac{e}{f}$, then $\frac{a}{b} < \frac{e}{f}$.

These properties only follow when the denominators are taken to be positive. Furthermore, the statement

$$\frac{a}{b} < \frac{c}{d} \quad \text{means the same as} \quad \frac{c}{d} > \frac{a}{b}.$$

Hence, among common fractions there exists a trichotomy: For two fractions $\frac{a}{b}$, $\frac{c}{d}$ one and only one of the following statements can be true:

$$\frac{a}{b} < \frac{c}{d}, \frac{a}{b} = \frac{c}{d}, \frac{a}{b} > \frac{c}{d}.$$

Indeed, among integers, either $ad < bc$ or $ad = bc$ or $ad > bc$. Because of the transitivity of equality and inequality, an ordering of the fractions becomes possible.

2.5.1 The Farey Sequence

An ordered sequence of non-negative reduced fractions between 0 and 1 whose denominators do not exceed the number N was discovered by Farey in 1816. The numerator and denominator of a reduced fraction have no common divider. The number N determines the *order* of the sequence. The low-order Farey sequences are

```
0/1                                                                1/1
                                     1/2
                1/3                                   2/3
       1/4                 2/5                 3/5                3/4
   1/5        2/7      3/8       3/7      4/7       5/8      5/7       4/5
1/6  2/9  3/11 3/10 4/11 3/13 5/12 4/9  5/9  7/12 8/13 7/11 7/10 8/11 7/9  5/6
```

Figure 2.16: *Farey sequence of rational fractions. Starting with the first row and reading from right to left 3/7 is identified as the 12th rational fraction.*

$$N = 1: \quad \frac{0}{1}, \frac{1}{1}$$

$$N = 2: \quad \frac{0}{1}, \frac{1}{2}, \frac{1}{1}$$

$$N = 3: \quad \frac{0}{1}, \frac{1}{3}, \frac{1}{2}, \frac{2}{3}, \frac{1}{1}$$

$$N = 4: \quad \frac{0}{1}, \frac{1}{4}, \frac{1}{3}, \frac{1}{2}, \frac{2}{3}, \frac{3}{4}, \frac{1}{1}$$

$$N = 5: \quad \frac{0}{1}, \frac{1}{5}, \frac{1}{4}, \frac{1}{3}, \frac{2}{5}, \frac{1}{2}, \frac{3}{5}, \frac{2}{3}, \frac{3}{4}, \frac{4}{5}, \frac{1}{1}$$

A generalized table that serves to order and enumerate all rational fractions, starting with 0/1 and 1/1, is obtained on generating each rational number from two that brace it from above, by separate addition of numerators and denominators of the bracing pair. The result is shown in Figure 2.16.

Farey established that two consecutive fractions h/k and l/m obey the relationship

$$hm = kl - 1 \quad \text{or} \quad \begin{vmatrix} h & l \\ k & m \end{vmatrix} = -1$$

This relationship, at the basis of (1.6), and other relationships were proved by Cauchy [4], using mathematical induction. Because of this relationship any Diophantine equation[6] of the form

$$ax + by = 1, \quad (a, b) = 1, \tag{2.12}$$

is solvable.

Proof Assume $0 < a < b$. Then, since $(a, b) = 1$, $\frac{a}{b}$ is a reduced fraction that occurs in some Farey sequence (e.g. of order b). Let $\frac{h}{k} < \frac{a}{b}$ be an adjacent pair, such that $ak - bh = 1$. Therefore, $x = k$, $y = -h$ solves the equation. As a corollary the equation

$$ax + by = c \tag{2.13}$$

[6] An equation satisfied by integers.

written in the form $a(x_0c) + b(y_0c) = c$, has the solution $x = x_0c$, $y = y_0c$. This result leads immediately to Euclid's lemma:
If a and b are coprime and $a|bc$ then $a|c$.

Let $d = (a, b)$ be the GCD of a and b. Then $(a/d, b/d) = 1$. If not, $(a/d, b/d) = \delta$ and hence $d.\delta$ divides both a and b. Because of the maximality of d it follows that $\delta = 1$. Using (2.12) integers x and y can be found such that

$$\frac{a}{d}x + \frac{b}{d}y = 1$$

and thus $ax + by = d$. The GCD of a and b has been shown to be a linear combination of a and b with integer coefficients. In terms of the LCM $[a, b]$ similar arguments can be used to show that $ab = (a, b)[a, b]$.

2.6 Modular Arithmetic

Modular, or Gaussian arithmetic is not unlike regular, or Euclidean arithmetic. Whereas Euclidean arithmetic operates on the infinite set of all integers, Gaussian (modulo) arithmetic works with sets of *residue classes*. In regular arithmetic $ab = 0$ is only possible with either a or b equal to zero, but not so in modular arithmetic that involves congruence relations.

2.6.1 Congruences

Let a, b, m be integers, m positive. It is said that a is congruent to b modulo m, $a \equiv b \pmod{m}$ if $m|(a-b)$. The congruence relation ($\equiv$) is an *equivalence* relation on the set of integers, and therefore is,

$$\text{Reflexive} : a \equiv a \pmod{m}$$

$$\text{Symmetric} : \text{if } a \equiv b \pmod{m} \text{ then } b \equiv a \pmod{m}$$

$$\text{Transitive} : \text{if } a \equiv b \pmod{m} \text{ and } b \equiv c \pmod{m} \text{ then } a \equiv c \pmod{m}$$

The condition $a \equiv a \pmod{m}$ simply means $m|(a - a)$.
If $a \equiv b \pmod{m}$, then $m|(a - b)$, so that $m|(b - a)$ and $b \equiv a \pmod{m}$. Finally, if $a \equiv a \pmod{m}$ and $b \equiv c \pmod{m}$ then $m|(a - b)$ and $m|(b - c)$ so $m|(a - b) + (b - c)$, $m|(a - c)$ and $a \equiv c \pmod{m}$. Thus congruence is an equivalence relation, and the set of integers is partitioned into disjoint classes. Any two integers in the same class are congruent to one another, and no two integers in distinct classes are congruent to one another.

According to (2.11) every integer leaves a remainder r, $0 \leq r < m$, on division by m. It can be shown that a and b are congruent modulo m iff they leave the same remainder on division by m. First suppose

$$a = qm + r, \quad b = q'm + r, \quad 0 \leq r < m$$

Then

$$a - b = (q - q')m, \quad \pm m|(a - b)$$

so that $m|(a - b)$. Conversely suppose $a \equiv b \pmod{m}$. Let

$$a = qm + r\,,\, b = q'm + r\,,\, 0 \leq r < m\,,\, 0 \leq r' < m$$

Then $(a - b) = (q - q')m + (r - r')$. Since $m|(a - b)$, $m|(r - r')$. But $-m < r - r' < m$, so $m \nmid (r - r')$, unless $r = r'$.

Congruence Classes

To estimate the number of congruence classes it is noted that the integers $0, 1, \ldots, m - 1$ lie in different classes. Given any integer some suitable multiple of m may be added or subtracted to arrive at one of the integers $0, 1, \ldots, m - 1$. Thus there are just m congruence classes modulo m, and the set of integers $0, 1, \ldots, m-1$ form a set of representations, one from each class.

The congruence classes of integers modulo m are said to provide a *partition* of the set of integers. The significance of this result is that, given two integers a and b, their congruence classes $[a]_m$ and $[b]_m$ are either equal or disjoint, summarized by the reciprocal implications ($\Leftrightarrow$),

$$a \equiv b \pmod{m} \Leftrightarrow [a]_m = [b]_m$$
$$a \not\equiv b \pmod{m} \leftrightarrow [a]_m \cap [b]_m = \varnothing$$

$\varnothing$ represents an empty set.

Arithmetic Operations

Congruences behave like equalities with respect to addition and multiplication, and obey the following rules:

(a) If $a \equiv b \pmod{m}$ then $b \equiv a \pmod{m}$.

(b) If $a \equiv b \pmod{m}$ and $b \equiv c \pmod{m}$, then $a \equiv c \pmod{m}$.

(c) If $a \equiv b \pmod{m}$, then $ka \equiv kb \pmod{m}$ for any integer k.

(d) If $a_i \equiv b_i \pmod{m}$ for $i = 1, 2, \dots, n$, then

$$a_1 + a_2 + \cdots + a_n \equiv b_1 + b_2 + \cdots + b_n \pmod{m}$$

$$a_1 a_2 \dots a_n \equiv b_1 b_2 \dots b_n \pmod{m}$$

To prove the additive rule for $a \equiv b \pmod{m}$ and $c \equiv d \pmod{m}$ it is noted that if $m|(a-b)$ and $m|(c-d)$ then $m|[(a-b)+(c-d)]$ or $m|[(a-c) - (b+d)]$. Hence $a + c \equiv b + d \pmod{m}$. The multiplication rule follows from the observation that $m|(a-b)c$ implies $ac \equiv bc \pmod{m}$ and that $m|b(c-d)$ implies $bc \equiv bd \pmod{m}$. Then $ac \equiv bd \pmod{m}$ follows through the transitivity of the congruence relation. The last part of (iv) is verified for $n = 2$ from which the general case follows by repeated application. By (iii) $a_1 a_2 \equiv b_1 a_2 \pmod{m}$, $b_1 a_2 \equiv b_1 b_2 \pmod{m}$ so that by (ii) $a_1 a_2 \equiv b_1 b_2 \pmod{m}$.

Unlike addition, subtraction and multiplication, congruences cannot be divided unrestrictedly. Compare

$$2 \equiv 12 \pmod{10} \qquad 1 \not\equiv 6 \pmod{10}$$

although

$$2 \equiv 24 \pmod{11} \qquad 1 \equiv 12 \pmod{11}$$

In fact, if $ab \equiv ac \pmod{m}$ and $(a, m) = 1$, then $b \equiv c \pmod{m}$. For if $m|(ab - ac)$ then $m|a(b-c)$ and it follows from Euclid's lemma that $m|(b-c)$. Hence

(e) If $ka \equiv kb \pmod{m}$, then $a \equiv b \pmod{\frac{m}{d}}$, where $d = (k, m)$. In particular, $a \equiv b \pmod{m}$ if k and m are coprime, $d = 1$.

A set of integers is a complete residue (or *remainder* system modulo m) iff it consists of exactly m integers, no two of which are congruent modulo m.

Theorem If $a_1, a_2, \dots, a_m$ form a complete residue system modulo m, and if $(a, m) = 1$, then $aa_1, aa_2, \dots, aa_m$ also form such a system.
For $aa_i \equiv aa_j \pmod{m}$, then $a_i \equiv a_j \pmod{m}$ by property (e).

Fermat's Theorem If p is a prime and $(a, p) = 1$, then $a^{p-1} \equiv 1 \pmod{p}$. The numbers $0, 1, 2, \dots, p-1$ form a complete residue system modulo p. Hence $0, a, 2a, \dots, (p-1)a$ do also, by the preceding theorem. Now each number on the list is congruent to exactly one on the other list. Omitting 0 from each list, since zeros correspond, it follows from (d) that

$$a.2a \dots (p-1)a \equiv 1.2 \dots (p-1) \pmod{p}$$

or

$$(p-1)!a^{p-1} \equiv (p-1)! \pmod p$$

Property (e) allows $(p-1)!$ to be divided out. Hence

$$a^{p-1} \equiv 1 \pmod p \tag{2.14}$$

Corollary If p is prime, then $a^p \equiv a \pmod p$, for any integer a.

Euler's Function

Whereas the number of congruence classes modulo m is equal to m, the number of classes with elements prime to m is designated $\varphi(m)$ and is called Euler's function. For a prime number p it is clear that $\varphi(p) = p - 1$. For $m = p^\alpha$, $\varphi(p^\alpha) = p^\alpha - p^{\alpha-1}$.

The Euler function $\varphi(n)$ has the property

$$n = \sum_{d|n} \varphi(d) \tag{2.15}$$

where the summation extends over all divisors d of n.

Proof Write down all fractions l/n of fixed denominator n, $0 \leq l/n \leq 1$, whether reduced or not:

$$\frac{1}{n}, \frac{2}{n}, \frac{3}{n}, \ldots, \frac{n-1}{n}, \frac{n}{n}$$

They are n in number. Some, as $1/n$ and $(n-1)/n$ are reduced in form. In the others common divisors of numerator and denominator may be cancelled. The resulting reduced denominators are then divisors d of n; for each divisor d of n all reduced proper fractions of denominator d will appear in the list. They are φ in number. Counting the fractions according to their reduced denominators then proves the theorem.

2.6.2 Higher Congruences

Solvability of the Diophantine equation $ax + by = c$, $(a, b) = 1$, implies solvability of the linear congruence $ax \equiv c \pmod m$, with $(a, m) = 1$, suggesting that congruences may be considered in much the same way as algebraic equations. Consider

$$f(x) = a_0x^n + a_1^{n-1} + \cdots + a_n \equiv 0 \pmod p, \quad (a_0, p) = 1 \tag{2.16}$$

where p is prime. Since $x^p \equiv x \pmod{p}$ for all x it is possible to eliminate all powers $x^p, x^{p-1}, x^{p-2}, \ldots,$ by replacing them by lower powers of x. It may therefore be assumed that $n < p$ in all of these congruences. Such a congruence need not have a solution. For example, $x^2 \equiv 3 \pmod{7}$ has no solution since $0^2 \equiv 0$, $1^2 \equiv 1$, $2^2 \equiv 4$, $3^2 \equiv 2$, $4^2 \equiv 2$, $5^2 \equiv 4$, and $6^2 \equiv 1 \pmod{7}$. Just as in the theory of algebraic equations, a congruence of degree n and prime number modulus p can have at most n solutions. This statement holds if (2.16) has no solutions. Suppose it has a solution x_1. Then

$$a_0 x_1^n + a_1 x_1^{n-1} + \cdots + a_n \equiv 0 \pmod{p}$$

Subtract this congruence from (2.16) to give

$$a_0(x^n - x_1^n) + a_1(x^{n-1} - x_1^{n-1}) + \cdots + a_{n-1}(x - x_1) \equiv 0 \pmod{p} \quad (2.17)$$

which any x satisfying (2.16) must also satisfy. The congruence (2.17), however, may be written as

$$(x - x_1) \cdot (a_0 x^{n-1} + b_1 x^{n-2} + \cdots + b_{n-1}) \equiv 0 \pmod{p} \quad (2.18)$$

where the b's are expressions obtained from x_1 and the a's. Since p divides the product in (2.18), it must divide one of the factors. Any x that satisfies (2.16) must therefore satisfy either $x - x_1 \equiv 0 \pmod{p}$ or

$$a_0 x^{n-1} + b_1 x^{n-2} + \cdots + b_{n-1} \equiv 0 \pmod{p}$$

The first alternative yields again x_1. The second may or may not yield a solution. That second congruence is of degree $(n-1)$, the highest coefficient is again a_0, and since the theory holds for the first-degree congruence $a_0 x + f_1 \equiv 0 \pmod{p}$, induction proves the

Theorem The number of solutions of the congruence (2.16) is at most n for any degree n.

Primitive Congruence Roots

A congruence of degree n modulo a prime number p cannot have more than n solutions modulo p. This maximal number can be attained as shown by the example

$$x^{p-1} - 1 \equiv 0 \pmod{p}$$

with solutions $x \equiv 1, 2, \ldots, (p-1)$.

It follows that if $d | p-1$ then $x^d - 1 \equiv 0 \pmod{p}$ has d solutions. From $p - 1 = md$ follows the identity

$$x^{p-1} - 1 = \left(x^d - 1\right)\left(x^{(m-1)d} + x^{(m-2)d} + \cdots + 1\right)$$

Now, the congruence $x^d - 1 \equiv 0 \pmod{p}$ has at most d solutions, and the congruence $x^{(m-1)d} + x^{(m-2))d} + \cdots + 1 \equiv 0 \pmod{p}$ has at most $(m-1)d$ solutions. If $x^d - 1 \equiv 0 \pmod{p}$ has less than d solutions, then $x^{p-1} - 1 \equiv 0 \pmod{p}$ would have less than $d + (m-1)d = p-1$ solutions, which is not the case. Thus $x^d - 1 \equiv 0 \pmod{p}$ does indeed have d solutions.

If $\delta | d$, then $x^\delta - 1$ divides $x^d - 1$ algebraically. Thus any solution of $x^\delta - 1 \equiv 0 \pmod{p}$ is also a solution of $x^d - 1 \equiv 0 \pmod{p}$. The solution x_0 of $x^{p-1} - 1 \equiv 0 \pmod{p}$ is said to *belong to the exponent* d if it is a solution of $x^d - 1 \equiv 0 \pmod{p}$, but is not a solution of $x^\delta - 1 \equiv 0 \pmod{p}$ for any $\delta < d$. It is also said that x_0 is a *primitive solution* of $x^d - 1 \equiv 0 \pmod{p}$. If a solution belongs to the exponent d, then necessarily $d|(p-1)$. For $e = (d, p-1)$, then $e = md + r(p-1)$ for suitable integers m and r, since the GCD of two numbers is a linear combination of these. Therefore $x^e \equiv (x^d)^m (x^{p-1})^r \equiv 1 \cdot 1 \equiv 1 \pmod{p}$. Since $e \leq d$, it follows from the minimality of d that $e = d$ and hence $d|(p-1)$. The solutions of $x^{p-1} - 1 \equiv 0 \pmod{p}$ may be separated into classes, each class containing those solutions that belong to the exponent d. Only the divisors d of $(p-1)$ need to be considered.

Let $\psi(d)$ denote the solutions that belong to the exponent d, that is the number of primitive solutions of $x^d - 1 \equiv 0 \pmod{p}$. It can be shown that $\psi(d) = \varphi(d)$, Euler's function.

Proof The statement is true for $d = 1, 2$. For $\varphi(1) = 1$, and the congruence $x - 1 \equiv 0 \pmod{p}$ has the unique solution $x \equiv 1$. Also $\varphi(2) = 1$, and the congruence $x^2 - 1 = (x-1)(x+2) \equiv 0 \pmod{p}$ has the unique primitive solution $x \equiv p-1$, to be proved by induction. Suppose $\psi(\delta) = \varphi(\delta)$ for all $\delta < d$. Since every solution of $x^d - 1 \equiv 0 \pmod{p}$ is a primitive solution of $x^\delta - 1 \pmod{p}$ if $\delta | d$, then

$$d = \sum_{\delta | d} \psi(\delta)$$

By induction, $\psi(\delta) = \varphi(\delta)$ for all divisors δ of d, except perhaps d itself, one has

$$d = \sum_{\delta | d} \varphi(\delta) + \psi(d) - \varphi(d)$$

By definition of Euler's function (2.15) however, $d = \sum_{\delta | d} \varphi(\delta)$. Thus $\psi(d) - \varphi(d) = 0$, as required.

As a corollary the primitive solutions of $x^d - 1 \equiv 0 \pmod{p}$ are seen to exist for all $d|(p-1)$. In particular, there exists a primitive solution to the congruence $x^{p-1} - 1 \equiv 0 \pmod{p}$ and the theorem insures that there exist

$\varphi(p-1)$ primitive solutions. A primitive solution of $x^{p-1} \equiv 0 \pmod{p}$ is called a primitive root modulo p.

The regular 17-gon

The theory of primitive congruence roots applies to the problem of cyclotomy: division of the circumference of a circle in equal parts. This was first achieved by Gauss and the most spectacular case, the regular heptadecagon is discussed in detail by Rademacher [4]; a condensed version of which is presented here. The geometrical procedure to construct the 17-gon by means of ruler and compass only, based on the analysis of Gauss, is described by Coxeter [22]. The regular 17-gon is considered as an object in the complex plane, as shown in Figure 2.17. The vertices are complex numbers inscribed in the unit circle about the origin and with one vertex at the point of the complex number 1. Any other vertex is at $z = x + iy$. Since $|z| = 1$ each $z_j = \cos\theta_j + i\sin\theta_j$, $j = 1, 2, \ldots, 16$. The vertices divide the circle into equal parts, so that $\theta_j = j\theta_1$, where $\theta_1 = (2\pi/17)$. The 17th vertex is again at 1 and hence, by (2.5)

$$1 = \cos 17\theta_1 + i\sin 17\theta_1 = (\cos\theta_1 + i\sin\theta_1)^{17} = z_1^{17}$$

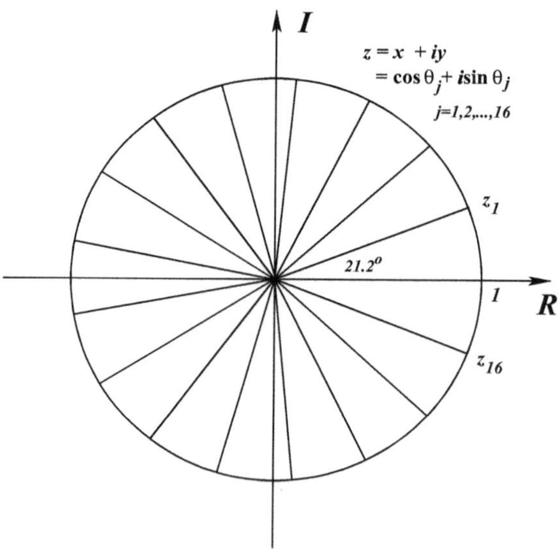

Figure 2.17: *Regular 17-gon in the complex plane.*

Table 2.5: *Table of* 10^n (mod 17) *residues.*

n	0	1	2	3	4	5	6	7	8	9	10	11	12	13	14	15
$10^n(17)$	1	10	15	14	4	6	9	5	16	7	2	3	13	11	8	12

Thus z_1 is a root of the equation $z^{17} - 1 = 0$. Since

$$z^{17} - 1 = (z-1)(z^{16} + z^{15} + \cdots + z + 1) \text{ and since } \quad z_1 \neq 1,$$

it follows that z_1 is a root of

$$z^{16} + z^{15} + \cdots + z + 1 = 0 \tag{2.19}$$

The problem reduces to finding a solution to this equation.

Since 17 is prime, there exist primitive roots modulo 17. For example, 10 is a primitive root (mod 17) since $10^{16} \equiv 1 \pmod{17}$, but no lower power of 10 is congruent to 1 modulo 17. Table 2.5 lists the values of $10^n \pmod{17}$ for specific powers of n. Each residue which is prime to 17 occurs once on the bottom row, which constitutes a cyclic group. When the exponents of (2.19) are ordered according to the order of the residue table, it becomes

$$z + z^{10} + x^{15} + z^{14} + \cdots + z^8 + z^{12} = 1 \tag{2.20}$$

where the exponents are successive powers of (mod 17).

The method introduced by Gauss breaks this equation up into two sums, called *periods*, by grouping together even and odd entries, respectively:

$$\eta_1 = z + z^{15} + z^4 + z^9 + x^{16} + z^2 + z^{13}$$
$$\eta^2 = z^{10} + z^{14} + z^6 + z^5 + z^7 + z^3 + z^{11} + z^{12}$$
$$\eta_1 + \eta_2 = -1$$

The desired solution consists of finding the 64 terms of the product $\eta_1\eta_2$. As a first step each element of η_1 is multiplied by the element of η_2 directly below it. The successive products are $zz^{10} = z^{11}$, $z^{15}z^{14} = z^{29}$, etc. After reducing the exponents modulo 17, the sum of the products

$$z^{11} + z^{12} + z^{10} + z^{14} + z^6 + z^5 + z^7 + z^3$$

corresponds again to η_2. The reason for this is that successive exponents of z in (2.19) differ by a factor of 10 (mod 17). Therefore successive exponents in η_1 and η_2 differ by a factor of 100 modulo 17. A set of products obtained by cyclic selection of factors from both η_1 and η_2 will always yield a sum modulo 17 corresponding to either η_1 or η_2. Eventually it is found that

$$\eta_1\eta_2 = 4\eta_1 + 4\eta_2 = -4$$

From these relations a quadratic equation with the roots η_1 and η_2 can be set up as

$$(y - \eta_1)(y - \eta_2) = 0$$

i.e.

$$y^2 + y - 4 = 0$$

It has roots[7]

$$\eta_1\,,\,\eta_2 = \frac{1}{2}\left(-1 \pm \sqrt{1+16}\right) = \frac{1}{2}\left(-1 \pm \sqrt{17}\right)$$

This procedure is continued by forming new periods such that

$$\eta_1' + \eta_2' = \eta_1 \qquad \eta_3' + \eta_4' = \eta_2$$

and $\eta_1'' + \eta_2'' = \eta_1'$, etc., ending up with the equation satisfied by z and z^{16},

$$z^2 + \eta_1'' z + 1 = 0$$

$$\eta_1'' = \frac{1}{8}\left\{-1 + \sqrt{17} + \sqrt{2(17 - \sqrt{17})}\right.$$
$$\left. + 2\sqrt{17 + 3\sqrt{17} - \sqrt{2(17 - \sqrt{17})} - 2\sqrt{2(17 - \sqrt{17})}}\right\}$$

which is a *real* quantity. Since $\eta_1'' = z + z^{16} = z + z^1 = 2\cos\theta$, where θ is the central angle of the 17-gon.

The success of the Gauss construction depends on the fact that $16 = 17 - 1$ is a power of 2. This fact allows the periods to be halved over and over, and so to reduce the solution of the cyclotomic equation of degree 16 to the solution of a sequence of quadratic equations. If p is a prime of the form $2^k + 1$, the method may be used to construct the regular p-gon with ruler and compass. For the number to be prime k must be a power of 2, say $k = 2^n$ and suitable primes would be of the form $2^{2^n} + 1$. Examples of such (Fermat) primes are

$$P_0 = 3 = 2^1 + 1$$

$$P_1 = 5 = 2^2 + 1$$

[7] In general, for a p-gon η_1 and η_2 are quadratic irrationalities and $\pm p$ appears under the radical. It will be shown later that this is a Gaussian sum.

$P_2 = 17 = 2^4 + 1$

$P_3 = 257 = 2^8 + 1$

$P_4 = 65537 = 2^{16} + 1$

$P_5 = 2^{32} + 1$ is not a prime, being divisible by 641.

2.6.3 Partitions and Equivalence Relations

The relation of congruence modulo m partitions the set of integers $\mathbb{Z}$ into m disjoint congruence classes. This is an example of a more general concept, known as an *equivalence relation*. An equivalence relation on a set corresponds to a partition of the set [16].

Partitions

The power of a set X, denoted by $\mathcal{P}(X)$, is the set of all subsets of X. A *partition* of the set X is a subset Π of $\mathcal{P}(X)$, i.e. a set of subsets of X, such that:

(a) the subsets in Π are non-empty, i.e. $A \in \Pi \Rightarrow A \neq \varnothing$,

(b) the subsets in Π are disjoint, i.e. $\forall A_1 A_2 \in \Pi(A_1 \neq A_2) \Rightarrow A_1 \cap A_2 = \varnothing$,

(c) the subsets in Π cover $X : \forall x \in X, \exists A \in \Pi, x \in A$. (For all x in X, for some subset A in Π, x is an element of A.)

The set of congruence classes modulo m

$$\mathbb{Z} = \{[a]_m | a \in \mathbb{Z}\} = \{[a]_m | a \in R_m$$

is a partition of $\mathbb{Z}$ into m subsets.

Two elements of a set that are in the same subset of the partition are said to be *related*, written as $a \sim b$, that constitutes an equivalence relation.

2.7 Periodic Arithmetic Functions

Gauss's construction shows that all cyclotomic equations

$$x^{p-1} + x^{p-2} + \cdots + x + 1 = 0\,,\ p \text{ prime}$$

can be solved by rational operations and successive extraction of roots. A solution of the algebraic equation

$$x^n - 1 = 0 \tag{2.21}$$

(compare Eq. 2.6) is called the nth root of unity or a root of unity of order n. The number 1 is a root of unity of any order, or a *trivial* root. Some roots of (2.21) are not roots of $x^k - 1 = 0$ with $k < n$. Such roots are called *primitive roots of unity.* Any root ζ of (2.21) is a primitive root of some equation $x^k - 1 = 0$, $0 < k \leq n$. In particular, any root of (2.21) is a primitive root of unity of order d, where $d|n$.

Let ζ_j, $j = 1, 2, \ldots, v$ be all the primitive roots of (2.21), to define the cyclotomic polynomial $F_n(x)$ of order n, as

$$F_n(x) = \prod_j (x - \zeta_j)$$

$F_n(x)$ will be a monic polynomial.[8] Evidently

$$x^n - 1 = \prod_{d|n} F_d(x) \tag{2.22}$$

For instance, $F_1(x) = x - 1$, $x^2 - 1 = F_1(x) \cdot F_2(x)$, $F_2(x) = x + 1$. It may be proved by induction that the cyclotomic polynomial $F_n(x)$ of order n is a monic polynomial of degree $\varphi(n)$ with integer coefficients. The theorem which holds for $n = 1, 2$ is next assumed valid for all $F_k(n)$, $k < n$. Now

$$x^n - 1 = F_n(x) \cdot \prod_{d(<n)|n} F_d(x) = F_n(x) \cdot G_n(x), \text{say}$$

But, because $d < n$, $G_n(x)$ is a product of monic polynomials with integer coefficients, hence it is also monic with integer coefficients. Then

$$F_n(x) = \frac{x^n - 1}{G_n(x)} \tag{2.23}$$

Since the divisor has highest coefficient 1 long division procuces only integer coefficients.

Let v be the degree of $F_n(x)$ and assume the degree $\varphi(d)$ for $F_d(x)$, $d < n$. The total number of classes modulo n, i.e.

$$n = v + \sum_{d(<n)|n} \varphi(d)$$

[8]Polynomial with leading coefficient, $a_n = 1$.

Since every solution of $x^n - 1 \equiv 0 \pmod{p}$ is a primitive solution of $x^d - 1 \equiv 0 \pmod{p}$

$$n = \sum_{d|n} \varphi(d)$$

By induction every solution of d is also a solution of n, except perhaps n itself,

$$n = v - \varphi(n) + \sum_{d|n} \varphi(d)$$

Since, by definition $n = \sum_{d|n} \varphi(d)$, the degree of $F_n(x)$ follows as $v = \varphi(n)$. By long division it follows that

$$F_3(x) = \frac{x^3 - 1}{F_1(x)} = x^2 + x + 1$$
$$F_4(x) = \frac{x^4 - 1}{F_1(x)F_2(x)} = x^2 + 1$$
$$F_5(x) = \frac{x^5 - 1}{F_1(x)} = x^4 + x^3 + x^2 + x + 1$$
$$F_6(x) = \frac{x_6 - 1}{F_1(x)F_2(x)F_3(x)} = \frac{x^6 - 1}{x^4 + x^3 - x - 1} = x^2 - x + 1$$

etc.

2.7.1 The Lagrange Resolvent

For any prime p, by (2.23)

$$x^p - 1 = F_1(x) \cdot F_p(x) = (x - 1)F_p(x)$$

i.e.

$$F_p(x) = x^{p-1} + x^{p-2} + \cdots + x + 1$$

Let ζ be a primitive root of unity of order p. Then $F_p(\zeta) = 0$ or $\zeta + \zeta^2 + \cdots + \zeta^r = -1$, where $r = p - 1$. Now take the primitive congruence root g modulo p. Then $g^r \equiv 1 \pmod{p}$ and no lower power of g can be congruent to 1 modulo p. The $r = p - 1$ numbers $g^0, g^1, \ldots, g^{r-1}$ are congruent modulo p to some permutation of the numbers $1, 2, \ldots, r$, and since the exponent of ζ counts only modulo p, the previous equation may be written

$$\zeta^{g^0} + \zeta^{g^1} + \cdots + \zeta^{g^{r-1}} = -1$$

This equation is written in the form of a Lagrange resolvent as

$$(1, \zeta) = -1$$

A Lagrange resolvent is defined in terms of an rth root of unity ρ, not necessarily primitive, as the sum

$$(\rho,\zeta) = \zeta + \rho\zeta^g + \rho^2\zeta^{g^2} + \cdots + \rho^{r-1}\zeta^{g^{r-1}}$$

Take a prime number $p \equiv 1 \pmod 4$ so that $r = p - 1 \equiv 0 \pmod 4$. The numbers ± 1, $\pm i$ are the 4th roots of unity and therefore also rth roots of unity. Four Lagrange resolvents are defined, in this notation, for the primitive root ζ of unity of order p:

$$\begin{aligned}
(1,\zeta) &= \zeta + \zeta^g + \zeta^{g^2} + \cdots + \zeta^{g^{r-1}} = -1\\
(-1,\zeta) &= \zeta - \zeta^g + \zeta^{g^2} - \cdots - \zeta^{g^{r-1}}\\
(1,\zeta) &= \zeta + i\zeta^g - \zeta^{g^2} - \cdots - i\zeta^{g^{r-1}}\\
(-i,\zeta) &= \zeta - i\zeta^g - \zeta^{g^2} + \cdots + i\zeta^{g^{r-1}}
\end{aligned}$$

2.7.2 Gaussian Sums

The sum of primitive roots of unity used in the theory of the regular 17-gon corresponds to $(1,\zeta) = -1$, broken down into the two sums

$$\begin{aligned}
\eta_1 &= \zeta^{g^0} + \zeta^{g^2} + \cdots + \zeta^{g^{r-2}}\\
\eta_2 &= \zeta^{g^1} + \zeta^{g^3} + \cdots + \zeta^{g^{r-1}}
\end{aligned}$$

where $r = p - 1$, an even number. In terms of Lagrange resolvents, therefore

$$\begin{aligned}
\eta_1 + \eta_2 &= (1,\zeta)\\
\eta_1 - \eta_2 &= (-1,\zeta)
\end{aligned}$$

The sum η_1 only shows exponents which are squares: $1 = g^0, g^2, \ldots, g^{r-2}$. The summands in η_2 show exponents which are not squares and not congruent to squares modulo p. Indeed, η_1 has all the summands which are congruent to a square modulo p, (except 0). For a square t^2, $p \nmid t$, there exists u such that

$$g^u \equiv t^2 \pmod p$$

An exponent v can be found so that

$$\begin{aligned}
t &\equiv g^v \pmod p\\
g^u &\equiv g^{2v} \pmod p
\end{aligned}$$

Since g is a primitive congruence root modulo p, this congruence implies $u \equiv 2v \pmod r$ which shows that u is even since r is even. The numbers that are congruent to a square modulo p are called *quadratic residues modulo* p, i.e. the numbers $1, g^2, g^4, \ldots, g^{r-2} \pmod{p}$; the others, i.e. $g, g^3, \ldots, g^{r-1}$ are called quadratic *nonresidues* modulo p. The number of quadratic residues is equal to the number of quadratic nonresidues, and that number is $r/2 = (p-1)/2$.

In $(-1, \zeta)$ the powers with quadratic residues as exponents have a plus sign, those with quadratic nonresidues a minus sign. This difference is reflected by the *Legendre symbol*, defined as

$$\left(\frac{m}{p}\right) = \begin{cases} +1 & \text{if } m \text{ is a quadratic residue modulo } p \\ -1 & \text{if } m \text{ is a quadratic nonresidue modulo } p \end{cases}$$

With $p \nmid m$, the special Lagrange resolvent

$$(-1, \zeta) = \sum_{m=1}^{p-1} \left(\frac{m}{p}\right) \zeta^m$$

is called a Gaussian sum, $G(\zeta)$. If also $\left(\frac{m}{p}\right) = 0$ for $p|m$, then, more generally

$$G(\zeta) = (-1, \zeta) = \sum_{m=0}^{p-1} \left(\frac{m}{p}\right) \zeta^m$$

This definition is further generalized by writing for any integer t

$$G(\zeta^t) = \sum_{m=0}^{p-1} \left(\frac{m}{p}\right) \zeta^{tm} \tag{2.24}$$

For $t \equiv 0 \pmod{p}$

$$G(\zeta^0) = \sum_{m=0}^{p-1} \left(\frac{m}{p}\right) = 0$$

since there are as many quadratic residues as nonresidues. For $t \not\equiv 0 \pmod{p}$ choose t' such that $tt' \equiv 1 \pmod{p}$. Put $tm \equiv m' \pmod{p}$ so that $m \equiv m't'$.

Since m' runs with m through a full residue system modulo p, from (2.24)

$$G(\zeta^t) \equiv \sum_{m' \pmod{p}} \left(\frac{m't'}{p}\right) \zeta^{m'} = \left(\frac{t'}{p}\right) \sum_{m'=0}^{p-1} \left(\frac{m'}{p}\right) \zeta^{m'}$$

or

$$G(\zeta^t) = \left(\frac{t}{p}\right) G(\zeta)$$

In this form $G(\zeta^t)$ is seen to be a periodic function in t of period p.

2.7.3 Finite Fourier Series

Periodic arithmetic functions such as $G(\zeta^t)$ can be expanded in finite Fourier series, in complete analogy to Fourier series in analysis. The expansion is based on the

Theorem Let $F(t)$ be a function defined for all integers t with the period m, and let η be a primitive mth root of unity. Then

$$F(t) = \sum_{u=0}^{m-1} a(u)\eta^{ut} \tag{2.25}$$

with

$$a(u) = \frac{1}{m}\sum_{t=0}^{m-1} F(t)\eta^{-tu} \tag{2.26}$$

Proof Formula (2.25) for $t = 0, 1, \ldots, m-1$ represents a system of m linear equations for m unknowns $a(u)$. To be solvable $a(u)$ must fulfill certain conditions. Multiply both sides of (2.25) by η^{-vt} and sum over t:

$$\begin{aligned}\sum_{t=0}^{m-1} F(t)\eta^{-vt} &= \sum_{t=0}^{m-1}\left(\sum_{u=0}^{m-1} a(u)\eta^{ut}\right)\eta^{-vt} \\ &= \sum_{u=0}^{m-1} a(u)\sum_{t=0}^{m-1}\eta^{(u-v)t} \\ &= m \cdot a(v)\,, \text{ i.e. } u = v\end{aligned}$$

It follows that $a(u)$, if it exists as a solution, is unique and can only have the form (2.26). That this $a(u)$ satisfies (2.25) can be seen by direct substitution

$$\begin{aligned}\sum_{u=0}^{m-1} a(u)\eta^{ut} &= \frac{1}{m}\sum_{u=0}^{m-1}\left(\sum_{s=0}^{m-1} F(s)\eta^{-us}\right)\eta^{ut} \\ &= \frac{1}{m}\sum_{s=0}^{m-1} F(s)\sum_{u=0}^{m-1}\eta^{u(t-s)} \\ &= F(t)\end{aligned}$$

which proofs the theorem for $s = t$.

The $a(u)$ may be called Fourier coefficients of the finite Fourier series (2.25). Because of the periodicity of $F(t)$ and $a(u)$ with the period m, the series in (2.25) and (2.26) may be taken over any complete residue system modulo m.

2.7.4 Periodic Functions

A function $f : \mathbb{R} \to \mathbb{C}$ is called periodic of period $L > 0$ if for every $x \in \mathbb{R}$, $f(x + L) = f(x)$ [30]. If f is periodic of period L, then the function $F(x) = f(Lx)$ is periodic of period 1. Moreover since $f(x) = F(x/L)$, it suffices to consider periodic functions of period 1 only, simply called *periodic.*

The functions $f(x) = \sin 2\pi x$, $f(x) = \cos 2\pi x$, and $f(x) = \exp(2\pi i x)$ are periodic. Further, every given function on the half-open interval $[0, 1)$ can be extended to a periodic function in a unique way. By definition the inner product $\langle .,. \rangle$ on a complex vector space V, is the map from $V \times V$ to $\mathbb{C}$ that satisfies the following conditions:

(a) For every $w \in V$ the map $v \mapsto \langle v, w \rangle$ is $\mathbb{C}$ linear

(b) $\langle v, w \rangle = \overline{\langle w, v \rangle}$

(c) $\langle .,. \rangle$ is positive definite, i.e. $\langle v, v \rangle \geq 0$; and $\langle v, v \rangle = 0$ implies $v = 0$

If f and g are periodic, so is $af + bg$ for $a, b \in \mathbb{C}$, so that the set of periodic functions forms a complex vector space, denoted by $C(\mathbb{R}/\mathbb{Z})$, the linear subspace of all continuous periodic functions $f : \mathbb{R} \to \mathbb{C}$.

For $k \in \mathbb{Z}$ the function $e_k = \exp(2\pi i k x)$ lies in $C(\mathbb{R}/\mathbb{Z})$. If $k, l \in \mathbb{Z}$ the inner product

$$\langle e_k, e_l \rangle = \begin{cases} 1 & if \quad k = l \\ 0 & if \quad k \neq l \end{cases}$$

It follows that, for varying k, the e_k give linearly independent vectors in the vector space $C(\mathbb{R}/\mathbb{Z})$. Finally, if

$$f(x) = \sum_{k=-n}^{n} c_k e_k(x)$$

for some coefficients $c_k \in \mathbb{C}$, then $c_k = \langle f, e_k \rangle$ for each k.

Proof If $k = l$, then

$$\langle e_k, e_l \rangle = \int_0^1 e^{2\pi i k x} e^{-2\pi i k x} \mathrm{d}x = \int_0^1 1 \mathrm{d}x = 1$$

Now let $k \neq l$ and set $m = k - l \neq 0$; then

$$\begin{aligned} \langle e_k, e_l \rangle &= \int_0^1 e^{2\pi i m x} \mathrm{d}x \\ &= \frac{1}{2\pi i m} e^{2\pi i m x} \Big|_0^1 \\ &= \frac{1}{2\pi i m} \Big(1 - 1 \Big) = 0 \end{aligned}$$

From this result the linear independence derives as follows. Suppose that

$$\lambda_{-n}e_{-n} + \lambda_{-n+1}e_{-n+1} + \cdots + \lambda_n e_n = 0$$

for some $n \in \mathbb{N}$ and coefficients $\lambda_k \in \mathbb{C}$. For the theorem to hold all coefficients λ_k must vanish. To show this, let k be an integer between $-n$ and n. Then

$$\begin{aligned} 0 &= \langle 0, e_k \rangle \\ &= \langle \lambda_{-n}e_{-n} + \cdots + \lambda_n e_n, e_k \rangle \\ &= \lambda_{-n}\langle e_{-n}, e_k \rangle + \cdots + \lambda_n \langle e_n, e_k \rangle \\ &= \lambda_k \end{aligned}$$

Thus the (e_k) are linearly independent, as claimed. In the same way it follows that $c_k = \langle f, e_k \rangle$ to complete the proof.

If the mapping $f : \mathbb{R} \to \mathbb{C}$ is periodic and integrable on the interval [0,1], the numbers

$$c_k(f) = \langle f, e_k \rangle = \int_0^1 f(x)e^{-2\pi ikx}\mathrm{d}x,\ k \in \mathbb{Z}$$

are called the Fourier coefficients of f. The series $\sum_{k=-\infty}^{\infty} c_k(f)e^{-2\pi ikx}$ is called the Fourier series of f.

The space $C(\mathbb{R}/\mathbb{Z})$ introduced as the space of continuous periodic functions on $\mathbb{R}$ has another interpretation [30]:
Assume the equivalence relation on $\mathbb{R}$:

$$x \sim y \Leftrightarrow x - y \in \mathbb{Z}$$

For $x \in \mathbb{R}$ its equivalence class is given by the restriction

$$[x] = x + \mathbb{Z} = \{x + k | k \in \mathbb{Z}\}$$

Let $\mathbb{R}/\mathbb{Z}$ be the set of all equivalence classes. This set can be identified by the half-open interval [0,1). It can also be identified with the unit torus

$$\mathbb{T} = \{x \in C : |z| = 1\}$$

since the map $e : \mathbb{R} \to \mathbb{T}$ that maps x to $e(x) = \exp(2\pi ix)$ gives a bijection between $\mathbb{R}/\mathbb{Z}$ and $\mathbb{T}$.

A sequence $[x_n]$ is said to converge to $[x] \in \mathbb{R}/\mathbb{Z}$ if there are representatives $x' \in \mathbb{R}$ and $x \in \mathbb{R}$ for the classes $[x_n]$ and $[x]$, such that the sequence (x'_n) converges to x' in $\mathbb{R}$. In the interval [0,1) this means that either x_n converges to x in the interval [0,1) or that $[x] = 0$ and the sequence x_n decomposes into two subsequences, one of which converges to 0 and the other to 1.

The best way to visualize $\mathbb{R}/\mathbb{Z}$ is as the real line rolled up by either identifying the integers or by using the map $\exp(2\pi ix)$ or by gluing the ends of the interval [0,1] together. It will be shown in the following that the periodicity of atomic matter arises from precisely such a closure in atomic number, neutron number or mass number.

Chapter 3

Periodic Table of the Elements

The periodic table of the chemical elements is the single most important concept in chemistry. It has evolved as a classification scheme through a number of well-defined stages. It has provided, at all stages, a detailed summary of all available knowledge pertaining to the physical and chemical properties of the elements and their compounds. More importantly, it reflects the current state of theoretical chemistry at any time. The classification was based at different times on the appearance of substances, chemical affinity, atomic weight, atomic number and the electronic configuration of atoms. The parallel development of a periodic table is traced in the following historical review, based to a large extent on the work of Partington [31].

3.1 Historical Development

The earliest concepts with a chemical flavour arose from philosophical discussions contrasting atomic and continuity models of matter. These theories were all swept aside by Aristotle (3rd century BC) who smothered further speculation with his pronouncement of primary matter on which specific forms could be impressed by the fundamental properties of being hot, cold, moist or dry, to various degrees. By combining these properties in pairs, as in Figure 3.1, the four elements, fire, air, water and earth are obtained. A fifth element, aether was added at a later stage. Early Egyptian arts of working with metals, glass and dyes inspired manipulations by magicians and alchemists in classical times. Despite a clear chemical content, these operations can hardly be recognized as chemistry in the same sense as today's science.

The origin of the modern science is usually traced back to the publication of Robert Boyle's *Sceptical Chymist* in 1661. The single most important feature of this work is the identification of primitive elements, of which mixed bodies are made up, and into which the mixed ones are ultimately resolved. Apart from signalling the beginning of chemistry, this book is certainly the first sensible classification, not only of chemical substances, but also of the chemical elements. Sixty years later (1718) Geoffroy the Elder published an Affinity Table, reproduced here in Figure 3.2. In each column any substance would displace any of those standing below it, from their compounds. In a series of acids, for example, sulphuric acid displaces hydrochloric and nitric acids from their salts, because it has a higher affinity for the base. Another 50 years later (1785) an English translation of Bergman's more detailed *Dissertation on Elective Attractions* was published, still without strict separation of elements and compounds.

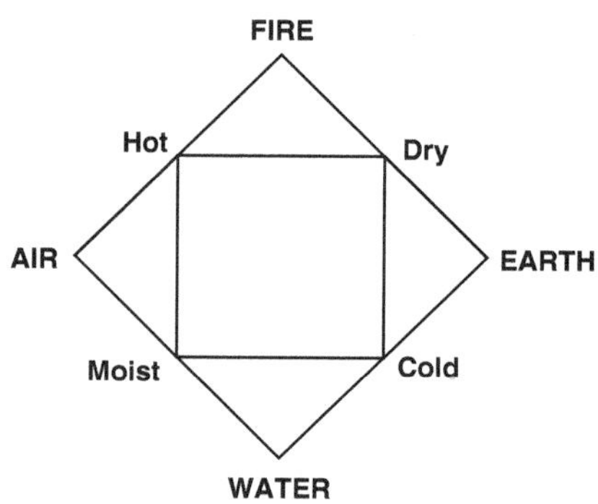

Figure 3.1: *Definition of the four elements by Aristotle.*

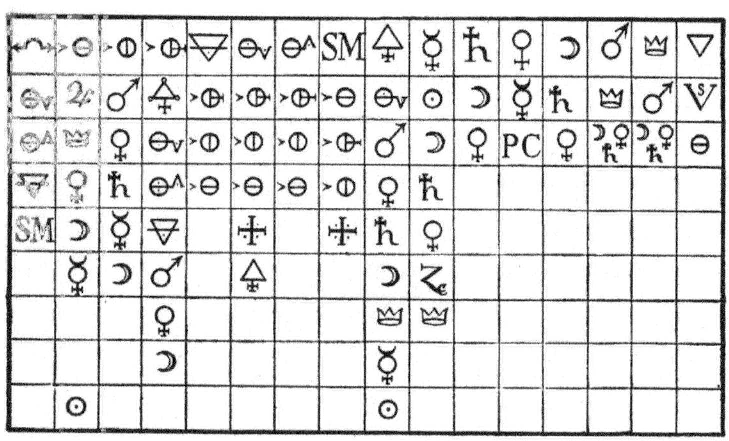

Figure 3.2: *Geoffroy's Affinity Table. Deciphering the table is left as an exercise to the reader.*

At about the same time (1789) in *Traite élémentaire de chimie*, Lavoisier published the first list of chemical elements based on the definition of Boyle. As he himself anticipated, a number of *earths* (lime, magnesia and alumina) and *alkalis* (potash and soda) included as elements, were later shown to be compounds. These are the oxides that cannot be decomposed by heating and chemical reduction. It is instructive, but not surprising, to note that *heat* was included in the list as an element. At the period in question chemical reaction was considered synonymous with the effects of heat or fire on ponderable matter. Central to chemistry was the theory of combustion, which has an interesting history of its own.

3.1.1 The Theory of Combustion

The alchemists used heat routinely to bring about chemical change. It was known to them that metals, other than gold and silver, when heated in open crucibles underwent chemical transformation into a *calx*. Since the calx is heavier than the metal before heating, it was natural to assume that the absorbed fire contributed the observed difference in weight. In alchemist theory the principle of combustibility was identified with the element sulphur, supposed to be a constituent, together with salt and mercury, of all matter. The element salt represented the fixed part that remained after calcination whilst mercury was the principle of metallicity, contained in all metals.[1]

Both this three-element theory of Paracelsus and Aristotle's four-element doctrine were overturned by Boyle on the basis of simple repeatable experiments to demonstrate that none of these seven elements could be extracted from metals, other than mercury itself, by any process. Boyle performed experiments to show that the calcination of tin, like the combustion of sulphur consumed an amount of gas from the atmosphere. To confuse the issue he also found that gunpowder caught fire when heated in the absence of air and continued to burn under water. Boyle's experiments were repeated by his erstwhile assistant, Robert Hooke who formulated the first rational theory of combustion (1665). In summary:

> A universal agent that supports combustion occurs in air, or fixed in saltpetre (nitre).

[1]The sought after transmutation of base metals into gold was taught to occur on altering the proportions of mercury and sulphur by adding the philosopher's stone.

He called the atmospheric component, *nitrous air*. The combustion behaviour of gunpowder was ascribed to its nitre content. John Mayow (1674) elaborated on Hooke's theory with supporting experiments and concluded that air consisted of nitrous air, which supports combustion and respiration, together with a second inert gas. Mayow never succeeded in isolating nitrous air although he merely had to heat nitre to melting in order to obtain that product:

$$2KNO_3 \rightarrow 2KNO_2 + O_2$$

This simple procedure could have changed the course of chemistry. In the event, Mayow's obviously valid theory lost ground, to be replaced by the totally speculative *phlogiston* theory of Becher and Stahl.

The Phlogiston Theory

Rather than reject the theories of Aristotle and Paracelsus, Becher (1669), in an effort to preserve their valid content, proposed some modification to bring them into line with new ideas of the day. It was argued that material bodies consisted of air, water and three earths; one inflammable (*terra pinguis*), the second *mercurial* and the third fusible or *vitreous*. These were the sulphur, mercury and salt of the alchemists, in more modern guise.

A refined version of the theory was published (1723) and popularized by another German professor, Georg Stahl. He used the name phlogiston for *terra pinguis*. When bodies burn phlogiston escapes. When the original bodies are recovered by reduction, phlogiston is replaced. Oil, wax and charcoal are rich in phlogiston, and may restore it to a burnt material. When zinc burns in air, phlogiston (ϕ) escapes[2]

$$\text{zinc} \rightarrow \text{calx} + \phi$$

[2]In modern terminolgy the oxidation/reduction, $Zn \rightleftharpoons ZnO$, is described in terms of electronic theory by the equations

$$Zn \rightarrow Zn^{++} + 2e$$

$$Zn^{++} + 2e \rightarrow Zn$$

not dissimilar from the phlogiston formulation.

$$P \rightleftharpoons P^{5+} + 5e$$

When the white calx of zinc is heated with charcoal zinc distils

$$\text{calx of zinc} + \phi \rightarrow \text{zinc}$$

Phosphorous burns to produce an acid and much heat

$$\text{phosphorous} \rightarrow \text{acid} + \phi$$

Heated with charcoal, phlogiston is absorbed and phosphorous is reproduced (see footnote).

Stahl's theory united many previously isolated facts and was widely accepted during the eighteenth century. The increase in weight during the calcination of metals was largely ignored and this fact was destined later to overturn the phlogiston theory. Instead of accepting the role of air in combustion processes, phlogistonists came up with the improbable assumption that phlogiston possessed negative weight.

The phlogiston theory is often held up to ridicule, but this is unfair both to the theory and the scientists who upheld it. When the opposing arguments finally came to a head both parties were too stubborn to consider compromise and one theory had to go. As suggested here in footnote material the phlogiston theory was re-invented a hundred years later as the modern electronic theory of oxidation and reduction. Maybe Stahl's thinking was too far ahead of his time, while his experimental acumen lagged too far behind.

A new theory of combustion was formulated after the discovery of oxygen and the pioneering work of Lavoisier, whose theory of combustion consisted of four half-truths:

1. Substances burn only in pure air (oxygen).
2. Non-metals such as sulphur, phosphorous and carbon, produce acids on combustion.
3. Metals produce calces (basic oxides) on absorption of oxygen.
4. Combustion is not due to the escape of phlogiston, but to chemical combination with oxygen.

One experimental observation with a ready explanation in phlogiston theory, could not be explained by Lavoisier: A metal like zinc dissolves in acid giving off inflammable air. A salt is left in solution, which on strong heating leaves the calx of the metal. The same salt is formed when the calx dissolves in acid, but no inflammable air is evolved. Lavoisier's theory failed to account for the inflammable air, which in the old theory could be identified with ϕ:

$$\text{metal(calx} + \phi) + \text{acid} \rightarrow \text{salt} + \phi$$

$$\text{calx} + \text{acid} \rightarrow \text{salt}$$

When it became clear (1781) that water consisted of oxygen and hydrogen (inflammable air), this last obstacle to Lavoisier's theory disappeared and from about 1785 the phlogiston theory was finally abandoned.

Although Lavoisier's theory of combustion, in some respects, was less comprehensive than phlogiston theory, it brought the importance of mass balance during chemical change to the fore. This new awareness of the conservation of matter dominated chemical research for the entire nineteenth century and, on the basis of Dalton's atomic theory, provided the experimental evidence that led to the formulation of the periodic law.

3.1.2 Atomic Theory

The law of conservation of mass is almost self-evident. Aristotle quoted the 5th century BC Greek philosopher Empedokles as saying:

> Nothing either comes into being or is destroyed, but things are merely transformed depending on the ratio of basic substances to one another.

Two thousand years later Lavoisier declared:

> Nothing can be created, and in every process there is just as much substance present before and after the process has taken place.

The new paradigm set in motion by Lavoisier's statement was soon to produce conspicuous results, embodied in the fundamental laws of chemical combination.

Laws of Chemical Combination

Within 10 years of Lavoisier's execution three basic laws had been established experimentally:

Law of Constant Proportions This law was stated by Proust in 1797:

> Elements combine in definite proportions by weight, so that the composition of a pure chemical compound is independent of the way in which it is prepared.

Law of Multiple Proportions This law was formulated by John Dalton in 1803, based on some theoretical speculation and later proven experimentally by Berzelius.

> When two elements combine to form more than one compound, the weights of one element which unite with identical weights of the other are in the ratio of small whole numbers.

Law of Equivalent Proportions Aspects of this law were already appreciated by Cavendish in 1766. The first clear statement of the law is probably due to Berthollet (1802):

> The weights of two (or more) substances which separately react chemically with identical weights of a third are also the weights which react with each other, or are related to them in the ratio of small whole numbers.

Dalton's Theory

The Greek philosophers Leukippos and Demokritos taught that matter is composed of small indivisible particles called atoms. Dalton (1803) developed this theory into a form that could explain the laws of chemical combination. Dalton's theory can be summarized in the form of three statements:

1. The chemical elements are composed of minute particles of matter called atoms, that remain undivided in all chemical changes. The atom is the smallest mass of an element which can take part in a chemical reaction.

2. Each kind of atom has a definite weight. Different elements have atoms differing in weight.

3. Atoms combine in simple numerical ratios.

The assumption that atoms are indestructible leads directly to the law of conservation of mass. Since compound atoms (molecules) in Dalton's theory are all alike and all atoms of the same element are identical, the law of constant proportions follows immediately.

If two elements combine in more than one ratio, any pair of different molecules are formed with compositions A_mB_n and A_kB_l, where m, n, k, l are small integers. The number of atoms A that combine with one B is m/n and k/l respectively, in the ratio ml/nk, i.e. a whole-number ratio. Hence the weights of B combining with identical weights of A are in the ratio of whole numbers that satisfies the law of equivalent proportions. Molecules formed by atomic pairs from the set A, B, C must have compositions A_mC_n, B_kC_l, A_pB_q, where m, n, k, l, p, q are small whole numbers. Hence p, q are either the same as m, k or whole multiples of them. This is the law of multiple proportions.

Dalton's theory does not address the problem of atomic weights directly. At best it allows the determination of relative atomic weights, but only if the atomic ratios in the final product of atomic interaction is known. Although it may be known that 8 parts of oxygen combine with 1 part of hydrogen, the relative weights cannot be deduced unless the exact atomic composition of water is known. Dalton actually considered this composition to be HO and determined the relative atomic weights of H and O to be 1 and 8, respectively.

Dalton represented atoms by circles with suitable identification marks, as shown by the following examples. The present chemical notation was finally

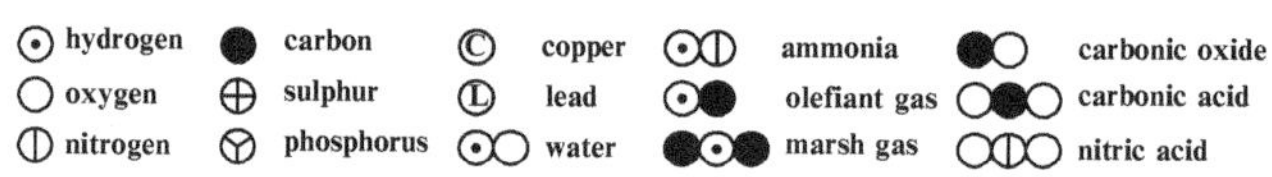

established by Berzelius (1813) and it soon replaced Dalton's symbols.

Not only does an atomic symbol identifies an element but it also represents an atom with the characteristic atomic weight of that element. The law of equivalent proportions and Dalton's theory allow the experimental determination of *equivalents* or equivalent weights of the elements.

Valency The *valency* of an element in a given compound is defined as the ratio of the atomic weight to the equivalent of the element in that compound. If E is the equivalent, A the atomic weight and n the valency of an element, $A = nE$. Since, by definition, one atom of hydrogen combines with a weight E of a given element, n atoms of hydrogen will combine with a weight A and therefore n is equal to the valency of the element in the resulting compound. In other words, the valency of an element is measured by the number of hydrogen atoms which unite with one atom of that element. Chlorine is univalent and may also be used instead of hydrogen to determinne the valencies of elements in compounds such as PCl_5. Oxygen is bivalent and may be used in the same way to fix valencies in compounds such as Cl_2O_7 and OsO_4.

Equivalents Once the hydrogen equivalent has been defined as 1 mass unit, the equivalents of other elements can be developed according to a tree structure on analyzing suitable binary compounds formed with other elements, by weight. Hydrogen is replaced from acids by many metals and by collecting the displaced gas of known density, metal equivalents are determined directly. The weight of metal displaced from a solution of its salts by other metals is used next to determine new equivalents. Once the equivalent of silver has been obtained, a known weight of silver nitrate in solution is treated with measured quantities of soluble metal chlorides to precitate the silver as AgCl and measure the equivalents of still other metals. Suitable reactions can always be found to expand the number of known equivalents and cross-check previous results until a self-consistent set has been created.

The exact equivalents of all known elements were determined and refined during the nineteenth century by many chemists, including Berzelius, Stas, Richards, Hönigschmid, and others. Without an independent measurement of valency however, atomic weights could not be obtained from these results, until Cannizaro demonstrated (1858) how to interpret the hypothesis of Avogadro to this effect.

The Hypothesis of Avogadro

The first quantitative studies of chemical reaction in the gas phase were conducted on mixtures of hydrogen and oxygen by Cavendish (1781) who established that the combining volumes were very nearly in the ratio 2:1. Investigations were extended to reaction between other gases by Gay-Lussac who formulated the general law:

> When gases take part in chemical changes the volumes of the reacting gases, and those of any gaseous products are in the ratio of small whole numbers.

All measurements were carried out under the same conditions of temperature and pressure.

The results of Gay-Lussac were interpreted by Berzelius and by Dalton to mean that equal volumes of elementary gases contain equal numbers of atoms and this led to a situation of conflict with Dalton's atomic theory. As an example, one volume of hydrogen reacts with one volume of chlorine to produce two volumes of hydrogen chloride. This result cannot possibly be interpreted to mean that one atom of hydrogen and one atom of chlorine produced two particles of the compound atom - Dalton's terminology. The issue was resolved by Amedeo Avogadro in 1811.

The Hypothesis Avogadro made the assumption that

> Equal volumes of all gases and vapours, under the same conditions of temperature and pressure, contain identical numbers of *molecules*. A molecule is the smallest mass of a substance capable of existing in the free state.

The essence of this assumption is that the particles of elementary gases are not atoms, but divisible molecules. It was shown that the volumetric relationships in the reaction between hydrogen and chlorine can be rationalized on the assumption that molecules of both hydrogen and chlorine consisted of two atoms each. The interaction may then be formulated as

$$H_2 + Cl_2 \rightarrow 2HCl$$

in line with the volumetric data.

Consistent application of Avogadro's hypothesis shows that whereas most metal vapours are monoatomic, most non-metals occur in the gas phase as diatomic molecules. Notable exceptions include O_3, P_4, As_4 and S_8.

Relative Density It follows from Avogadro's hypothesis and Dalton's theory that the weights of equal volumes of gases or vapours, at the same temperature and pressure, are in the ratio of their molecular weights. On defining the relative density of a gas or a vapour as the ratio

$$\frac{\text{Weight of a given volume of the gas}}{\text{Weight of an equal volume of hydrogen}}$$

at the same temperature and pressure, Avogadro's hypothesis shows that the relative density is also equal to the ratio

$$\frac{\text{Weight of one molecule of the gas or vapour}}{\text{Weight of one molecule of hydrogen}}$$

since equal volumes contain identical numbers of molecules.

On the hydrogen equivalent scale the molecular weight of hydrogen gas is 2, and the relative density of a gas reduces to (molecular weight)/2. Hence

$$\text{Molecular weight} = (\text{relative density}) \times 2$$

Another consequence of Avogadro's hypothesis is that a gram-molecular weight (*mol*) of any gas at S.T.P[3] occupies the same volume, known as the *molar volume*. If M is the molecular weight of a gas of density D (in gl^{-1} at STP)

$$\frac{M}{D} = V_m$$

[3]Standard Temperature and Pressure is defined as 0°C and 1 atmosphere pressure.

Oxygen with $M = 32$ has $D = 1.429$ gl^{-1}, hence $V_m = 22.4$ l. According to Avogadro V_m has this same value for all gases. Furthermore, the number of molecules in a mol is the same for all gases, and is known as Avogadro's number N_A, It has been measured by a variety of methods, giving values in general good agreement, now accepted as $N_A = 6.022 \times 10^{23}$ mol^{-1}.

Cannizaro's Principle

For many years, during the first half of the nineteenth century, there appears to have been a stand-off between the research camps using atomic theory and gas laws respectively. Whereas Dalton confused molecules with atoms, Avogadro used the hydrogen molecule rather than the hydrogen atom as molecular weight standard. Once this confusion had been cleared up by Cannizaro in 1858 the determination of absolute atomic weights became possible by systematic use of Avogadro's hypothesis. A typical procedure starts with the determination of molecular weights, by vapour density measurement, of a number of volatile compounds of a given element. The weight of the element under study, in 1 mol of each selected compound, is next established by chemical analysis. The smallest weight so established is interpreted as the gram-atomic weight of that element.

Cannizaro's principle may be summarized to state:

> The atomic weight of an element is the smallest weight of that element contained in a molecular weight of its compounds.

It is not sufficient to apply Cannizaro's principle to one or two compounds of an element, even if the vapour of the element itself is included. The chances of a meaningful result improve as a variety of more compounds are included in the test battery. Since vapour density measurements are not of the highest accuracy, the molecular weights obtained by this method are approximate and the derived atomic weights are in need of refinement by other methods.

3.1.3 Measurement of Atomic Weights

Cannizaro's insight showed how to convert the accumulated data on chemical equivalents into atomic weights. In view of the large number of elements and chemical reactions involved in these measurements it was imperative to strive for the highest possible accuracy in order to obtain a self-consistent set of reliable atomic weights. Two of the most diligent scientists in this pursuit were the Belgian chemist Jean Stas and Theodore Richards at Harvard.

The type of challenge these scientists had to face is summarized well in the words of Richards:

> Every substance must be assumed to be impure, every reaction must be assumed to be incomplete, every method of measurement must be assumed to contain some constant error, until proof to the contrary can be obtained. As little as possible must be taken for granted.

Atomic weights, by definition are relative weights. It is essential therefore that independent workers should agree on a standard reference. At various times atomic weight scales were based on H = 1, H = 2, O = 1, O = 10, O = 16 and O = 100. In the second half of the nineteenth century the oxygen 16 scale was generally accepted.

Long before final agreement on the atomic weights of different elements was possible many regularities between atomic weights and chemical properties were noticed. One observation, more than anything else, that highlighted clear parallels between atomic weight and chemical properties was the discovery of *isomorphism*, and the use of this concept in the determination of atomic weights.

Isomorphism

The Abbé René Haüy, the founder of crystallography postulated (1822) two fundamental axioms:

1. Identity of crystalline form (except in the regular system) implies identity of chemical composition

and conversely,

2. difference in crystalline form implies difference in chemical composition

One of the early known exceptions was calcium carbonate that crystallizes either as hexagonal calcite, or orthorhombic aragonite. The alums were found to have the same crystalline form despite different chemical composition.

On the basis of similarities between crystals such as $Na_2HPO_4{\cdot}12H_2O$ and $Na_2HAsPO_4{\cdot}12H_2O$ Mitscherlich proposed (1819) a variation on the axioms, based on the property of isomorphism, or the tendency of different substances to crystallize in the same form. Since many analogous compounds of phosphorous and arsenic were found to be *isomorphous*, this name was applied to the elements themselves. *Isomorphous elements* were later defined more

Table 3.1: *Elements of the same group are known to form isomorphous compounds. Several elements occur in more than one group.*

ISOMORPHOUS ELEMENTS

I	Cl, Br, I, F; Mn (compare $KMnO_4$ and $KClO_4$)
II	S, Se; Cr, Mn, Te (in compounds K_2RO_4)
III	As, Sb, Bi; Te(element); P, V (salts); N, P (organic bases)
IV	K, Na, Cs, Rb, Li; Tl, Ag
V	Ca, Sr, Ba, Pb; Fe, Zn, Mg; Ni, Co, Cu
VI	Al, Fe, Cr, Mn
VII	Cu, Ag in lower oxides, Au
VIII	Pt, Ir, Pd, Rh, Ru, Os; Au, Fe, Ni, Sn, Te
IX	C, Si, Ti, Zr, Sn; Fe,Ti
X	Mo, W, Cr

generally as elements that form isomorphous compounds with the same elements or radicals, and can replace one another in compounds without change in crystalline form. The free elements need not have the same crystalline form, although this possibility is not excluded. In most, but not all cases, isomorphism was found to occur among elements that are chemically alike.[4]

An important consequence of isomorphic relationships is the possibility of classifying the elements into groups, such that the members of each group can replace one another in their compounds without significant change in crystallographic form. The groups of isomorphous elements shown in Table 3.1 reveal chemical similarities in many cases. The grouping of elements according to Table 3.1 must have provided powerful impetus to encourage development of a general classification of the elements to reflect their chemical similarities. In that sense tables of this kind may be considered as precursors of the periodic table.

The idea of isomorphism also made an important contribution to the determination of atomic weights. Vapour densities could in general be measured only for compounds of non-metals and another way had to be found to establish the atomic weights of metals. Isomorphism provided a way of linking up the atomic weights of metals and non-metals.

Analysis of a pair of isomorphous salts gives the relative atomic weights of the two distinct atoms directly. For example, the atomic weight of selenium

[4]Accurate crystallographic work of a later era has shown that isomorphous crystals have the same symmetry without having identical unit cells.

was established in this way in 1828, following the discovery of the isomorphism of sodium sulphate and selenate, and also the corresponding silver salts. The atomic weight of a metal may also be linked to that of a non-metal. Potassium permanganate is isomorphous with potassium perchlorate, giving the atomic weight of Mn assuming that of Cl to be known. As another example, the manganate, chromate, sulphate and selenate of potassium are all isomorphous, thus connecting the atomic weights of Mn, Cr and Se to that of S. Finally a whole series of divalent metals form double sulphates of the general type $M_2^IM^{II}(SO_4)_2{\cdot}6H_2O$ where M^I is an alkali, or other monovalent metal and M^{II} may be Mg, Mn, Fe, Co, Ni, Cu and Zn, the atomic weights of which may be linked through Mn to those of S and Cl.

Another interesting case of isomorphism led to the correction of the atomic weight of vanadium. Berzelius described the following four minerals by the formulae shown:

Apatite	$3Ca_3(PO_4)_2{\cdot}CaF_2$
Pyromorphite	$3Pb_3(PO_4)_2{\cdot}PbCl_2$
Mimetite	$3Pb_3(AsO_4)_2{\cdot}PbCl_2$
Vanadinite	$3Pb_3V_2O_6{\cdot}PbCl_2$

Since the four minerals are isomorphous Roscoe re-investigated the vanadium compound and concluded that Berzelius must have mistaken the oxide VO for V and so assigned the wrong atomic weight. The corrected formula $3P_3(VO_4)_2{\cdot}PbCl_2$ established the correct atomic weight for V.

Atomic weight, as the most fundamental property of an element that could be measured accurately, soon became an important parameter, in addition to chemical properties, for the classification of the elements, culminating in the formulation of the periodic law.

3.1.4 The Periodic Law

The publication of affinity tables, designed to impose some order amongst the increasing number of elements identified since the time of Boyle, were mentioned before. That was one of many attempts to classify chemical entities. The general aim of any classification is to group together those entities that resemble each other and to separate those which differ. The best classification will be that which brings together things which resemble each other in the greatest number of ways. The elements were grouped at various times into metals and non-metals, into acidic and basic, electronegative and electropositive. They were classed according to valency and according to many other properties.

In all chemical change one property was found to remain unaltered and the most satisfactory system of classification was based on this property, the atomic weight of the elements. Early efforts to achieve such classification suffered from unreliable experiments and inconsistent results, reported by different schools. As methodologies improved and consensus was reached on theories of atoms and molecules, the classifications became more convincing.

Credit for the first successful classification of the elements in terms of atomic weight usually goes to Döbereiner (*c.*1820) who noticed regularities in the atomic weights of chemically related elements and formulated the *law of triads*. In groups of three related elements the atomic weights are either much the same (e.g. iron, cobalt, nickel) or else exhibit a constant difference when arranged in sequence. Three triads with their atomic weights and differences are listed in Table 3.2.

Several other workers (Lenssen, Dumas, Gmelin, Van Pettenkofer, Cooke) elaborated on the triad theme to formulate families of related elements with numerical relationships, according to simple arithmetic rules, between their atomic weights. By 1858 the five families shown in Table 3.3 had been recognized.

In 1862 De Chancourtois proposed a classification, based on revised atomic weights according to Cannizaro. He plotted atomic weights on the surface of

Table 3.2: *Examples of Döbereiner triads of elements. The atomic weight of each middle element is approximately the mean of the atomic weights of the extreme elements.*

Calcium		Strontium		Barium
40		87		137
	47		50	
Chlorine		Bromine		Iodine
35.5		80		127
	44.5		47	
Sulphur		Selenium		Tellurium
32		79		128
	47		49	

Table 3.3: *Early classification of elements with recognizable periodic features.*

	Mg	Ca	Sr	Ba	
Li	Na	K			
	F	Cl	Br	I	
	O	S	Se	Te	
	N	P	As	Sb	Bi

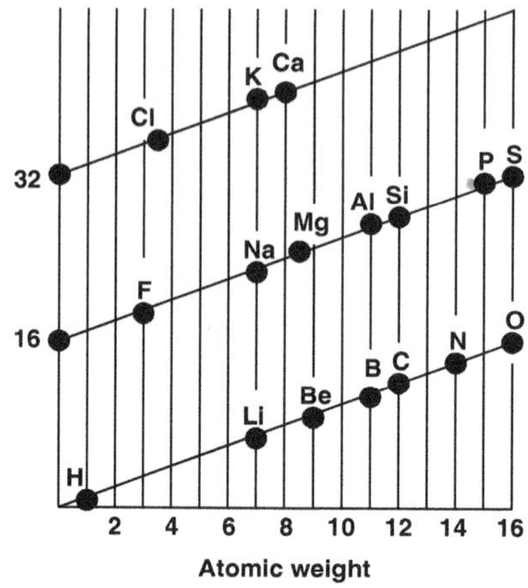

Figure 3.3: *Helical arrangement of elements with increasing atomic weight.*

Table 3.4: *Illustration of the law of octaves as proposed by Newlands.*

1 H	2 Li	3 Be	4 B	5 C	6 N	7 O	
8 F	9 Na	10 Mg	11 Al	12 Si	13 P	14 S	
15 Cl	16 K	17 Ca	19 Cr	18 Ti	20 Mn	21 Fe	etc.

a regular cylinder with a circumference of 16 units, equivalent to the atomic weight of oxygen. The resulting helical curve is shown, opened up, in Figure 3.3. It positions related elements close to a set of straight lines. He concluded that *the properties of the elements are the properties of numbers.* This conclusion summarizes the main theme of the present monograph rather well.

A new system of classification, proposed by Newlands in 1865, was obtained by arranging the elements in order of increasing atomic weight (Table 3.4). The position of each element in the sequence was assigned an *ordinal number*, starting from 1 for H. He noted that, in this sequence, every eighth element is a kind of repetition of the first, like the eighth note in an octave of music, and he called this the *law of octaves.* In such a table, elements of the same family tend to occur in the same column. Like the helix of De Chancourtois the law of octaves breaks down after Ca.

Periodic Classifications

In 1869, quite independently and in apparent ignorance of previous work Dmitri Mendeléeff in Russia and Lothar Meyer in Germany proposed a comprehensive classification of the elements according to a scheme that became known as the periodic law.

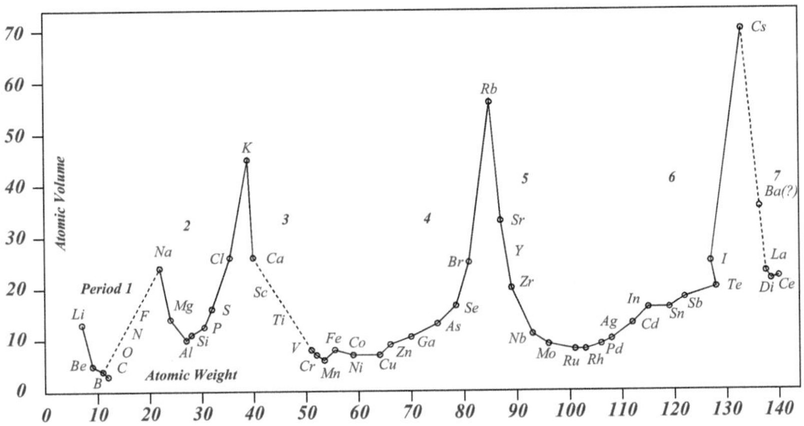

Figure 3.4: *Lothar Meyer's atomic volume curve, redrawn.*

Lothar Meyer Lothar Meyer went beyond mere classification of the elements as a function of atomic weight. He provided graphical evidence that elemental periodicity was more than an illusion, as hinted by some contemporaries.[5] He calculated atomic volumes in milliliter on dividing gram-atomic weights by elemental density in units of gram per cubic centimeter ($V = M/d$), and plotted this quantity as a function of atomic weight. A reconstruction of his original plot is shown in Figure 3.4. It is noticed immediately that the family of alkali metals signal periodic limits to an almost smoothly varying function over the intervening elements. Between Li and Cs there appears to be two short periods, followed by two long ones. Lothar Meyer himself considered each long period to consist of two short ones centred at Ni and Pd respectively, and he saw the beginnings of a seventh period beyond Cs.

Apart from providing dramatic illustration of periodicity amongst elements, Lothar Meyer formulated a periodic classification, almost indistinguishable from that of Mendeléeff. His work was, admittedly published somewhat later than that of his Russian rival. It still does not explain why Mendeléeff managed to get almost sole credit for the first formulation of the periodic law. What made the difference probably was Mendeléeff's audacity to predict the existence of undiscovered elements on the basis of gaps in his table.

[5]When Newlands discussed the law of octaves at a meeting of the London Chemical Society in 1866 he was asked if he had ever examined the elements according to their initial letters [32].

Mendeléeff The name Mendeléeff became a household word after publication of his periodic law and periodic table, shown in Table 3.5.

The following quotation [33] describes the first announcement of the periodic law in Mendeléeff's own words:

> (1) The elements, if arranged according to their atomic weights, exhibit an evident *periodicity* of properties. (2) Elements which are similar as regards their chemical properties have atomic weights which are either of nearly the same value (platinum, iridium, osmium) or which increase regularly (*e.g.* potassium, rubidium, cæsium). (3) The arrangement of the elements or of groups of elements in the order of their atomic weights, corresponds with their so-called *valencies.* (4) The elements, which are the most widely diffused in nature, have *small* atomic weights, and all the elements of small atomic weight are characterised by their sharply-defined properties. They are therefore typical elements.

Table 3.5: *The 1891 version of Mendeléeff's Periodic Table (Redrawn from [33]).*

Distribution of the Elements in Groups and Series

Group	I.	II.	III.	IV.	V.	VI.	VII.	VIII.			
Series 1 · ·	· **H**	—	—	RH_4	RH_3	RH_2	RH				
" 2 · ·	**Li** ·	**Be** ·	**B** ·	**C** ·	**N** ·	**O** ·	**F** ·	Hydrogen compounds			
" 3 · ·	· **Na**	· Mg	· Al	· Si	· P	· S	· Cl				
" 4 · ·	K ·	Ca ·	Sc ·	Ti ·	V ·	Cr ·	Mn ·	Fe	Co	Ni	Cu
" 5 · ·	· (Cu)	· Zn	· Ga	· Ge	· As	· Se	· Br				
" 6 · ·	Rb ·	Sr ·	Y ·	Zr ·	Nb ·	Mo ·	— ·	Ru	Rh	Pd	Ag
" 7 · ·	· (Ag)	· Cd	· In	· Sn	· Sb	· Te	· I				
" 8 · ·	Cs ·	Ba ·	La ·	Ce ·	Di? ·	— ·	— ·	—	—	—	—
" 9 · ·	· —	· —	· —	· —	· —	· —	· —				
" 10 · ·	— ·	— ·	Yb ·	— ·	Ta ·	W ·	— ·	Os	Ir	Pt	Au
" 11 · ·	· (Au)	· Hg	· Tl	· Pb	· Bi	· —	· —				
" 12 · ·	— ·	— ·	— ·	·	— ·	U ·	— ·				
	R_2O —	R_2O_2 RO	R_2O_3 —	R_2O_4 RO_2	R_2O_5 —	R_2O_6 RO_3	R_2O_7 —	Higher oxides. RO_4			

> (5) The *magnitude* of the atomic weight determines the character of an element. (6) The discovery of many yet unknown elements may be expected. For instance, elements analogous to aluminium and silicon, whose atomic weights would be between 65 and 75. (7) The atomic weight of an element may sometimes be amended by aid of a knowledge of those of the contiguous elements. Thus the combining weight of tellurium must lie between 123 and 126, and cannot be 128. (8) Certain characteristic properties of the elements can be foretold from their atomic weights.

In the 1891 version of the table shown here, the gaps in the original, on the basis of which the elements eka-boron, eka-aluminium and eka-silicon had been predicted with remarkable accuracy, had already been filled after discovery of Sc, Ga and Ge.

Mendeléeff's table takes a form still in current use, which arranges the elements horizontally in order of their atomic weights and vertically according to common properties. Although Mendeléeff's achievement is not in dispute his statement [33] about Lothar Meyer's contribution rings less than generous:

> With regard to the work of Prof. Lothar Meyer respecting the periodic law (Notes 12 and 13), it is evident, judging from the method of investigation, and of his statement (Liebig's Annalen Supt. Band **7**, 1870, 354), at the very commencement of which he cites my paper of 1869 above mentioned, that he took the periodic law in the same form that it was given by me.

Fact is that the atomic volume curve focuses the periodicity of many other elemental properties much better than any matrix tabulation.

In his Faraday lecture of June 4, 1889, Mendeléeff made the following two statements [33]:

> (1) The periodic law has shown that our chemical individuals (*atoms*) display a harmonic periodicity of properties, dependent on their masses. Now natural science has long been accustomed to deal with periodicities observed in nature, to seize them with the vice of mathematical analysis, to submit them to the rasp of experiment. And these instruments of scientific thought would surely, long since, have mastered the problem connected with the chemical elements, were it not for a new feature which was brought to light by the periodic law, and which gave a peculiar and original character to the periodic function. . . . The periods of

> the elements have a character very different from those which are so simply represented by geometers. They correspond to points, to numbers, to sudden changes of the masses, and not to a continuous evolution.
>
> (2) In the theory of numbers only do we find problems analogous to ours, and two attempts at expressing the atomic weights of the elements by algebraic formulæ seem to be deserving of attention, although neither of them can be considered as a complete theory, nor as promising finally to solve the problem of the periodic law. ... What we can at present only be certain of is this: that attempts like the(se) two must be repeated and multiplied, because the periodic law has clearly shown that the masses of the atoms increase abruptly, by steps, which are clearly connected in some way with Dalton's law of multiple proportions.

Apart from the work of Reynolds, described below, we are not aware of any serious effort to follow up on these suggestions, until now.

The Zero Group

It is ironic that Mendeléeff who gained recognition for predicting the discovery of new elements on the basis of gaps in his table, failed to spot the largest series of gaps, now filled by the noble gases of group 0. The discovery of argon in the atmosphere by Raleigh and Ramsey in 1894 came as a complete surprise to the chemical world. No logical place could be found in the periodic table for the new element with atomic weight slightly higher than that of potassium. Helium gas was discovered a few years later and it was predicted by Thomsen [34] that there was room for six inert gases to occupy the space between neighbouring halogens and alkali metals in the periodic table. He proposed the long form of the periodic table, shown in Figure 3.5, and later adopted by Bohr as a basis of the electronic configuration of the atoms.

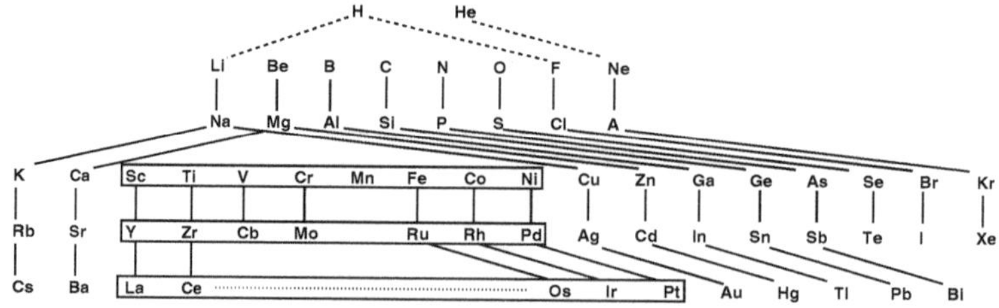

Figure 3.5: *Long form of the periodic table of Thomsen and Bohr.*

Discovery of the inert gases, to which a valency of zero was assigned, stimulated many theories to link the relative periodic position of an element to its valency, the inert gases supplying a bridge between electronegative and electropositive elements.

3.1.5 Interlude

The classical periodic table attained its final form with the accommodation of the inert gases, and represented a self-consistent logical construct to which most important concepts of chemistry could be reduced. During the final period of its reign it was used to make adjustments to atomic-weight measurements and direct the search for undiscovered elements.

Reynolds A final effort to find a simple numerical basis for the periodic classification was made by Reynolds [35], who likened elemental periodicity to the stationary waves in a knotted string, such as are set up when one end of a light cord is vibrated with a large tuning fork. Part of the proposed wave structure is shown in Figure 3.6 as a function of atomic weight.

Discrepancies

When all major anomalies had been removed from the periodic table, a few niggling defects persisted and some fundamental questions remained unresolved. One of the most serious problems was the inverted positions of pairs of elements: K and Ar, Co and Ni, Te and I. There was the difficulty to allocate positions to the rare earth elements. A remarkable difficulty was the position of hydrogen, which belonged with neither the alkali metals nor the halogens, but could fit with either group. Like the related problem presented by Prout's hypothesis these difficulties remained unresolved depite numerous efforts.

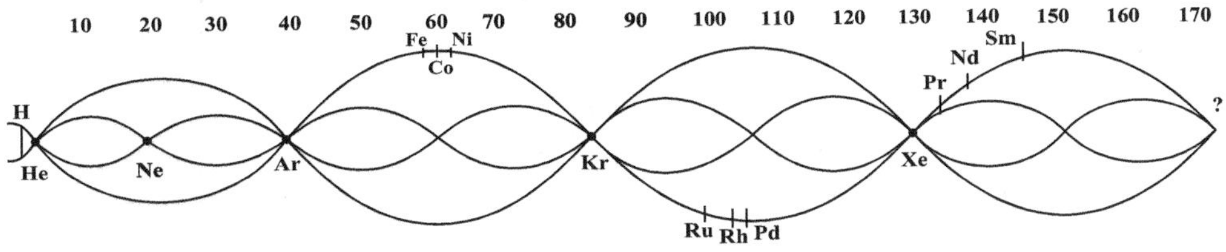

Figure 3.6: *Periodic table in the form of harmonic waves, as a function of atomic weight, proposed by Reynolds [35]. Elements are located on the bold curves.*

Prout's Hypothesis Prout's hypothesis, published anonymously in 1815, was based on the observation that the atomic weights of most atoms are whole multiples of the atomic weight of hydrogen. The simplest explanation of this observation is that *the elements are different aggregates of the atoms of primordial hydrogen.* Despite many results in support of the hypothesis there were notable exceptions, such as chlorine with an atomic weight of 35.5 on the $H = 1$ scale. Interest in the hypothesis flared up again some 40 years later when some deviant atomic weights were remeasured by Stas, who found more of them close to whole numbers. After an exhaustive search however, even Stas had to admit (1865) that, based on experimental evidence, Prout's law was an illusion. Discovery of the periodic law, which soon after demonstrated a clear relationship amongst elements, based on atomic weight, once again seemed to imply a common origin of all matter.

Marignac, discoverer of the elements Yb and Gd, tried (1860) to rescue Prout's hypothesis by suggesting that small variations of atomic composition could be responsible for deviations from whole number values. The same theme was taken up by Crookes in 1888 as a result of his experiments on electric discharge through gases. He postulated that a given element could be a mixture of several meta-elements which, despite identical chemical properties, could have different atomic weights. Crookes had the same experience as Stahl, Prout, Newlands and Marignac before him, his views being dismissed as fantasy by his peers.

End of the Line

Towards the end of the nineteenth century, at about the time that the periodic table gained consensus shape, a number of unexpected phenomena that revealed an internal structure of atoms, were discovered. In less than two decades new concepts such as *line spectra, cathode rays, radioactivity, X-radiation, positive rays, quantum effects, Rutherford scattering* and *mass spectrometry* became commonplace in physics and chemistry. These discoveries had an immediate impact on the conceptual basis of the periodic table and dictated its future development.

Most dramatic was the discovery of radioactivity (1896) by Becquerel, who observed that uranium salts spontaneously emitted radiation which affected photographic plates. It soon became clear that the radiation occurs when an atom decays into an atom of another element. By 1910 enough evidence had accumulated to cause the collapse of the atomic-weight basis of the periodic table. Study of the radio elements revealed that when radiation in the form of an α-particle is emitted, an element shifts back its position in the periodic table by two places, but when a β-particle is ejected, it goes forward one place. From this rule it followed that more than one type of atom may end up at the same periodic site. The end product of three radioactive elements uranium, actinium and thorium were found to be three chemically equivalent types of lead atom, called isotopes, with atomic weights of 206, 207 and 208, respectively. The message was clear: atomic weight is not a unique descriptor of an element and measured atomic weights may well be the average of several isotopes.

By 1918 *isotopy* had been shown to be common amongst most elements and by the technique of mass spectrometry, devised by Aston, the atomic weights of different isotopes could be measured individually and accurately. Aston found all atomic weights measured by his technique to be whole numbers on the oxygen standard. He drew the conclusion that Prout's hypothesis of a century ago, may, after all prove to be correct. Radioactivity signalled the end of the line for the periodic table of Lothar Meyer and Mendeléeff.

3.1.6 Atomic Structure

The first inkling that an atom has internal structure came with the discovery of the electron.

The Electron

The existence of electrons was first inferred from electrochemical measurements, summarized by Faraday's law of electrolysis:

> a given quantity of electricity always liberates the same mass of a given substance, proportional to the equivalent weight of that substance, i.e.
>
> $$m = \frac{MQ}{nF} \tag{3.1}$$

Here m is the mass of an element of atomic weight M liberated at an electrode by the passage of an amount Q of electricity. A constant amount F of electricity is always associated with one equivalent of electrochemical reaction. The integer n was interpreted as the number of elementary charges on an *ion* or charged atom, in solution. It was proposed by Stoney (1891) that this natural unit of electricity be called an *electron* and taken to be the quantity of electricity that must pass through an electrolytic solution in order to liberate one atom of a monovalent substance. Since one *Faraday* (9.6484×10^4 C) liberates one mole of a monovalent substance that contains an Avogadro number (6.0221×10^{23}) of atoms, an electron of charge is given by $e = F/N_A$.

Final discovery of the electron is credited to J.J. Thomson (1900) who observed free electrons in an evacuated discharge tube and measured the ratio of charge to mass, e/m_e, using an instrument as described graphically in Figure 3.7. Displacements of the beam, measured in perpendicularly applied electric and magnetic fields of known strengths, were assumed to be proportional to charge and inversely proportional to the mass (or momentum) of an electron and their ratio calculated directly. The charge of the electron was finally measured independently by Millikan (1909) who balanced the gravitational force on charged oil droplets by an applied electric field.

Currently accepted values of the charge and mass of an electron are

$$e = 1.6022 \times 10^{-19}\text{C}$$
$$m_e = 9.1094 \times 10^{-31}\text{kg}$$

If the mass of an electron is considered to be of purely electrostatic origin, it may be assumed that the potential field energy of an electron confined to a sphere of radius r_{el} is equal to the relativistic energy of the rest mass m_0. Hence,

$$E_{pot} = \frac{e^2}{4\pi\epsilon_0 r_{el}} = m_0 c^2$$
$$r_{el} = \frac{e^2}{4\pi\epsilon_0 m_0 c^2} = 2.8 \times 10^{-15}\text{m} \tag{3.2}$$

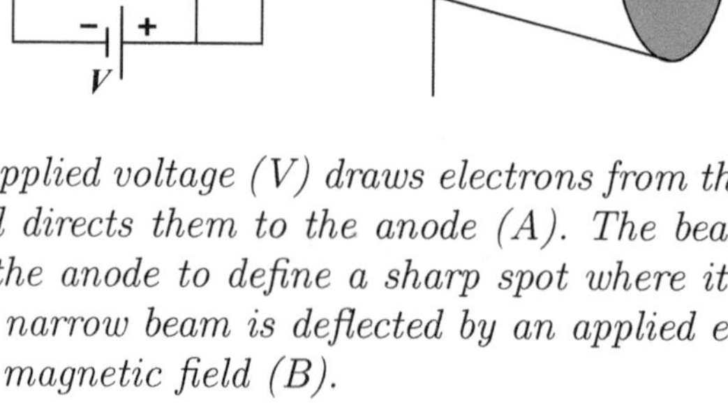

Figure 3.7: *An applied voltage (V) draws electrons from the negatively charged cathode (C) and directs them to the anode (A). The beam is passed through narrow slits in the anode to define a sharp spot where it hits the fluorescent screen (S). The narrow beam is deflected by an applied electric field (E) and a perpendicular magnetic field (B).*

known as the *classical radius* of the electron.[6] From the known atomic weight of hydrogen and Avogadro's number the mass of a hydrogen atom was known to be

$$m_H = \frac{1.008}{6.024 \times 10^{26}} = 1.673 \times 10^{-27}\mathrm{kg}$$

The ratio $m_H/m_e = 1837$ shows that the mass of an electron is very small compared to that of H. It therefore seemed plausible to assume that an atom consisted of a heavy, positively charged sphere of matter with small, light, regularly spaced negatively charged electrons embedded therein, to give a neutral whole.

Radioactivity

Another discovery soon led to observations that showed this plum-pudding model to be seriously in error. Some of the heavy elements were found to spontaneously emit a mysterious type of radiation that indicated atoms of these elements to be intrinsically unstable. Such *radioactive* elements were found to emit three types of radiation, classified as α, β and γ rays respectively. The γ-rays could be identified as high-energy radiation, β-rays as identical to electrons and α-rays turned out to be doubly ionized He atoms, i.e. He^{2+}. It was α-rays, used in scattering experiments that caused a rethink of the simple atomic model of the day.

In 1910 Rutherford, and others, studied the scattering of a beam of α-particles when passed through thin metal films. Most of the α-particles passed through the metal films with little or no deviation. Surprisingly however, some of the α-particles were scattered through large angles and even bounced back from the metal film. To account for the observed scattering it was argued that high-angle scattering resulted from near collisions between α-particles and dense, positively charged atomic cores.

Nuclear Model of the Atom

Scattering of an α-particle with positive charge $2e$, approaching a positive point charge Ze with velocity v_0 can be simulated in terms of Coulomb's law that specifies the repulsive force between these charged bodies at a vector distance $\boldsymbol{r}$ apart, as

$$\boldsymbol{F} = \frac{2Ze^2}{4\pi\epsilon_0 r^2}\frac{\boldsymbol{r}}{r} \tag{3.3}$$

[6]The actual size of an electron however, is a function of its environment [36] and in the hydrogen atom it may be of the same size as the entire atom.

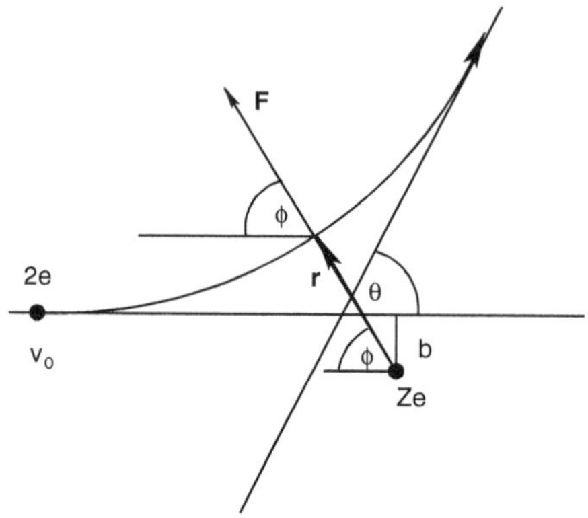

Figure 3.8: *Diagram to illustrate the scattering of an α-particle of charge* 2e *by a fixed nucleus of charge Ze, according to Coulomb's law.*

The scattering is shown diagrammatically in Figure 3.8 in terms of the *impact parameter* b, which is the distance of closest approach of the α-particle to the target nucleus, assuming that no deflection occurs. A simple derivation [37] shows that

$$b = \frac{k}{mv_0^2}\cot(\theta/2) \tag{3.4}$$

$$\mathrm{d}b = -\frac{k}{2mv_0^2}\frac{1}{\sin^2(\theta/2)}\mathrm{d}\theta \tag{3.5}$$

$$k = \frac{2Ze^2}{4\pi\epsilon_0}$$

The differential $\mathrm{d}b$ quantifies the range $\mathrm{d}\theta$ in scattering angle at which particles are observed. The angle θ decreases with increasing b. Integration, to allow for rotational symmetry of the scattering process, produces the differential cross section. The surface element on the unit sphere (Figure 3.9) is called a *solid angle.* If the detector subtends a solid angle $\mathrm{d}\Omega$ the number of particles observed at angle θ obeys the scattering formula in terms of the differential cross section

$$\frac{\mathrm{d}\sigma}{\mathrm{d}\Omega} = \frac{Z^2e^4}{(4\pi\epsilon_0)^2m^2v_0^4\sin^4(\theta/2)} \tag{3.6}$$

This formula was tested experimentally. Keeping the solid angle $\mathrm{d}\Omega$ constant, the $\sin^4(\theta/2)$ law is reproduced exactly in the counting rate of scattered α-particles. With particle energies up to 5 MeV and scattering angles of 150°, corresponding to an impact parameter of 6×10^{-15}m, no deviations from the

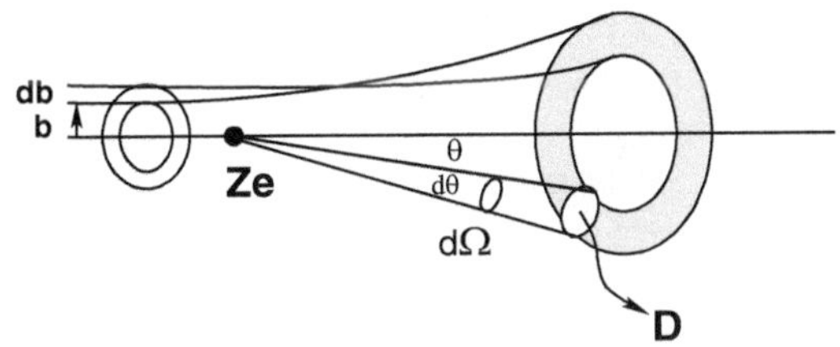

Figure 3.9: *Solid angle of scattering defined. The detector D subtends a solid angle $d\Omega$ at the scattering centre.*

Rutherford formula are observed. The number of charges on the nucleus Z was fitted empirically and found to agree with the numerical position of the target material in the periodic table.

For near-central collisions, the scattering of fast particles at large θ or small b, deviations from the Rutherford formula are observed. Under these circumstances Coulomb's law is apparently no longer obeyed. The α-particles approach the nuclei so closely that another, short-range interaction force, the nuclear force, becomes effective. The final conclusion is that virtually all of the atomic mass is concentrated in a nucleus of radius $R \simeq 10^{-15}$m.

For very large impact parameters (small deflection angles), the Rutherford formula is likewise no longer exactly valid, because now the Coulomb potential of the nucleus is screened by the extranuclear electrons. These effects occur for $b > 10^{-10}$m.

The radius of the nucleus is defined (Figure 3.10) as the distance at which the effect of the nuclear potential is comparable to that of the Coulomb potential. For nuclei of mass number A it is found empirically that

$$\begin{aligned} R &\simeq (1.3 \pm 0.1)A^{1/3} \times 10^{-15}\text{m} \\ &= r_0 A^{1/3} \end{aligned}$$

e.g. $R(^{12}\text{C}) = 2.7 \times 10^{-15}$m $\quad R(^{208}\text{Pb}) = 7.1 \times 10^{-15}$m

These radii are comparable to the classical radius of an electron. The relationship between nuclear mass and nuclear radius implies that the density of nuclear matter is constant and independent of the size of the nucleus. If the nucleus is assumed to be made up of A identical *nucleons*, this number is proportional to the nuclear volume. On the atomic mass scale each nucleon has mass $1/N_A$. The nuclear density calculates as

$$\begin{aligned} \rho_n &= \frac{M_n}{V_n} = \frac{(A/N_A)}{\frac{4}{3}\pi(r_0 A^{-1/3})^3} \\ &= \frac{3}{4\pi r_0^3 N_A} \approx 2 \times 10^{17}\text{kg m}^{-3} \end{aligned} \tag{3.7}$$

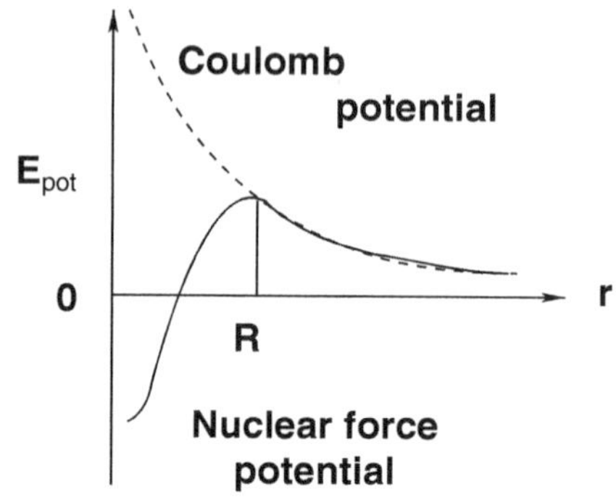

Figure 3.10: *Schematic diagram to compare Coulomb and nuclear potentials as a function of the radial distance from the nuclear centre.*

The number density of nucleons in the nucleus

$$N_n = \rho_n N_A = \frac{3}{4\pi r_0^3} \simeq 1.4 \times 10^{44}\text{m}^{-3} \tag{3.8}$$

It was postulated by Rutherford on the basis of (3.6) that an atom had a dense positively charged *nucleus*, surrounded by a diffuse cloud of negatively charged electrons. Suggestions that electrons revolve about the nucleus like planets around the sun, were difficult to defend in view of the fact that accelerated charges were known to radiate energy. Although electrostatic attraction to the positive nucleus would be balanced by the centrifugal force on the orbiting electron, loss of radiant energy would continually disturb this balance.

3.1.7 Atomic Number

The numerical position of an element in the periodic table has already been mentioned, with hindsight, as a measure of the positive charge, concentrated at the atomic nucleus. This identification was first proposed by Van den Broek [38]. Atomic order, rather than atomic weight, was stated to provide an improved guide to the solution of elemental problems. Thus, the correct sequence of elements in the periodic table depends on atomic order rather than atomic weight. It was suggested that, if the elements were arranged in increasing order of number of electrons surrounding the positive nucleus of the neutral atom, the numerical position of each element in this sequence, to be called its *atomic number*, corresponds to the proper position occupied by this element in the periodic sequence. At the time, atomic numbers could be assigned for elements up to about 57, where Mendeléeff's table also ran into problems to include the recently discovered lanthanides.

These proposals were not only in line with Rutherford's atomic model, but also corrected the patently wrong sequence for the pairs Ar, K and Co, Ni, according to atomic-weight classification. The significance of atomic number was confirmed soon after through an analysis of the characteristic X-ray spectra of the elements.

X-radiation

X-rays were discovered by Röntgen in 1895 as the radiation emitted when a beam of fast cathode rays strikes a solid target in an evacuated tube (Figure 3.11). Since accelerated charges are known to be a source of electromagnetic radiation it could be argued that when a beam of electrons is stopped at a solid anode radiation should result. Radiation produced by such a mechanism is aptly named in German as *Bremsstrahlung*, (brake radiation).

The wavelength of x-rays was first measured in one of the crucial experiments of physics, proposed by Von Laue (1912). If theoretical speculation about the nature of crystals as made up of periodically repeating elementary building blocks, or unit cells, were right, it was argued that such unit cells should be of dimensions comparable to the conjectured wavelength of x-rays. Indeed, crystals were found to act as a three-dimensional diffraction grating for x-rays and the positive outcome of the first experiment vindicated two theories at once: x-rays consist of waves and crystals of regular building blocks. Wavelengths in the region of 13–48 pm could be measured.

The nature of the emitted X-radiation depends on the accelerating voltage and the composition of the anode target material. In the case of pure anode material the radiation consists of a continuous component, called *white radiation*, which is Bremsstrahlung, with a superimposed line spectrum (Figure 3.12).

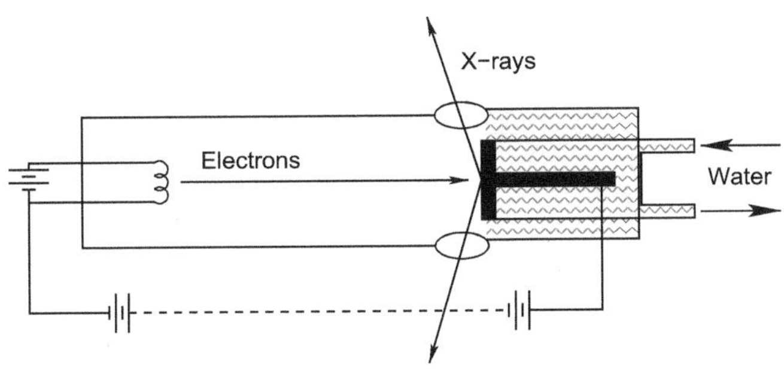

Figure 3.11: *Schematic diagram of a typical x-ray tube consisting of an electron gun and a water-cooled anode in an evacuated metal tube. The transparent windows are made of beryllium.*

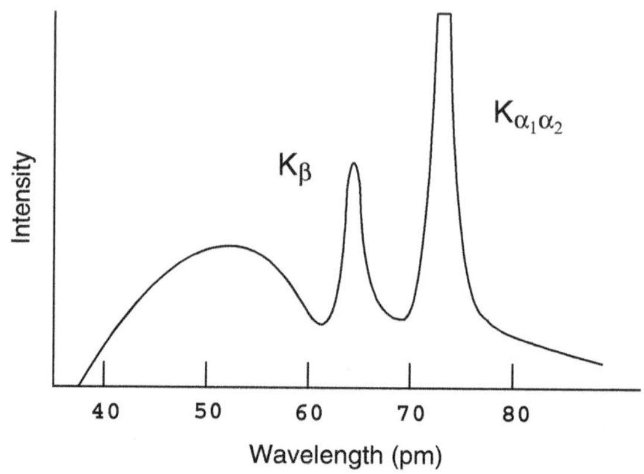

Figure 3.12: *Part of the Mo x-ray emission spectrum.*

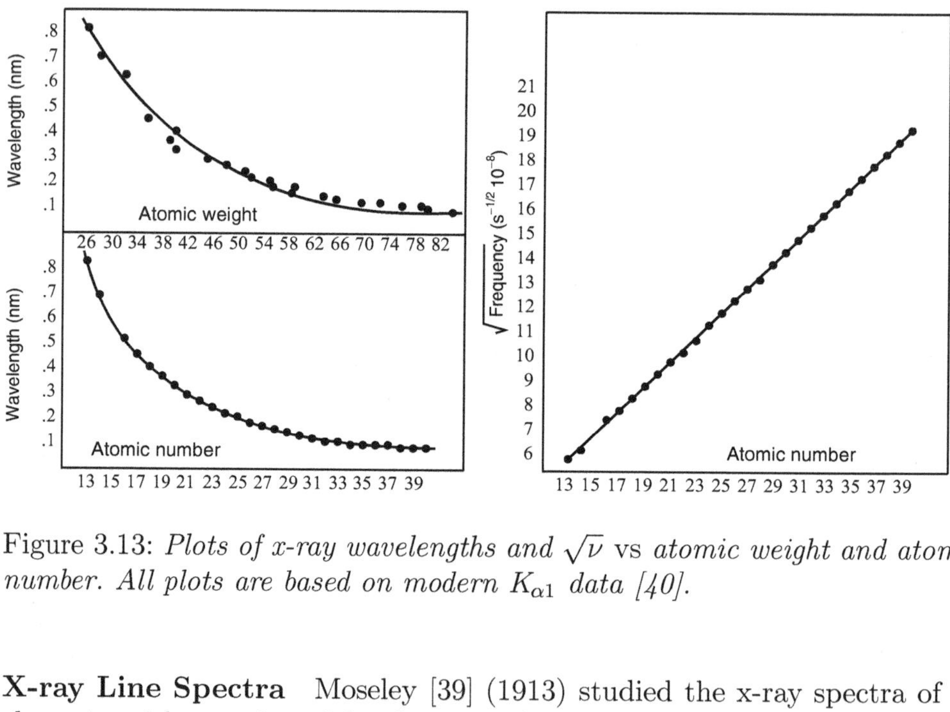

Figure 3.13: *Plots of x-ray wavelengths and* $\sqrt{\nu}$ vs *atomic weight and atomic number. All plots are based on modern* $K_{\alpha 1}$ *data [40].*

X-ray Line Spectra Moseley [39] (1913) studied the x-ray spectra of 39 elements with atomic weights between Al and Au. All of the spectra were remarkably alike and consisted of groups of lines, referred to as K, L, M, etc. It was noticed that the wavelengths of corresponding lines from each group, for different elements, varied in a regular way with the atomic weights of the target material. This relationship, as demonstrated graphically in Figure 3.13, is not mathematically precise. The relationship between wavelength and atomic number, shown for comparison in the same figure, is obviously a better fit. The final frame in Figure 3.13 shows the square root of characteristic frequencies plotted against atomic number. This linear relationship must obviously be of fundamental importance, which only became apparent after publication of Bohr's quantum model of the atom later in the same year. It is remarkable how unrelated work of Rutherford, Van den Broek, Von Laue and Bohr, from four different countries, happened at about the same time. Together, but not in isolation, these researches established a new paradigm for the refinement of the periodic table of the elements.

3.2 Theoretical Development

One of the unsolved problems of late nineteenth century physics with a direct bearing on the arrangement of electrons in atoms and which became of decisive importance in explaining the structure of the periodic table was the understanding of atomic line spectra.

3.2.1 Atomic Line Spectra

It has been known since the days of Kirchhoff and Bunsen (*c.*1859) that each element emits a characteristic line spectrum when in the incandescent state. This observation was interpreted to mean that atoms cannot be structureless, but an interpretation of the spectra in terms of atomic structure only became possible when definite relationships between the spectral lines of any one element came to light.

When hydrogen gas is excited in an electric discharge tube, the emitted light does not occur as a continuous spectrum, but as a mixture of components with sharply defined frequencies. A schematic drawing of such a photographically recorded spectrum is shown in Figure 3.14. In 1885, Balmer pointed out that a simple relation existed between the different lines in this spectrum of hydrogen. The spectrum, illustrated in Figure 3.14 consists of a series of lines known as H_α, H_β, H_γ, H_δ, etc., extending from the red into the ultraviolet region. Balmer observed that the wavenumber of each line of the series could be represented by the formula

$$\bar{\nu} = \frac{1}{\lambda} = R\left(\frac{1}{2^2} - \frac{1}{n^2}\right) \tag{3.9}$$

where R is a constant and $n = 3, 4, 5, 6$, etc. As $n \rightarrow \infty$ the limiting wavenumber, or *head of the series* is approached at $\bar{\nu} = R/4$. This simple numerical

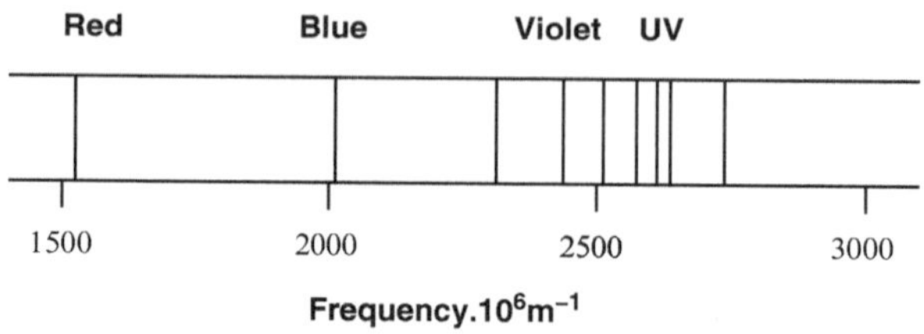

Figure 3.14: *First eight lines in the Balmer series of H emission lines,* $n = 3 \to 10$ *in* $\nu = Rc\left(\frac{1}{2^2} - \frac{1}{n^2}\right)$. *The line on the right marks the head of the series.*

relationship is probably the most important empirical formula ever discovered. It represents the first instance of an atomic property characterized by integers, which is now recognized as fundamental to the development of modern quantum theory. The significance of the discovery was apparently lost on contemporary scientists who dismissed it as a numerical oddity without physical meaning.[7]

Further regularities in line spectra were soon discovered by Rydberg, Ritz, Lyman, Paschen and others, culminating in the general formula

$$\nu = Rc\left(\frac{1}{m^2} - \frac{1}{n^2}\right) \tag{3.10}$$

that summarized all known series of hydrogen line spectra. The constant R, named *Rydberg's constant* has the numerical value $R = 1.0974 \times 10^7$ m^{-1}, $c = 3 \times 10^8$ ms^{-1} is the speed of light and m and n are integers, such that $n > m = 1, 2, 3 \ldots$, each value of m defining another spectral series.

The mysterious integers that were found to characterize atomic line spectra found a ready explanation in terms of the quantum theory of radiation first mooted by Max Planck in 1901.

3.2.2 Quantum Theory

The quantum theory as first used by Planck is nothing but the first atomic theory of energy. Planck's problem dealt with the radiation emitted by hot objects and the dependence of wavelength on the temperature of the emitter. It is well known that the colour inside a furnace changes from red,

[7]Numerical relationships between planetary orbits are still treated with the same contempt as mere coincidence by astrophysicists.

through yellow, to white as the temperature increases. A closed furnace with a small orifice that allows radiation to escape without disturbing the interior equilibrium has the peculiar name *black body.* A black body is any cavity that contains radiation in thermal equilibrium with the walls of the cavity. The material of the walls emits and absorbs thermal radiation at all wavelengths. It was shown experimentally and theoretically that the energy radiated by a black body depends on the temperature. The spectral distribution of this radiation is of a universal nature and the results of typical measurements are shown schematically in Figure 3.15. The fact that the observed spectral distribution could not be simulated correctly by any theoretical thermodynamic model developed into a major irritant for the physics community.

The total radiated energy and radiation density were known from Stefan's law,

$$\rho = aT^4 \tag{3.11}$$

but the theoretical derivation of the radiation density and hence the value of a, by integrating the average energy contributions $\bar{\varepsilon}$ over all vibrational modes, i.e.

$$\rho(\nu)\mathrm{d}\nu = \frac{8\pi\bar{\varepsilon}\nu^2\mathrm{d}\nu}{c^3} \tag{3.12}$$

was unsuccessful. The problem consists of finding the correct form of $\bar{\varepsilon}$.

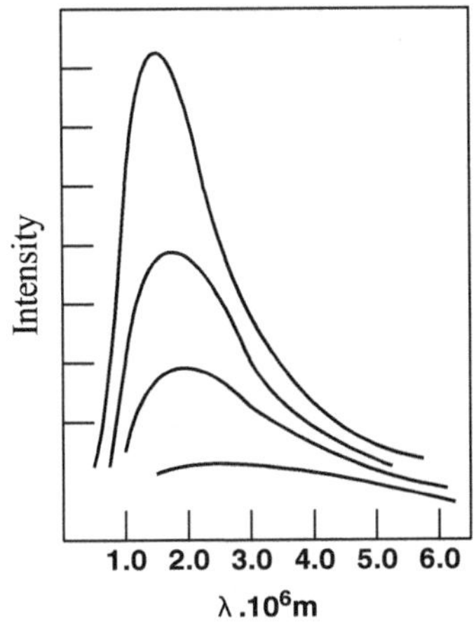

Figure 3.15: *Spectral distribution of black-body radiation at different temperatures.*

The brute force attempt by Raleigh and Jeans to integrate over all energy levels in the Boltzmann distribution

$$\bar{\varepsilon} = \frac{\sum_0^\infty \varepsilon_n e^{-\varepsilon_n/kT}}{\sum_0^\infty e^{-\varepsilon_n/kT}} \tag{3.13}$$

based on the assumption that energy changes continuously, produced the form $\varepsilon = kT$ and the density distribution

$$\rho(\nu)\mathrm{d}\nu = \frac{8\pi kT\nu^2\mathrm{d}\nu}{c^3} \tag{3.14}$$

in terms of Boltzmann's constant k.

This formula fits the experimental curve only at long wavelength. At short wavelength

$$\lim_{\lambda\to 0}\rho(\lambda) = \lim_{\nu\to\infty}\rho(\nu) = \infty$$

It predicts that thermal equilibrium can be sustained only by endless absorption of thermal energy, and conversion into radiation energy of wavelengths approaching zero. This result is known as the ultraviolet catastrophe.

In another attempt to find the correct solution Wien proposed the empirical formula

$$\rho(\nu)\mathrm{d}\nu = \frac{8\pi k\beta}{c^3}e^{-\beta\nu/T}\nu^3\mathrm{d}\nu \tag{3.15}$$

that gives an excellent fit at high frequencies, but fails at low frequency and high temperature.

By writing $x = \nu/T$ both the Raleigh and Wien formulae can be cast into the form

$$\rho(\nu)\mathrm{d}\nu = \frac{8\pi}{c^3}F(x)\nu^3\mathrm{d}\nu$$

such that

$$F(x) = \begin{cases} k\beta e^{-\beta x} & \text{(Wien)} \\ k/x & \text{(Raleigh)} \end{cases}$$

Planck, who eventually solved the problem tried to find a fundamental basis for the form

$$F(x) = k\beta\left(e^{\beta x} - 1\right)^{-1}$$

that converts to the Wien function at large x and to the Raleigh function at small x.[8]

[8]The exponential function

$$e^x = 1 + x + (x^2/2!) + (x^3/3!) + \ldots$$

For small x it reduces to $e^x \simeq 1 + x$

To obtain the correct form of β, consistent with experiment, some drastic assumptions are necessary. Most important, and based on the fact that integration of the energy function leads to the ultraviolet catastrophe, it is necessary to assume that the energy does not change continuously and hence that the infinite summations must be carried out directly. To achieve that, summation over an elementary energy pulse ε, such that $n\varepsilon = \varepsilon_n$, is required. Since the distribution of ε_n must predict smaller contributions to the radiation density at high frequency, it is necessary that the factor $\exp(-\varepsilon_n/kT)$ should diminish. This happens when $\varepsilon \propto \nu$, i.e. $\varepsilon_n = nh\nu$. The proportionality factor h is known as Planck's universal constant. The average energy is calculated by evaluation[9] of the resulting expression:

$$\bar{\varepsilon} = \frac{h\nu \sum_{n=0}^{\infty} n \exp(-nh\nu/kT)}{\sum_{n=0}^{\infty} \exp(-nh\nu/kT)}$$

A function of the required form is obtained, provided $\beta = h/k$, to give the radiation density

$$\rho(\nu)\mathrm{d}\nu = \frac{8\pi h}{c^3}\left(e^{h\nu/kT} - 1\right)^{-1} \nu^3 \mathrm{d}\nu \qquad (3.17)$$

This formula is consistent with the experimental distribution and on integration it leads to Stefan's law, i.e.

$$\rho = \left[\frac{8\pi^5 k^4}{15h^3c^3}\right] T^4$$

[9] The numerator, written in the form

$$\begin{aligned} e^{-x} + e^{-2x} &+ e^{-3x} + \dots \\ + e^{-2x} &+ e^{-3x} + \dots \\ &+ e^{-3x} + \dots \\ &\qquad + \dots \end{aligned}$$

rearranges to

$$\begin{aligned} &\left(e^{x} - 1\right)^{-1} + e^{-x}\left(e^{x} - 1\right)^{-1} + e^{-2x}\left(e^{x} - 1\right)^{-1} + \dots \\ &= \left(e^{x} - 1\right)^{-1}\left(1 - e^{-x}\right)^{-1} \end{aligned}$$

The denominator has the same form as the second term and hence

$$\bar{\varepsilon} = h\nu\left(e^{h\nu/kT} - 1\right)^{-1} \qquad (3.16)$$

The essential problem highlighted by Planck's analysis is that the equipartition principle ($\bar{\varepsilon} = kT$) does not hold for vibrational states and that a quantum (atomic) theory is required for radiation energy. The value of Planck's constant is independent of the composition of the black body and has dimensions of action = energy × time. $h = 6.6254 \times 10^{-34}$ Js.

Returning to the Rydberg equation (3.10), multiplied by h it becomes an energy equation

$$\begin{aligned} h\nu &= Rch\left(\frac{1}{m^2} - \frac{1}{n^2}\right) \\ &= E_m - E_n \end{aligned} \tag{3.18}$$

where E_m and E_n correspond to the energy of the system before and after the emission of the energy quantum $h\nu$. Furthermore, apart from an arbitrary constant

$$E_n = -Rch/n^2 \tag{3.19}$$

On the basis of these results Bohr developed a theory to account for line spectra in terms of atomic structure.

3.2.3 The Bohr Model

As the basis of his atomic model Bohr accepted the frequency condition of (3.18).

$$\nu = \frac{1}{h}\left(E_n - E_m\right)$$

to mean that extranuclear electrons in atoms are confined to discrete energy levels described by positive integers, n and m. The allowed energies correspond to a series of *stationary states* of the electron and transition between two stationary states occurs with the emission or absorption of a monochromatic quantum of radiation of frequency $\nu = \Delta E/h$. While the electron remains in a stationary state n its energy remains constant (3.19). In the case of a hydrogen atom a single electron is associated with a proton at the nucleus and to avoid the electron being pulled into the nucleus it is assumed to be accelerated by the Coulombic attraction of $e^2/4\pi\epsilon_0 r^2$ to stabilize a circular orbit of radius r, about the nucleus. If the electron orbits the nucleus at constant velocity $v = 2\pi r\omega$ and frequency of revolution ω, mechanical stability requires

$$\begin{aligned} \frac{mv^2}{r} &= mr(2\pi\omega)^2 = \frac{e^2}{4\pi\epsilon_0 r^2} \\ \omega^2 &= \frac{e^2}{16\pi^3\epsilon_0 mr^3} \end{aligned} \tag{3.20}$$

Contrary to classical theory that requires an accelerated electron to radiate energy at the same frequency, $\nu = \omega$, Bohr postulated a quantum condition that prevents radiation by electrons in stationary states. The kinetic energy of the orbiting electron

$$\frac{1}{2}mv^2 = \frac{1}{2} \cdot 4\pi^2 mr^2\omega^2 = \frac{e^2}{8\pi\epsilon_0 r}$$

The potential energy

$$E_p = \int_\infty^r \frac{e^2}{4\pi\epsilon_0 r^2}\mathrm{d}r = -\frac{e^2}{4\pi\epsilon_0 r}$$

The total energy

$$E = \frac{1}{4\pi\epsilon_0}\left[\frac{e^2}{2r} - \frac{e^2}{r}\right] = -\frac{e^2}{8\pi\epsilon_0 r}$$

Removal of the electron from the atom requires an ionization energy $W = -E$. The radius of the orbit and the frequency of revolution become

$$r = \frac{e^2}{8\pi\epsilon_0 W}, \quad \omega = \left(\frac{32\epsilon_0^2 W^3}{me^4}\right)^{\frac{1}{2}} \tag{3.21}$$

There are no grounds to assume that the classical frequency of rotation should be simply related to the quantum-mechanical frequency of radiation generated by transition between (radiationless) stationary states. By analogy with Planck's analysis of black-body radiation however, Bohr could argue that such a relation should exist at low frequency. The classical Raleigh-Jeans equation gives the same result as the quantum-mechanical Planck distribution at very low frequency. Bohr postulated this *Principle of Correspondence* between classical and quantum theories to be generally valid, and in the case of the hydrogen atom to apply at large values of n. For a transition between neighbouring orbits the frequency of radiation emitted is

$$\nu = Rc\left\{\frac{1}{n^2} - \frac{1}{(n+1)^2}\right\}$$

which, for large n reduces to $2Rc/n^3$, and under these conditions the correspondence principle implies that $\nu = \omega_n$, i.e.

$$\frac{2Rc}{n^3} = \left(\frac{32\epsilon_0^2 W^3}{me^4}\right)^{\frac{1}{2}}$$

This expression is combined with that for orbital energy, $W = Rch/n^2$, such that

$$\frac{2Rc}{n^3} = \frac{2}{nh}\left(\frac{Rch}{n^2}\right) = \frac{2W}{nh}$$

and

$$\left(\frac{2W}{nh}\right)^2 = \frac{32\epsilon_0^2 W^3}{me^4}$$

$$E_n = -W = -\frac{1}{n^2}\cdot\frac{me^4}{8\epsilon_0^2 h^2} \tag{3.22}$$

The calculated value of

$$Rc = \frac{n^2 W}{h} = \frac{me^4}{8\epsilon_0^2 h^3} = 3.29 \times 10^{15}\text{s}^{-1}$$

agrees with the experimental value of Rydberg's constant. The electronic energy is *quantized* in terms of the *quantum number* n and the orbital radius

$$r_n = \frac{n^2 h^2 \epsilon_0}{\pi m e^2} \tag{3.23}$$

The square of the angular momentum

$$L^2 = (pr)^2 = E_k(2mr^2) = \frac{e^2}{8\pi\epsilon_0 r}\cdot 2mr^2 = \frac{2me^2 r}{8\pi\epsilon_0} = \left(\frac{nh}{2\pi}\right)^2$$

The angular momentum, like the energy, is therefore quantized, in units of $\hbar = h/2\pi$.

The origin of the hydrogen line spectrum is neatly accounted for in terms of the transitions indicated on Figure 3.16. Despite the convincing explanation of hydrogen line spectra, the planetary model proposed by Bohr could never be extended to describe the electronic configuration of atoms other than hydrogen, or to provide an acceptable model for chemical bonding. On the other hand, regularities in the K_α x-ray emission spectra observed by Moseley, e.g.

$$\nu = Rc(Z-1)^2\left(\frac{1}{1^2} - \frac{1}{2^2}\right) \tag{3.24}$$

where Z is atomic number, seem to follow the Bohr model rather well. It seems reasonable to interpret (3.24) to define the observed lines as due to the transition of an electron from an orbit of quantum number 2 to one of quantum number 1 for a nuclear charge of $Z - 1$. The equation works less well for the L and M series that involve transitions between higher electronic levels.

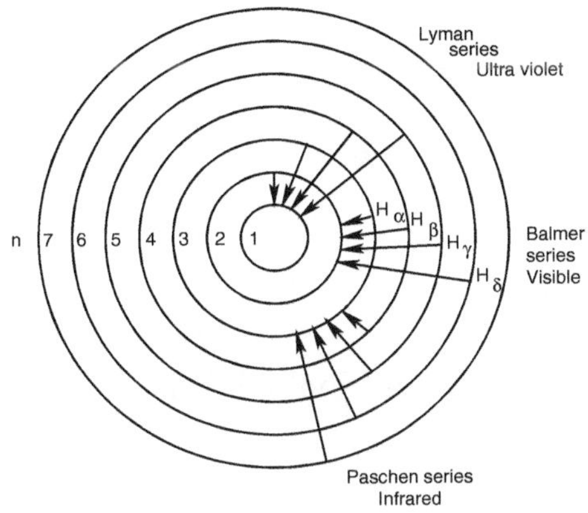

Figure 3.16: *Diagram to show how the line spectrum of hydrogen is generated by electronic transitions between energy levels of quantum number n.*

Any relationship with the periodic table claimed on the basis of the Bohr model remained conjectural because there is no reliable estimate of the number of electrons that may co-exist at a given Bohr level. It is primarily for this reason that the Bohr model was never really accepted by the chemistry community. An alternative, based on traditional thinking on chemical bonding, valency and electron affinity, although static rather than dynamic, was no less effective than the Bohr model to account for the ultimate stability of the atom. From a modern perspective, the Bohr model is essentially classical with a quantum gloss, to be accepted in good faith as a valid explanation of stationary states. In the long term however, it stimulated enquiry that produced the more successful wave-mechanical model of the atom. During the interim the periodic table acquired its final form, still in general use today, on the basis of the static atomic model of Lewis, Langmuir and Bury.

3.2.4 Static Model of the Atom

Efforts to correlate observed variation of chemical parameters with known periodicities of the elements, independent of any atomic model, resulted in a scheme that could be interpreted to define the most likely relative positions of electrons with respect to an atomic nucleus and one another. This scheme was based on the planetary Bohr model, ignoring questions of stability. The simplified model, proposed by Kossel, considered the extranuclear electrons to be situated in concentric shells. In passing along the elemental sequence, each fresh atom contains one electron more than its predecesser. The opening of each fresh period in the periodic table was assumed to correspond to placement of the first electron in a new shell. The maximum number of electrons accommodated in any shell was considered fixed by numbers in the series

$$N = 2(1 + 2^2 + 2^2 + 3^2 + 3^2 + 4^2 + \dots) \tag{3.25}$$

proposed by Rydberg [41].

In response, Bohr is quoted [42], to state:

> This interpretation of the atomic number may be said to signify an important step toward the resolution of a problem which for a long time has been one of the boldest dreams of natural science, namely, to build up an understanding of the regularities of nature upon the consideration of pure numbers

A more general scheme was proposed at about the same time by Lewis.

The Lewis Model

The scheme, first proposed by Lewis [43], accepted the nuclear structure of the atom and that the number of electrons in each atom corresponds to the ordinal position (2.2) of the element in the periodic table. To account for the chemical inertness of the elements of group 0, a special stability, associated with either an electron pair or a group of eight electrons, was postulated.

In terms of this scheme the helium atom has two electrons close to the atomic nucleus. Neon consists of a pair of inner electrons (as in helium) inside a cube of eight electrons. Argon has the same arrangement of ten inner electrons, surrounded by eight additional electrons at the corners of a larger external cube. The electrons of all other atoms have the same arrangement as atoms with atomic number one unit less, plus an additional electron, placed according to the sequence defined for the inert gases. In chemical reactions all atoms tend to either take up or give up electrons so as to resemble either helium or one of the other elements of the inert group. Thus, lithium, which contains one electron outside the helium pair, tends to give up this extra electron and hence is electropositive in its chemical properties, while fluorine that has only one electron short of the neon cube, is highly electronegative. When lithium reacts with fluorine an electron is transferred to create the Li^+F^- pair. This tendency of atoms to acquire an outermost shell of eight electrons is satisfied not only by transfer, but also by sharing of electron pairs between atoms to form *covalent bonds*. It was emphasized by Lewis that shared electrons always occur in pairs and that the rule of two was even more fundamental than the rule of eight.

The Langmuir Model

The ideas of Lewis were extended by Langmuir [44] to all elements of the periodic table. It was stated:

> In attempting to determine the arrangement of electrons in all atoms, we must be guided by the numbers of electrons which make up the atoms of the inert gases; in other words by the atomic numbers of these elements, namely, helium 2, neon 10, argon 18, krypton 36, xenon 54, and niton 86.

These are the numbers of the series defined by Rydberg [41]. Presumably therefore, there must be something about an arrangement of 18 or 32 electrons that makes it just as stable as the cube in the case of neon and argon. Langmuir proposed that electrons in atoms are distributed through a series of concentric shells, and, in the case of the inert gases, that the electrons in each shell are arranged symmetrically about a plane through the nucleus. Consecutive shells contain 2, 8, 18 and 32 cells. The innermost cells can accommodate only one electron while each of the other cells can hold two electrons. The factor 2 in Rydberg's series is accounted for by the assumption of the symmetry plane and repetition of the square factors leads to the assumption of two electrons per cell. On the basis of these postulates Langmuir derives the electronic structures of different elements shown in Table 3.6. The elements are arranged so that elements with electrons in the same layer occur in the same column. A periodic table based on this proposed electronic distribution is in almost perfect register with the modern long form of the table. Moving the two sets of elements shown enclosed in stippled boxes, as indicated by vertical arrows, reproduces the modern table.

The Bury Model

Langmuir's theory was modified by Bury [45], assuming that instead of 2, 8, 18, or 32 electrons in the outer layer, the maximum number of electrons in this layer cannot exceed 8. It was argued that the number of electrons in a shell can exceed 8, only in inner shells, when an accumulation of electrons has already commenced in an outer layer. On the basis of this postulate an arrangement of inert gas electronic configuration in keeping with known chemical properties was derived as shown in Table 3.7. Bury's modification does not affect the general layout of the periodic table and like Langmuir's scheme it predicts the correct classification of the chemical elements. It has the additional merit of reflecting electronic structures of individual elements, well in line with modern experimental results.

Table 3.6: *Arrangement of electrons in atoms, according to Langmuir (1919).*

N:	1	3	11	19	37	55	87
Layer:	I	IIa	IIb	IIIa	IIIb	IVa	IVb
Group I.....	H 1	Li 1	Na 1	K 1	Rb 1	Cs 1	— 1
II....		Be 2	Mg 2	Ca 2	Sr 2	Ba 2	Ra 1
III...		B 3	Al 3	Sc 3	Y 3	La 3	Ac 3
IV...		C 4	Si 4	Ti 4	Zr 4	Ce 4	Th 4
						Pr 5	
						Nd 6	
						— 7	
						Sm 8	
						Eu 9	
						Gd 10	......Rare Earths
						Tb 11	
						Dy 12	
						Ho 13	
						Er 14	
						— 15	
						Yb 16	
						Lu 17	
						Hf 18	
V....				V 5	Cb 5	Ta 19	Ux_2 5
VI...				Cr 6	Mo 6	W 20	U 6
VII..				Mn 7	— 7	— 21	
				Fe 8	Ru 8	Os 22	
VIII.				Co 9	Rh 9	Ir 23	
				Ni 10	Pd 10	Pt 24	
I.....				Cu 11	Ag 11	Au 25	
II....				Zn 12	Cd 12	Hg 26	
III...				Ga 13	In 13	Tl 27	
IV...				Ge 14	Sn 14	Pb 28	
V....		N 5	P 5	As 15	Sb 15	Bi 29	
VI...		O 6	S 6	Se 16	Te 16	RaF 30	
VII..		F 7	Cl 7	Br 17	I 17	— 31	
0.....	He 2	Ne 8	Ar 8	Kr 18	Xe 18	Rn 32	

Finally it was possible to reconcile the chemical table with the Bohr-Sommerfeld atomic model, which introduced additional quantum numbers ad hoc to describe individual elliptical orbits for electrons, and to be in line with spectroscopic results.

Table 3.7: *Electronic structure of rare-gas atoms in terms of Bury's proposals.*

Element	Shell						
	K	L	M	N	O	P	Q
He	2						
Ne	2	8					
Ar	2	8	8				
Kr	2	8	18	8			
Xe	2	8	18	18	8		
Rn	2	8	18	32	18	8	

3.2.5 The Sommerfeld Model

The first modification to the Bohr model of the atom, by Sommerfeld [46] was the introduction of elliptical orbits, in addition to the circular. To specify an elliptic orbit it was necessary to introduce an *azimuthal* quantum number n_a together with a *radial* quantum number n_r, such that the principal quantum number $n = n_a + n_r$. To keep the new system consistent with spectroscopic observation it was necessary to specify a number of selection criteria that restrict possible electronic transitions between orbital energy levels, e.g. (i) When n changes by unity, the transition is restricted to occur between circular orbits; (ii) In all other cases, $\Delta n_r = \pm 1$.

Classification of spectral lines on the basis of general appearance and behaviour, showed that an emission spectrum could be divided up into four series, called *principal, sharp, diffuse* and *fundamental.* To address this issue Sommerfeld argued that a third quantum number was needed to describe electronic orbits in three-dimensional space. The discovery [47] that an electron has intrinsic angular momentum, called *spin*, neccessitated the introduction of a fourth quantum number. The physical significance of the four quantum numbers was interpreted to be:

n principal quantum number that defines the energy
l defines the angular momentum
m_l defines the direction of the angular momentum
m_s the spin quantum number

A fifth selection principle was provided by Pauli's *exclusion principle*, which states that two electrons on the same atom may not have four quantum numbers the same.

In terms of the four quantum numbers it was possible to associate the appearance of the four spectral series with specific changes in values of l during transition. On this basis orbital electrons were characterized as s, p, d, f, for $l = 0$, 1, 2, 3, respectively.

It is not too difficult to show the parallel between the Sommerfeld and Bury schemes, either of which can serve as a basis to interpret the periodic table, although the order of orbital stability remains an empirical result. The basic common assumption is that electrons in an atom occupy the available energy levels in the order of increasing principal quantum number. To first approximation, all electrons with common n therefore have the same orbital energy. Because of mutual repulsion between electrons however, electronic energies at a given level also vary as a function of angular momentum and quantum number l. For a given value of l, the quantum number m_l can have integral values in the range $-l \leq m_l \leq l$. For the sub-level defined by $l = 0$ there is only a single allowed value of $m_l = 0$, known as the s-state. For $l = 1$, $m_l = -1, 0, 1$, the sub-level has a *degeneracy* of 3, and is known as a p-state. For d and f states, $l = 2, 3$, the sub-levels are five- and seven-fold degenerate respectively. The exclusion principle dictates that a maximum of two electrons can be accommodated for each allowed value of m_l. It follows that s, p, d and f sub-levels can accommodate maxima of 2, 6, 10 and 14 electrons respectively. The restrictions on the quantum number l therefore imply that for $n = 1$, only one s sub-level is allowed. For $n = 2$, s and p sub-levels are allowed. For $n = 3$ there are s, p and d sub-levels, and for $n = 4$ there are s, p, d and f sub-levels. The accepted notation to specify the electronic distribution at a given energy level is of the form nl^x, where l indicates the sub-level specification and x the number of electrons at that sub-level. The occupation of energy levels and sub-levels does not strictly follow the sequence dictated by increasing n and l, but an empirical scheme follows directly from Table 3.7. As an example, the electronic configuration of Kr is defined by $1s^2 2s^2 2p^6 3s^2 3p^6 3d^{10} 4s^2 4p^6$.

An improved theoretical model that leads to quantitative relationships between the electronic configurations that feature in the periodic classification was provided by the wave-mechanical description of hydrogen.

3.2.6 Wave-mechanical Atomic Model

Quantum-mechanical description of the hydrogen atom without ad hoc assumptions was made possible by De Broglie's postulate that any particle or object with linear momentum p has an associated wavelength $\lambda = h/p$ that accounts for its wavelike properties. Wave mechanics, as an alternative to particle mechanics, is formulated by incorporation of the quantum conditions of Planck ($E = h\nu$) and De Broglie into a classical equation of wave motion.

Wave Motion and Wave Mechanics

The simplest discription of a wave is as a disturbance which is periodic in space and time. Periodicity in space means that it repeats at regular intervals, known as the *wavelength*, λ. Periodicity in time means that it moves past a fixed point at a steady rate characterised by the *period* τ, which counts the number of crests passing per unit time. By definition then, the velocity of the wave $c = \lambda/\tau$. It is customary to use the reciprocals of wavelength $1/\lambda = \bar{\nu}$, known as the *wave number* and $1/\tau = \nu$ known as the *frequency*, in some applications. The *amplitude* A is the maximum value of the disturbance. In the case of a sine wave that repeats at intervals of 2π, the wave disturbance can be described as

$$z = A\sin\left(\frac{x}{\lambda}\cdot 2\pi - \frac{t}{\tau}\cdot 2\pi\right) = A\sin 2\pi(\bar{\nu}x - \nu t)$$

A more general formulation is

$$z = A\sin 2\pi\left(\frac{x}{\lambda} - t\nu\right) + B\cos 2\pi\left(\frac{x}{\lambda} - t\nu\right)$$

In terms of the identity $\exp(i\theta) = \cos\theta + i\sin\theta$, this formulation is equivalent to

$$z = A\exp 2\pi i\left(\frac{x}{\lambda} - t\nu\right)$$

In order to impart wave properties to material particles, momentum and energy defined according to the postulates of De Broglie and Planck

$$p = h/\lambda \qquad E = h\nu$$

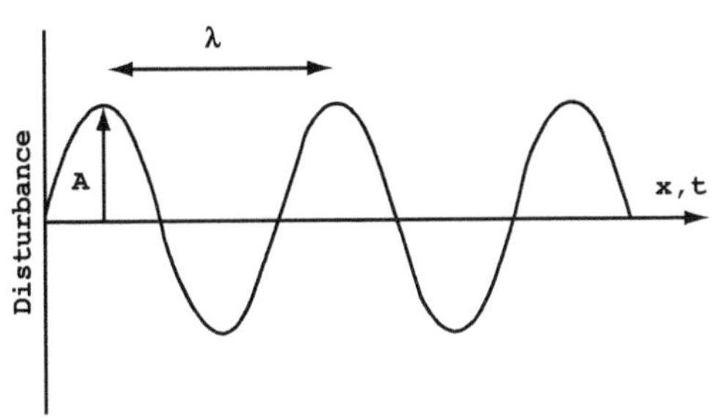

Figure 3.17: *Wave motion, periodic in space and time.*

are introduced into the wave equation instead of the normal wave variables, to define a *wave function*

$$\begin{aligned}\Psi &= A\exp 2\pi i\left(\frac{x}{\lambda}-\frac{t}{\tau}\right)\\ &= A\exp 2\pi i\left(\frac{px}{h}-\frac{Et}{h}\right)\end{aligned}$$

To examine how this wave function describes the position of a particle it is necessary to determine how Ψ changes as a function of x and t. The first derivatives are obtained as

$$\frac{\partial\Psi}{\partial x}=\Psi\left(\frac{2\pi ip}{h}\right)\quad\text{and}\quad\frac{\partial\Psi}{\partial t}=\Psi\left(-\frac{2\pi iE}{h}\right)$$

using the prescription

$$\frac{d}{dx}\left(e^u\right)=e^u\frac{du}{dx}$$

The derivative expressions rearrange into

$$p\psi=\frac{h}{2\pi i}\frac{d}{dx}(\psi)\quad\text{and}\quad E\Psi=\frac{ih}{2\pi}\frac{\partial}{\partial t}(\Psi)$$

The lower case symbol ψ is used to denote a time-independent wave function. In each of these equations the product of a variable that represents a dynamical observable and a wave function, equals a differential operator that operates on the wave function. It is said that to each quantum-mechanical observable there corresponds an operator,

$$p\rightarrow\frac{h}{2\pi i}\frac{d}{dx}\tag{3.26}$$

$$E\rightarrow\frac{ih}{2\pi}\frac{d}{dt}\tag{3.27}$$

The operator that corresponds to the total energy is known as the Hamiltonian operator, H. The corresponding wave equation reads

$$H\Psi=\frac{ih}{2\pi}\frac{\partial\Psi}{\partial t}=E\Psi$$

In another formulation the classical Hamiltonian

$$H=E=T+V=\frac{(mv)^2}{2m}+V=\frac{p^2}{2m}+V$$

By substituting the momentum operator this expression is transformed into the wave mechanical form

$$H\psi = \frac{1}{2m}\frac{h}{2\pi i}\frac{d}{dx}\left(\frac{h}{2\pi i}\frac{d\psi}{dx}\right) + V\psi = E\psi$$

i.e.

$$-\frac{h^2}{8\pi^2 m}\cdot\frac{d^2\psi}{dx^2} = (E - V)\psi$$

which rearranges into the more familiar forms

$$\frac{d^2\psi}{dx^2} + \frac{8\pi^2 m}{h^2}(E - V)\psi = 0$$

$$\frac{d^2\psi}{dx^2} + \frac{2m}{\hbar^2}(E - V)\psi = 0$$

The variable h-bar, $\hbar = h/2\pi$. The wave equation in one of these forms is known as Schrödinger's equation in one dimension. The first term is the kinetic energy in operator form and V is the classical potential energy. Schrödinger's equation is a second-order differential equation and therefore it has an infinite number of solutions. The physical interpretation of ψ however, implies that some of the mathematical solutions are physically unacceptable. When these solutions are rejected it means that certain values of E are not allowed, which introduces the idea of *quantized* energy.

Wave-mechanical Model of the Hydrogen Atom

An electron associated with the nucleus of a hydrogen moves under the influence of a potential

$$V = -\frac{e^2}{4\pi\epsilon_0 r}$$

where r corresponds to the separation between the positive and negative charge centres. Like the particle in a circular Bohr orbit the electron is expected to have quantized energy and angular momentum. The origin of the latter can be visualized in terms of De Broglie's postulate by assuming a standing wave stabilized by a cyclic boundary condition, as shown in Figure 3.18(a). To avoid destructive interference it is necessary that an integral number of wavelengths span the orbit and hence

$$n\lambda = 2\pi r = \frac{nh}{p} \quad \text{hence} \quad pr = \frac{nh}{2\pi}$$

This condition, first proposed by Bohr, leads directly to the quantum conditions of the Bohr model and accounts for observed line spectra. It is an

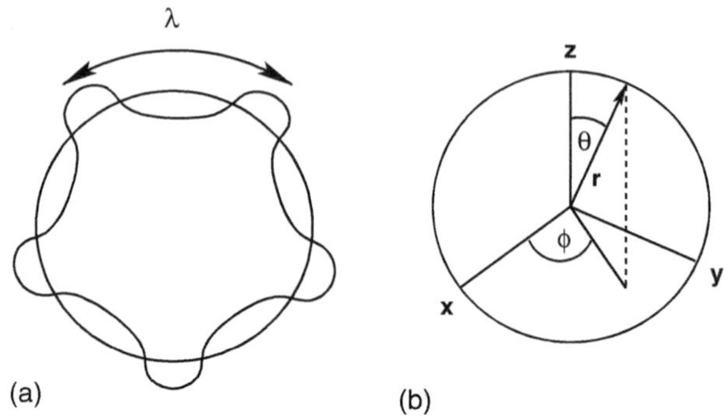

Figure 3.18: *(a) Orbital electron in the hydrogen atom represented by a standing De Broglie wave. (b) Diagram to show the relationship between Cartesian (x,y,z) and spherical polar (r, θ, ϕ) coordinates, in terms of the vector* $\boldsymbol{r}$.

improvement on the Bohr model in the sense that an accelerated charge is not invoked directly, although a mechanically stable orbit is still assumed. The pure quantum-mechanical description does not suffer from this defect and is provided by solution of Schrödinger's equation with Coulomb potential.

Schrödinger's equation in three dimensions reads

$$\nabla^2\psi = \frac{d^2\psi}{dx^2} + \frac{d^2\psi}{dy^2} + \frac{d^2\psi}{dz^2} + \frac{8\pi^2 m}{h^2}(E - V)\psi = 0$$

Since cartesian coordinates are not appropriate to handle problems in spherical symmetry it is necessary to transform to the more appropriate spherical polar coordinates r, θ and ϕ as defined by Figure 3.18(b) and the transformation equations:

$$\begin{aligned} x &= r\sin\theta\cos\phi \\ y &= r\sin\theta\sin\phi \\ z &= r\cos\theta \end{aligned}$$

An expression for the Laplacian operator ∇^2 can be obtained directly by forming the appropriate derivatives. The result, given without proof is

$$\nabla^2 = \frac{\partial^2}{\partial r^2} + \frac{2}{r}\frac{\partial}{\partial r} + \frac{1}{r^2}\Lambda^2$$

The term in Λ contains the total angular dependence, independant of r. The radial wave equation, without angle dependence is

$$\frac{d^2\psi}{dr^2} + \frac{2}{r}\frac{d\psi}{dr} + \frac{2m}{\hbar^2}\left(E + \frac{e^2}{4\pi\epsilon_0 r}\right)\psi = 0 \qquad (3.28)$$

For large r the equation becomes

$$\frac{d^2\psi}{dr^2} + \frac{2mE\psi}{\hbar^2} = 0$$

This equation is familiar in the form $k^2 = 2mE/\hbar^2 < 0$, with solutions $\psi = \exp(-kr)$. Substitution of this solution into (3.28) gives

$$\frac{d\psi}{dr} = -ke^{-kr} = -k\psi$$

$$\frac{d^2\psi}{dr^2} = -k\frac{d\psi}{dr} = k^2\psi$$

and hence

$$k^2 - \frac{2k}{r} + \frac{2mE}{\hbar^2} + \frac{me^2}{2\pi\epsilon_0\hbar^2 r} = 0$$

Since this equation is valid for all r the respective sums that are dependent and independent of r should individually be equal to zero, i.e.

$$k^2 + \frac{2mE}{\hbar^2} = 0 \quad \text{and} \quad -2k + \frac{me^2}{2\pi\epsilon_0\hbar^2} = 0$$

Thus

$$k = \frac{me^2}{4\pi\epsilon_0\hbar^2} = \left(\frac{1}{r_1}\right)_{Bohr} \quad \text{and} \quad E = -\frac{k^2\hbar^2}{2m} = (E_1)_{Bohr}$$

Rigorous solution of the three-dimensional Schrödinger equation

$$\nabla^2\psi + \frac{2m}{\hbar^2}\left(E + \frac{e^2}{4\pi\epsilon_0 r}\right)\psi = 0$$

is achieved by separation of variables, writing

$$\psi(r,\theta,\phi) = R(r)\cdot Y(\theta,\phi)$$

The radial equation has acceptable solutions only for integral values of the principal quantum number n, related to the total energy by an equation

$$E = -\frac{\text{constant}}{n^2} = -\frac{E_1}{n^2}, \quad n = 1, 2, \ldots$$

$$E_n = -\left(\frac{e^4 m_e}{32\pi^2\epsilon_0^2\hbar^2}\right)\left(\frac{1}{n^2}\right)$$

$$E_n - E_m = \left(\frac{e^4 m_e}{8\epsilon_0^2 h^2}\right)\left(\frac{1}{m^2} - \frac{1}{n^2}\right)$$

The angular dependent part of the wave function likewise has solutions characterised by integral values of the two quantum numbers l and m_l, known as the angular momentum and magnetic quantum numbers respectively,

$$l = 0, 1, 2, \ldots, (n-1)$$

$$m_l = l, l-1, \ldots, -l, \quad 2l+1 \,\text{values}$$

More complicated atoms have complicated wave functions. However, since all atoms are centred around a nucleus it follows that one-electron wave functions will all have the same angular dependence in all atoms. It means that the hydrogen wave functions $Y(\theta, \phi)$ will separate out of all atomic wave functions. These one-electron wave functions, called *orbitals* will therefore be the same for all atoms and for this reason the different s, p, d, f orbitals, derived from the H solution can be assigned to all other atoms. Their shapes are determined by the values of the quantum number l. The negative values of the energy indicate that the electron is bound to the nucleus. To uncouple it from the nucleus it is necessary to promote the electron to the level $E = 0$, $n = \infty$. This ionization requires an amount of energy $-E_1 = 22 \times 10E^{-19}$J. The experimentally measured value agrees precisely with the theoretical value.

Electronic states of wave functions with common n, but different values of l and m_l have the same energy and are said to be *degenerate*. Wave functions with $l = 0, 1, 2, 3$ describe electronic states with different angular momenta and are referred to as s, p, d, f states. The degeneracy of these states depend on allowed values of m_l and therefore amount to 1, 3, 5, 7 respectively.

Electron Spin

Solutions of Schrödinger's equation for the hydrogen electron predict that different states at the same energy level are distinguished by the quantum numbers l and m_l, which quantify the angular momentum of the electron in quantities of

$$L^2 = L_x^2 + L_y^2 + L_z^2 = l(l+1)\hbar^2 \quad \text{(in total)} \quad L_z = m_l\hbar \quad \text{(in projection)}$$

For any given l there are $2l+1$ different values of m_l, from $-l$ to l. These predictions can be tested in Stern-Gerlach experiments. Electrons interact with a magnetic field by virtue of their angular momenta. An electric charge (electron) only has a magnetic moment when it rotates, or has non-zero angular momentum, and it can only interact with a magnetic field if it has a magnetic moment. In a typical SG experiment a beam of evaporated metal atoms is passed between the poles of a magnet. When a beam of silver atoms is used

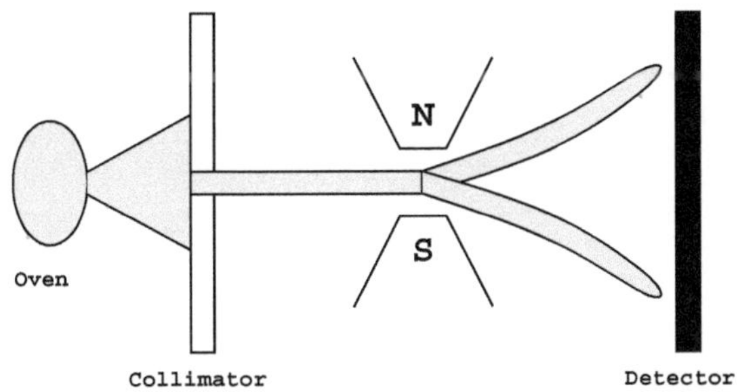

Figure 3.19: *Schematic illustration of a beam of silver atoms split into two when passed through an inhomogeneous magnetic field.*

an unexpected result is obtained. As shown in Figure 3.19 the magnetic field splits the beam into two components. All but one of the 47 electrons in a silver atom are part of a spherically symmetrical core, with no net angular momentum. The observed magnetic effect must therefore be linked to the single valence electron, which however, has $l = 0$ and hence zero angular momentum, as well.

To account for the magnetic interaction it was necessary to postulate an intrinsic magnetic moment arising from electron spin. If the electron were like a classical spinning object the magnetic moment μ_z in the field direction would assume all values between $\pm\boldsymbol{\mu}$. A continuous beam of atoms, evenly spread around the direction of the incoming beam, is predicted to emerge from the SG magnet; some atoms displaced in the direction of the field and some in the opposite direction. What is observed instead is that the beam is split into two distinct components. The observation implies that the component S_z of the electron spin $\boldsymbol{S}$ has only two possible values, S_z^+ (up) and S_z^- (down). The two possible values of S_z are multiples of some fundamental unit of angular momentum; numerically $S_z = \pm\hbar/2$.

The same SG arrangement can be used to measure orbital angular momentum. Atoms with different m_l will respond differently to the magnetic field and the beam is split into $2l + 1$ bundles. The quantum number l can hence be determined by counting the number of spots on the detecting screen, one for each value of m_l. A typical outcome is shown in Figure 3.20.

Angular momentum is a strictly conserved quantity in classical mechanics, and the two new levels observed with silver atoms suggest that there is an additional contribution of $s\hbar$ to the angular momentum, L_z, through a quantum number s, such that $2s + 1 = 2$, i.e. $s = 1/2$. As before, a quantum number m_s, with possible values between $-s$ and $+s$ describes the additional

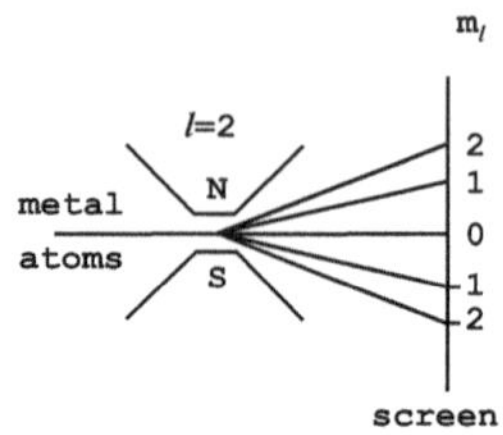

Figure 3.20: *Stern-Gerlach experiment to demonstrate the resolution of electronic orbital angular momentum states in an inhomogeneous magnetic field.*

angular momentum in projection, $S_z = \pm(1/2)\hbar$. The quantum-mechanically conserved quantity becomes

$$J = L + S$$

only part of which (L) is accounted for by Schrödinger's equation.

Schrödinger's Equation and Spin The non-appearance of electron spin as a solution of Schrödinger's equation shows that in some sense the equation is incomplete. The Schrödinger equation for a free electron consists of a time and a space part

$$i\hbar\frac{\partial\Phi}{\partial t} = -\frac{\hbar^2}{2m}\nabla^2\Phi \tag{3.29}$$

or rearranged into

$$\left(2mi\hbar\frac{\partial}{\partial t} + \hbar^2\nabla^2\right)\Phi = 0$$

it appears as a combination of one linear and one squared operator, $\hat{E}$ and $\hat{p}^2$, where

$$\hat{E} \leftarrow i\hbar\frac{\partial}{\partial t} \quad \text{and} \quad \hat{p} \leftarrow -i\hbar\nabla$$

A linear Schrödinger operator would have the advantage of a wave equation which is linear in both space and time derivatives. The most general equation with the required form is

$$S = A\hat{E} + \mathbf{B}\cdot\hat{\mathbf{p}} + C = 0$$

such that $S^2 = 2m\hat{E} - \hat{p}^2$, i.e.

$$(A\hat{E} + \mathbf{B}\cdot\hat{\mathbf{p}} + C)^2 = A^2(\hat{E}^2) + 2AB(\hat{E}\hat{p}) + 2AC(\hat{E}) + 2BC(\hat{p}) + B^2(\hat{p}^2) + C^2$$

To match the Schrödinger formulation, $(2m\hat{E} - \hat{p}^2)$, it is necessary that

$$A^2 = AB = BC = C^2 = 0$$

$$2AC = 2m$$

$$B^2 = \pm 1$$

These conditions are clearly impossible if A, B and C are complex numbers, but could be satisfied by square matrices. Typical examples are the Pauli matrices:

$$\sigma_1 = \begin{pmatrix} 0 & 1 \\ 1 & 0 \end{pmatrix} \quad \sigma_2 = \begin{pmatrix} 0 & -i \\ i & 0 \end{pmatrix} \quad \sigma_3 = \begin{pmatrix} 1 & 0 \\ 0 & -1 \end{pmatrix}$$

and in particular

$$B_i = \begin{pmatrix} \sigma_i & O \\ O & \sigma_i \end{pmatrix}$$

where O is the 2×2 null matrix, and

$$A = \begin{pmatrix} O & O \\ I & O \end{pmatrix} \quad C = \begin{pmatrix} O & 2mI \\ O & O \end{pmatrix}$$

where I is the 2×2 unit matrix. In terms of these matrices the Schrödinger equation becomes

$$E\begin{pmatrix} O & O \\ I & O \end{pmatrix}\begin{pmatrix} \psi \\ \chi \end{pmatrix} + \begin{pmatrix} \sigma_i & O \\ O & \sigma_i \end{pmatrix} p_i \begin{pmatrix} \psi \\ \chi \end{pmatrix} + \begin{pmatrix} O & 2mI \\ O & O \end{pmatrix}\begin{pmatrix} \psi \\ \chi \end{pmatrix} = 0$$

i.e.

$$E\begin{pmatrix} O \\ \psi \end{pmatrix} + (\sigma \cdot p)\begin{pmatrix} \psi \\ \chi \end{pmatrix} + 2m\begin{pmatrix} \chi \\ O \end{pmatrix} = 0$$

It is required that

$$\Phi = \begin{pmatrix} \psi \\ \chi \end{pmatrix}$$

is a two-component vector in order to allow matrix multiplication. By adding the three terms the equation becomes

$$\begin{pmatrix} \sigma.p & 2m \\ E & \sigma.p \end{pmatrix}\begin{pmatrix} \psi \\ \chi \end{pmatrix} = O$$

which is

$$(\sigma \cdot p)\psi + 2m\chi = 0$$

$$E\psi + (\sigma \cdot p)\chi = 0$$

It follows that ψ and χ are not linearly independent,

$$\chi = -\frac{(\sigma \cdot p)\psi}{2m}$$

whereby

$$E\psi = \frac{(\sigma \cdot p)^2\psi}{2m}$$

Since $\sigma^2 = 1$, this is simply the Schrödinger equation

$$(2mE - p^2)\psi = 0$$

$$\Longrightarrow \left(2mi\hbar\frac{\partial}{\partial t} + \hbar^2\nabla^2\right)\psi = 0$$

This may well appear not to produce anything new until the electron is examined in an external magnetic field. The wave equation then takes the form

$$i\hbar\frac{\partial\psi}{\partial t} = -\frac{\hbar^2}{2m}\left(\nabla - \frac{ie\mathbf{A}}{\hbar c}\right)^2\psi + \frac{\hbar e}{2mc}(\sigma \cdot \mathbf{B})\psi + V\psi$$

which is the Pauli equation, first obtained empirically by the addition of an extra operator to the Schrödinger Hamiltonian to account for electron spin. The importance of this equation lies in the term that reflects an intrinsic magnetic moment

$$\mu = \frac{\hbar e}{2mc}\sigma$$

which is exactly as required to explain the fine structure of electronic spectra in terms of a spin angular momentum of $\sigma/2$ and a Landé factor of 2. The matrix representation of the spin operator requires the spin state of a particle to be represented by row vectors, commonly interpreted as spin 'up' or 'down'. An arbitrary state function Ψ must be represented as a superposition of spin up and spin down states

$$|\uparrow\rangle = \begin{pmatrix}1\\0\end{pmatrix}, \quad |\downarrow\rangle = \begin{pmatrix}0\\1\end{pmatrix}$$

$$\Psi = \psi_+ \,|\uparrow\rangle + \psi_- \,|\downarrow\rangle = \begin{pmatrix}\psi_+\\\psi_-\end{pmatrix}$$

This two-component wave function is known as a spinor.

Exclusion Principle To understand how electrons with spin interact, it is useful to examine a system consisting of two electrons, such as the helium atom. Let this two-electron system be described by the wave function, in space coordinates, $\Phi(r)$. If the electrons are interchanged the wave function will in general be different, (Φ'), but since the electrons are identical (ignoring spin) the energy of the system will not be affected. The wave functions therefore belong to degenerate levels.

Suppose that the effect of interchanging of electrons is correctly described by some operator, π, *i.e.* $\pi\Phi = \Phi'$. By repeating the operation the situation is reversed and the system returns to its initial state, i.e.

$$\pi\Phi' = \pi^2\Phi = \Phi$$

which shows that $\pi = \pm 1$. The wave function is said to be either symmetrical $(\pi = 1)$ or anti-symmetrical with respect to the interchange of electron coordinates. Now let $\Phi_a(r_1)$ and $\Phi_b(r_2)$ be one-electron wave functions for electrons a and b at coordinates r_1 and r_2 respectively. The combined function can then be formulated in terms of the product functions

$$\psi_1 = \Phi_a(r_1)\Phi_b(r_2) \quad \text{and} \quad \psi_2 = \Phi_a(r_2)\Phi_b(r_1)$$

or a linear combination of these. It is noted that the functions

$$\psi_\pm = \frac{1}{\sqrt{2}}(\psi_1 \pm \psi_2)$$

correctly describe wave functions that are either symmetrical or anti-symmetrical with respect to electron interchange. The factor $1/\sqrt{2}$ ensures normalization.

For two identical particles $(\Phi_a = \Phi_b)$ it is noted that ψ_+ approaches a maximum at $r_1 = r_2$, whereas ψ_- approaches zero. This means that two electrons in the same state would tend to stay together if the wave function is symmetrical and to avoid each other when the wave function is anti-symmetrical under exchange. It follows that the function $\psi(r)$ is automatically symmetrical and that the anti-symmetric function vanishes identically. To arrive at the correct formulation of the ground state of the helium atom it is necessary to also take into account the effect of spin, represented by the functions α and β. There are four possibilities, according to the electrons having the same spin, either up or down:

$$\alpha(1)\alpha(2) \quad \text{or} \quad \beta(1)\beta(2)$$

and states with the spins opposed:

$$-\frac{1}{\sqrt{2}}\{\alpha(1)\beta(2) \pm \beta(1)\alpha(2)\}$$

It is reasonable to expect that each of these spin states could occur in combination with the ground-state space function $\psi_+(r)$ to yield four different levels at the ground state. However, for the helium atom only one ground-state function can be identified experimentally and it is significant to note that only one of the spin functions is anti-symmetrical, i.e.

$$\psi_-(s) = \frac{1}{\sqrt{2}}\{\alpha(1)\beta(2) - \beta(1)\alpha(2)\}$$

Since $\psi_+(r)$ is symmetrical, this suggests that the total wave function $\psi = \psi_+(r)\psi_-(s)$, and anti-symmetrical. There is no theoretical ground for this conclusion, which is a purely empirical result based on a variety of experimental measurements. However, it seems to apply everywhere and to represent a law of Nature, stating that systems consisting of more than one particle of half-integral spin are always represented by anti-symmetric wave functions.

To ensure that many-electron wave functions are anti-symmetrical, each electron must be in a different quantum state. This result is known as Pauli's exclusion principle, which states that no two electrons in a many-electron system can have all quantum numbers the same. In the case of atoms it is noted that since there are only two quantum states of the spin, no more than two electrons can have the same set of orbital quantum numbers.

3.2.7 Aufbau Procedure

The currently most popular form of the periodic table is shown in Figure 3.21. The numbering of the groups from 1 to 18 is considered best to emphasize the relationship between the periodic position of an element and the electronic configuration of its atoms.

The energy levels of a hydrogen electron, according to the Schrödinger solution increase in the order

$$1s < 2s < 2p < 3s < 3p < 3d < 4s < 4p < 4d < 4f < 5s < 5p < 5d < 5f, \text{etc.}$$

The exclusion principle restricts the number of electrons per energy level to a pair with opposed spins. An empirical rule, proposed by Hund, requires that degenerate levels be singly occupied by electrons of parallel spin, before pairing occurs. It is commonly accepted that the electronic configuration of any element, up to the second most energetic electron, is identical to that of the element at the preceeding position in the periodic classification. The final electron goes into the lowest unoccupied level. Application of these rules to derive the electronic configuration of any element, starting from hydrogen, is known as the Aufbau procedure (building up procedure).

1	2	3	4	5	6	7	8	9	10	11	12	13	14	15	16	17	18
Ia	IIa	IIIb	IVb	Vb	VIb	VIIb	VIII			Ib	IIb	IIIa	IVa	Va	VIa	VIIa	0
1 H																	2 He
3 Li	4 Be											5 B	6 C	7 N	8 O	9 F	10 Ne
11 Na	12 Mg											13 Al	14 Si	15 P	16 S	17 Cl	18 Ar
19 K	20 Ca	21 Sc	22 Ti	23 V	24 Cr	25 Mn	26 Fe	27 Co	28 Ni	29 Cu	30 Zn	31 Ga	32 Ge	33 As	34 Se	35 Br	36 Kr
37 Rb	38 Sr	39 Y	40 Zr	41 Nb	42 Mo	43 Tc	44 Ru	45 Rh	46 Pd	47 Ag	48 Cd	49 In	50 Sn	51 Sb	52 Te	53 I	54 Xe
55 Cs	56 Ba	71 Lu	72 Hf	73 Ta	74 W	75 Re	76 Os	77 Ir	78 Pt	79 Au	80 Hg	81 Tl	82 Pb	83 Bi	84 Po	85 At	86 Rn
87 Fr	88 Ra	103 Lr	104 Db	105 Jl	106 Rf	107 Bh	108 Hn	109 Mt									

57 La	58 Ce	59 Pr	60 Nd	61 Pm	62 Sm	63 Eu	64 Gd	65 Tb	66 Dy	67 Ho	68 Er	69 Tm	70 Yb
89 Ac	90 Th	91 Pa	92 U	93 Np	94 Pu	95 Am	96 Cm	97 Bk	98 Cf	99 Es	100 Fm	101 Md	102 No

Figure 3.21: *Modern version of the periodic table. The most recent convention is to number groups from 1 to 18.*

The scheme works well in terms of Schrödinger's solution, up to element 18. Instead of the $3d$ level however, the final electron of K goes to $4s$. Undergraduate teachers explain that interelectronic effects like penetration, screening and repulsion, not operative in hydrogen, are responsible for switching the levels $ns < (n-1)d < (n-2)f$, that seems to account for the block structure of the periodic table in terms of a modified Aufbau sequence. The s, p, and d blocks thereby consist of groups 1–2, 3–12, and 13–18, respectively. The f block fits between groups 2 and 3 and is placed separately for printing economy only.

Spectroscopically measured electronic configurations confirm the broad classification, but not the detail. The transition series (d block), for instance, do not end in group 12. The coinage metals of group 11 all have outer configurations of $d^{10}s^1$, suggesting completion of d levels in group 10. However, the configurations Ni($3d^8 4s^2$), Pd($4d^{10}$) and Pt($5d^9 6s^1$) are all different. Still, groups 11 and 12 clearly fit better with the s block.

The number of electrons in a closed shell of principal quantum number n is given by the sum $2\sum_{l=0}^{n-1}(2l+1) = 2n^2$, which is related to but not identical with Rydberg's empirical series (3.25). According to the Schrödinger formula, closed shell arrangements occur at atomic numbers $Z = 2, 10, 28, 60$, etc. These are not the numbers, 2, 10, 18, 36, etc. at which elemental periods come to a close according to the empirical rules of Langmuir and Bury.

3.3 Conclusion

The importance of the periodic table of the elements in the development of 20th century chemistry cannot be overemphasized, despite the fact that it led to a complete misreading of the chemical properties of the 'inert' gases [48]. Systematic inorganic chemistry without a periodic table is impossible to imagine. Of even more importance is the development of quantum theory, in direct response to the electronic structure of atoms as revealed by the periodic classification. The belief of overzealous commentators that the periodic table has been reduced to a by-product of quantum theory, is belied by the facts.

Despite serious discrepancies between the Schrödinger model and the observed periodic table, it is pretended that a complete theory of the periodic table exists. To quote one authority [49]:

> We imagine the bare nucleus of atomic number Z, and then feed into the orbitals Z electrons in succession. The order of occupation is
>
> $$1s2s2p3s3p4s3d4p5s4d5p6s\ldots$$
>
> and each orbital can accommodate two electrons. This order of occupation is approximately the order of energies of the individual orbitals, because, in general, the lower the energy of the orbital, the lower the total energy of the atom as a whole when that orbital is occupied. However, there are complicating effects arising from electron-electron repulsions that are important when the orbitals have similar energies (such as $4s$ and $3d$ orbitals near Ca and Sc), and we must take special care... electron-electron repulsions are comparable to the energy difference between the $4s$ and $3d$ orbitals, and a simple analysis no longer works.

The less simple analysis that works, is never presented, because it does not exist.

It must be concluded that a periodic system based on the energy spectrum of the hydrogen electron, according to Schrödinger's solution, cannot be reconciled with the observed periodic table of the elements. It fails because it ignores all interaction with the environment and no amount of patchwork is ever going to rescue this situation. It is a myth that chemistry derives from quantum theory. More fundamental than both is the periodic table that reduces the properties of matter to a number basis, which is revealed only peripherally in the differential equations of quantum theory.

The failure of quantum theory to account for the details of nuclear structure relates to a more elaborate number system involved there. Sporadic efforts to classify sub-atomic species into periodic groups have not succeeded so far and the resolution of the problem lies in the future.

Chapter 4

Structure of Atomic Nuclei

4.1 Introduction

The search for a nuclear structure started with the first observation of radioactivity. As described in the previous chapter, three types of natural radioactivity are observed.

In α-decay, a helium nucleus is emitted from the radioactive nuclide with a change in both atomic and mass numbers:

$$(Z, A) \rightarrow (Z-2, A-4) + {}_2^4\text{He}$$

In β-decay, a negative electron is emitted where upon the nuclear charge increases by one unit, without change in mass number:

$$(Z, A) \rightarrow (Z+1, A) + \beta^-$$

In γ-decay an electromagnetic photon is emitted without change in atomic or mass number:

$$(Z, A)^* \rightarrow (Z, A) + h\nu$$

The naturally occurring radio elements occur in series that represent the decay steps, starting from a parent nucleus and ending with a non-radioactive, stable nuclide. The atomic and mass numbers of any member of a radioactive series are related to those of a precursor by the equations

$$A = A_0 - 4N_\alpha$$

$$Z = Z_0 - 2N_\alpha + N_\beta$$

where N_α and N_β are the numbers of α and β particles emitted. These expressions obey the mass-number relationship $A = 4n + m$, that gives rise to four series, defined by $m = 0, 1, 2, 3$, known as the $4n$, $4n + 1$, $4n + 2$ and $4n + 3$ series. The starting and end members of the four series are

$$
\begin{aligned}
{}^{232}\mathrm{Th} &\rightarrow {}^{208}\mathrm{Pb} \\
{}^{237}\mathrm{Np} &\rightarrow {}^{209}\mathrm{Bi} \\
{}^{238}\mathrm{U} &\rightarrow {}^{206}\mathrm{Pb} \\
{}^{235}\mathrm{U} &\rightarrow {}^{207}\mathrm{Pb}
\end{aligned}
$$

The $4n+1$ series does not occur in Nature, and starts from an artificially produced nuclide. The remaining three series account for most naturally occurring radioactive nuclides, not formed by cosmic radiation. Those unaccounted for are $4n\left({}^{40}_{19}\mathrm{K}(\beta),\ {}^{144}_{60}\mathrm{Nd}(\alpha),\ {}^{176}_{71}\mathrm{Lu}(\beta),\ {}^{192}_{78}\mathrm{Pt}(\alpha)\right)$, $4n+2\left({}^{50}_{23}\mathrm{V}(\mathrm{EC}),\ {}^{138}_{57}\mathrm{La}(\beta)\right.$, $4n+3\left({}^{87}_{37}\mathrm{Rb}(\beta),\ {}^{115}_{49}\mathrm{In}(\beta),\ {}^{147}_{62}\mathrm{Sm}(\alpha),\ {}^{187}_{75}\mathrm{Re}(\beta)\right)$. It is noticable how the $4n+1$ series is, once more, not represented. The symbol EC designates *electron capture*, that occurs when an atomic nucleus absorbs an orbital electron to reduce its nuclear charge by one unit.

The energetics of α-decay is commonly discussed as a tunnelling phenomenon that assumes the particle confined to a box of nuclear dimensions. However, there is no evidence that free α-particles occur in atomic nuclei and it would be meaningless to consider their potential energy in this region. The energy of the total nucleus is reduced by the process that creates an α-particle.

The same argument applies to β decay. The fact that electrons are emitted does not imply their free existence in the nucleus. The fusion between proton and electron in the nucleus produces a new species, $p + e \rightarrow n$, with the disappearance of their individual identities. The only problem with this formulation is that all of the particles involved here have half-integral spins. The neutrino was invented to ensure the conservation of angular momentum.

4.2 Mass and Binding Energy

The discovery of isotopes cleared up the problem of fractional atomic weights and showed that the mass of each atom is close to an integral multiple of a basic atomic mass unit, which is the approximate mass of a nucleon. By international agreement the atomic mass unit is defined as 1/12 of the mass of ^{12}C. The energy equivalent is 931.481 MeV.

The atomic mass unit is about 0.8% smaller than the average mass of a nucleon. This means that an energy equivalent to 0.8% of their mass is needed to free nucleons from the nucleus. The energy needed to split up a nucleus completely into free protons and neutrons is called its *binding* energy.

In practice it is the binding energies of atoms rather than of smaller nuclei that are measured. The binding energy (E_B) therefore contains a contribution from electron mass, i.e.

$$E_B = c^2(Zm_H + Nm_n - M) \tag{4.1}$$

where m_H and m_n are the masses of hydrogen and a neutron, respectively. $N = A - Z$. The electron mass contributions to atomic mass M and to Zm_H cancel.

Nuclear mass data are obtained mainly by mass spectrometry, including measurement of mass differences between species with equal mass number, such as $^{16}O^+$ and $(^{12}C\,^1H_4)^+$. The relationship $E = mc^2$ has been tested exhaustively in terms of such measurements and its validity confirmed experimentally. Many tabulations of mass excess (*ME*) and binding energy of nuclides fail to emphasize that nuclear binding energy must be calculated from mass excess on the ^{1}H rather than ^{12}C mass scale. The convertion of *ME* on ^{12}C scale to binding energy therefore consists of

$$B_E = -ME + (Zm_p + Nm_n)$$

with $m_p = 7288.696$ keV, $m_n = 8071.596$ keV. A plot of binding energy per nucleon [50] is shown in Figure 4.1.

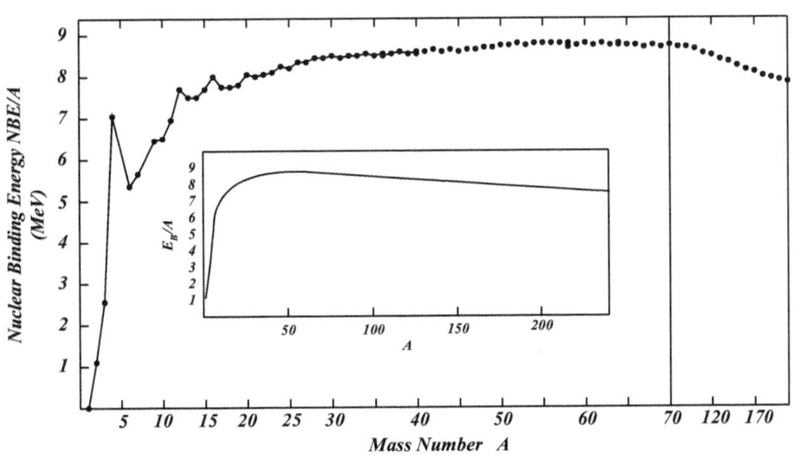

Figure 4.1: *Binding energy per nucleon (MeV) shown as a function of mass number. Note the change of scale at A = 70. The small inset shows the semi-empirical binding-energy curve that fits experimental values surprisingly well.*

The graph displays three important features. The fact that atomic masses are close to integral multiples of the atomic mass unit suggests that the binding energy per nucleon should be independent of nuclear size. This inference is well supported by observation, with a mean of about 8 MeV per nucleon for all atoms.

There is a maximum around $A = 60$, with a slow decrease as A increases above 60. The existence of the maximum shows that energy would be released in either the *fusion* of two light elements to form a heavier one or in the *fission* of a heavy element to form two lighter elements.

There are secondary peaks at the nuclei ${}^{4}_{2}\mathrm{He}$, ${}^{12}_{6}\mathrm{C}$, ${}^{16}_{8}\mathrm{O}$, ${}^{20}_{10}\mathrm{Ne}$, and ${}^{24}_{12}\mathrm{Mg}$, made up of integer numbers of α-particles. The most prominent of these is that of ${}^{4}\mathrm{He}$ itself. ${}^{8}_{4}\mathrm{Be}$ decays very rapidly into two α-particles.

4.2.1 Models of the Nucleus

Attempts to rationalize the shape of the binding-energy curve led to the formulation of three simple nuclear models and a semi-empirical formula that reproduces the binding curve rather well. The physical interpretation of the various terms in the semi-empirical formula are based on different models, suggesting that the three models represent different aspects of the nuclear structure and that each of them gains prominence over a limited mass range. Each of these will be discussed in turn.

Alpha-particle Model

The secondary maxima observed in the binding curve may be interpreted to mean that the relevant nuclei might be treated as groups of α-particles. The relatively high stability of these nuclei, and the large binding energy of ${}^{4}\mathrm{He}$, make it plausible to assume that pairs of protons and neutrons become associated together inside the nucleus, for short periods as α-particles.

The average binding energy per nucleon is about 8 MeV, and that for an α-particle is 7 MeV. So, it seems plausible that a nucleus could consist of α-particles bound together with energies of about 1 MeV per nucleon. About 7/8th of the binding energy per nucleon would then be accounted for if the nucleons were grouped into α-particles, with relatively weak interaction between α-particles. Quantitative details of the model have not been calculated.

The Fermi Gas Model

In another approach the nucleons are regarded as noninteracting Fermi particles, like the sea of electrons in a metal. The size of the well is based on the nuclear radius derived from the Rutherford model of the atom,

$$R = r_0 A^{\frac{1}{3}} \tag{4.2}$$

The spherical potential well of volume $4\pi r_0^3 A/3$ and depth sufficient to accommodate all nucleons to a Fermi level at -8 Mev with respect to the exterior is assumed to accommodate all nucleons. The energy levels of protons and neutrons are alike, but not identical, because of a Coulomb force that affects only protons. Both protons and neutrons obey the exclusion principle, so that filling of the lowest and subsequent pairs of energy levels corresponds to the formation of α-particles. The saturation of each pair of levels corresponds with the formation of the next α-particle. After a level is filled the next nucleon must go into a higher level which reduces the average binding energy per nucleon and a dip occurs in the E_A/A curve.

This α-particle effect becomes less visible when the proton and neutron levels get out of step and the nth proton level coincides with the $n-1$ neutron level. This mismatch becomes apparent as $A > 40$. Schematic nucleon levels are shown in Figure 4.2.

Because of electrostatic repulsion proton levels are increasingly pushed upwards and heavy nuclei contain more neutrons than protons. All that remains of the α-particle effect is pairing, which causes nuclei with an even number of protons and an even number of neutrons to be more tightly bound than those with an odd number of each.

The Liquid-drop Model

Because of the short range of the nuclear force, nucleons only interact with their closest neighbours, but because of large vibrations interaction between nucleons is more like the interaction in a liquid rather than a crystal. In terms of this model the nucleus is held together by surface tension, working against Coulomb repulsion between protons, that tends to disrupt the nucleus. The model has been particularly useful in visualizing nuclear fission. Whereas the disruptive Coulomb repulsion increases with A, the stabilizing surface effects decrease. Sufficiently heavy nuclei may therefore be predicted to spontaneously fracture into smaller fragments and less heavy nuclei should split if excited into the proper kind of oscillation. In many cases such excitation is brought about by the absorption of a slow neutron. The shape of the binding-energy curve explains the release of energy when the heavy nucleus splits into two roughly equal parts, as is the case for ^{235}U.

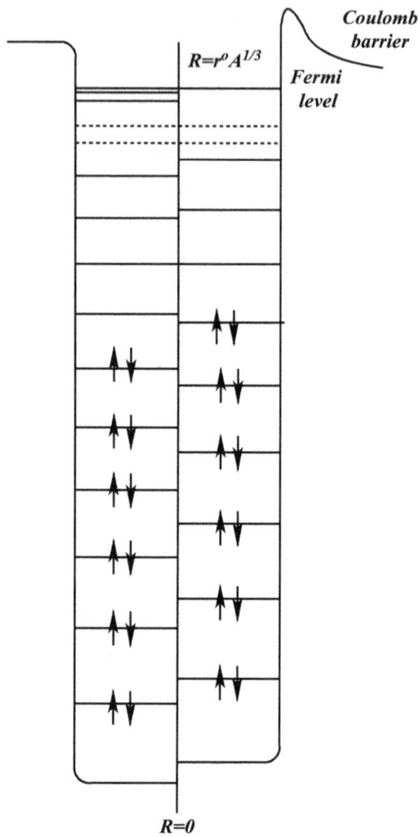

Figure 4.2: *Nuclear proton and neutron energy levels to show the low-energy forms corresponding to integral numbers of α-particles in the nucleus.*

4.2.2 The Semi-empirical Mass Formula

A formula that matches nuclear binding energies quite well was proposed by Von Weizsäcker (1935), in the form

$$E_B = a_1 A - a_2 A^{2/3} - \frac{a_3(N-Z)^2}{A} - a_4 Z^2 A^{-1/3} \pm \delta(A) \tag{4.3}$$

where a_1, a_2, a_3, and a_4 are adjustable parameters and $\delta(A)$ is a function of A chosen to account for the pairing effect. Combined with (4.1) a formula for the mass of each nucleus is obtained as:

$$M = Zm_H + Nm_n - \frac{E_B}{c^2}$$

Each adjustable parameter has a well-defined physical interpretation, with values adjusted to agree with experimental observation.

The first two terms of (4.3) are based on the liquid-drop model. The number of nearest neighbours does not depend on the total number of nucleons and the binding energy per nucleon is therefore the same in all nuclei. The contribution to the total binding energy is simply proportional to the mass number. This is not a good assumption for light nuclei that contain a small number of nucleons.

The contribution made by nucleons in the surface of the nucleus to the binding energy, must clearly be less than of those in the interior. A nucleus of radius R has surface area of $4\pi R^2 = 4\pi r_o^2 A^{2/3}$. Hence the correction term $-a_2 A^{2/3}$.

The third term is based on the Fermi gas model of the nucleus. Because neutron and proton levels are almost equal, there is a tendency, if electrostatic interaction is ignored, for the binding energy to be greater when $N = Z$. To understand the effect of a neutron excess on this term it is supposed that a situation of balance ($N = Z$) is disturbed by conversion of protons into neutrons, and *vice versa.* The number so converted will be $N - Z$. For each conversion the binding energy decreases by an amount which is proportional to the fractional difference between the highest occupied neutron and proton levels, $(N - Z)/A$.

The electrostatic effect is taken into account by the next term. The energy of interaction between two point charges $E_e \propto q_1 q_2/r$. For the charge distribution represented by Z protons in a sphere of radius $r_o A^{-1/3}$ it is assumed that additional protons are added one by one to those already smeared out over the nucleus. Since there are $Z(Z-1)/2$ pairs of protons the work required to bring the protons together is

$$W = \frac{Z(Z-1)e^2}{8\pi\epsilon_0 (r)_{av}}$$

If the protons are uniformly distributed throughout a nucleus of radius R, $(r)_{av}$ is proportional to R and hence to $A^{1/3}$, so that the Coulombic interaction term may be written as $-a_4 Z^2 A^{-1/3}$.

The term $\pm\delta(A)$ allows for the pairing effect. The positive (negative) sign applies for nuclei with both N and Z even (odd). When A is odd this term is zero.

The binding energy curve calculated from parameters that best fit the observed masses is shown in Figure 4.1.

4.2.3 Nuclear Stability

The cornerstone of all classical atomic theories was the permanence of the atom. The first indication that this atomic axiom was wrong, came with the discovery of radioactivity. It was finally destroyed by the detection of cosmic rays, that revealed atomic instability on a cosmic scale. That the illusion of atomic permanence persisted for so long is probably related to the fact that the naturally occurring radioactive nuclei are comparatively rare and of high atomic mass. The more common elements occur in a narrow band within certain limits of neutron to proton ratios, reasonably well defined empirically, but theoretically poorly understood. The common mapping to define the stability of naturally occuring nuclides, known as a Segré chart is shown in Figure 4.3.

Although the region of stability is not precisely delineated, three general regions of instability are easily identified. Nuclides with $(A - Z) : Z$ ratios below the upper limit for stable nuclides, decay by either positron emission and/or electron capture:

$$^{a}_{p}\mathrm{X} - \beta^{+} \rightarrow {}^{\;\;\;a}_{p-1}\mathrm{Y}$$

Those with ratios larger than the field of stability decay by β emission:

$$^{a}_{p}\mathrm{X} - \beta^{-} \rightarrow {}^{\;\;\;a}_{p+1}\mathrm{Y}$$

Nuclides with $Z > 83$ decay in a sequence of steps, but mainly by α emission:

$$^{a}_{p}\mathrm{X} - \alpha \rightarrow {}^{a-4}_{p-2}\mathrm{Y}$$

A few α-unstable nuclides with $Z < 83$ are known, but they have not been categorized systematically. The effect of β^{+} decay is to increase the ratio $(A - Z)/Z$, while β^{-} decay decreases this ratio. The effect of α decay is to increase $(A - Z)/Z$ and decrease Z. In all cases the radioactive decay serves to produce a daughter nucleus closer to the region of stability.

A number of empirical rules for the distribution of stable nuclides, isotopes and isotones (nuclides with the same number of neutrons) emerge from Figure 4.3.

1. For even Z there are always at least two values of $N = A - Z$ which give stable isotopes, with the exception of $^{9}_{4}\mathrm{Be}$.

2. For odd Z there are never more than two stable isotopes.

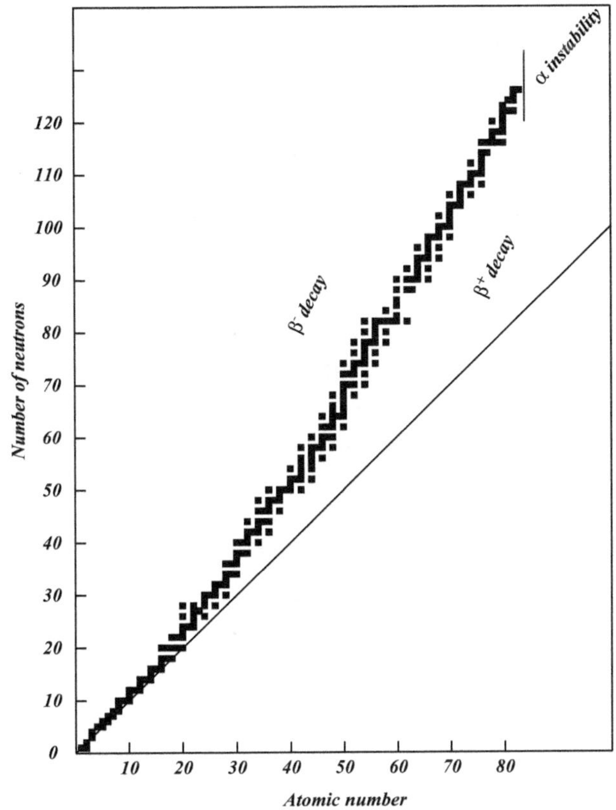

Figure 4.3: *The stable nuclides and their composition in terms of protons and neutrons. The straight line represents the ratio* $(A - Z)/Z = 1$.

3. For even N there are always at least two values of Z which give stable isotones, except $N = 2$ and 4.

4. For odd N there are never more than two stable isotones.

The only odd$-Z$ elements that form stable nuclides with the same number of protons and neutrons are H, Li, B, and N. For large Z the only stable nuclides with equal proton and neutron numbers are of even Z.

It is clear from these rules that the behaviour of neutrons and protons in an atomic nucleus is essentially the same. The missing elements $Z = 43$ and 61 are the analogues of the missing isotones N = 19, 21, etc. The large number of stable isotopes of even-Z elements is the analogue of the large number of stable even-N isotones. The number of even-Z isotopes and even-N isotones reach local maxima for Z or $N = 20, 28, 50$, and 82, referred to as magic numbers. These observations support the view that protons and neutrons are two forms of a more fundamental particle, the nucleon.

For heavy nuclei there is an increasing excess of neutrons over protons, also described as an asymmetry factor that has an effect on the stability of a nucleus. An empirical formulation of this energy term

$$E_{as} \propto \frac{(A-2Z)^2}{A}$$

has been discussed many times, e.g. [51]. The asymmetry term expresses the fact that, in a quantized system of neutrons and protons, any excess neutrons will be pushed up to levels occupied by neutrons only (compare Figure 3.2). The resulting nuclei are therefore not stabilized by the α-particle effect. If the $(A-2Z)$ excess neutrons are regarded as producing a deficit of binding energy, the fraction of nuclear volume so affected is $(A-2Z)/A$. The total deficit is proportional to the product of the two factors

$$E_{as} = -a_a \frac{(A-2Z)^2}{A}$$

where a_a is to be evaluated empirically. The Coulomb energy,

$$E_c = -a_c \frac{Z^2}{A^{\frac{1}{3}}}$$

is affected by the asymmetry factor such that

$$\Delta E = E_c - E_a = a_a \frac{(A-2Z)^2}{A} - a_c \frac{Z^2}{A^{\frac{1}{3}}}$$

For each value of A there is a minimum, hence

$$\left(\frac{\partial(\Delta E)}{\partial Z}\right)_A = 0$$

i.e.

$$4a_a \frac{(A-2Z)}{A} - 2a_c \frac{Z}{A^{\frac{1}{3}}} = 0$$

Hence

$$\frac{a_a}{a_c} = \frac{1}{2} \frac{ZA^{\frac{2}{3}}}{(A-2Z)}$$

This ratio has been determined empirically equal to 32, such that the neutron excess

$$A - 2Z = \frac{1}{2} \frac{ZA^{\frac{2}{3}}}{32}$$

With the rough rule $A \simeq 2Z$ this expression reduces to

$$A - 2Z \simeq \frac{A^{\frac{5}{3}}}{128} = 0.0078A^{\frac{5}{3}}$$

Alternatively,

$$A - 2Z = 0.025Z^{\frac{5}{3}} \tag{4.4}$$

Cosmic Rays

It was observed early in the previous century that experiments involving electroscopes consistently registered some residual leak, caused by ionizing radiation. The major cause of this leakage could be traced to radioactive impurities in the earth. Placing ionization chambers in high-altitude balloons however, showed that although the ionization decreased with altitude, it reached a minimum at about 2,000 m and then started to increase steadily with height. It was concluded that some form of penetrating radiation that comes to the earth from outside was partially absorbed by the atmosphere and therefore becomes more intense at high altitudes.

It is known today that cosmic-ray flux almost certainly fills the Milky Way and corresponds to an energy density in interstellar space of about 1 MeV m^{-3} ($10^{-13}Jm^{-3}$) [52]. Most cosmic rays consist of high-energy protons, α-particles and heavier nuclei up to an atomic number 26 (iron), although much heavier particles are also observed. The mean energy of cosmic ray protons is about 2 GeV, but cosmic-ray particle energies up to 10^{11} GeV (30 J) have been measured.

Primary cosmic rays consist almost entirely of fully ionized atomic nuclei moving at relativistic speeds. In the atmosphere these particles have frequent collisions with stationary nuclei, struck with such violence that nucleons may become dislodged, leaving the residual nucleus in a highly excited state. A great many *spallation* fragments may evaporate from the excited nucleus as fast-moving singly-charged particles, several slower and multiply-charged fragments, together with nucleons, mesons and other exotic particles.

A fundamental discovery arising from cosmic ray research was the detection in 1932 of a new particle. It was noticed that among cosmic ray tracks in an expansion chamber were some, identical to those produced by an electron, but deflected by a magnetic field in a direction opposite to that expected for a negatively charged particle. The tracks appeared to be formed by positive electrons, now called *positrons*. When hard x-rays are absorbed by matter the quantum of radiation is sometimes completely absorbed by an atom which simultaneously emits a positron and an electron. This process is known as pair production. If the absorption takes place in a cloud chamber a paired track

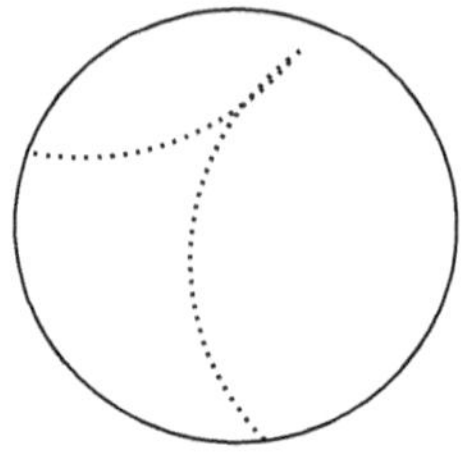

Figure 4.4: *Schematic drawing of pair production of electron and positron in a cloud chamber and deflected by an applied magnetic field.*

may be observed to come from the absorbing atom in a magnetic field. Similar pairs, shown schematically in Figure 4.4, are sometimes seen in cosmic-ray photographs.

The discovery of heavy electrons now called *mesons*, has a striking similarity to the discovery of the positron, since the existence of both types was predicted theoretically, before being recognized during cosmic-ray studies. Mesons and positrons will be discussed in more detail later on.

4.2.4 Nuclear Synthesis and Abundance

A central, but elusive aim of nuclear research for almost a century has been the understanding of nuclear synthesis, particularly in the hands of astrophysicists. The dearth of experimental data to drive such enquiry has often resulted in the bending of synthesis theory to follow current thinking in cosmogony, rather than the other way around. This situation has not changed materially since the publication of two comprehensive reviews of the problem in 1950 [53, 54], after the decline of Lemaître's, but before the advent of Gamow's big-bang theories. This period provided an opportunity to consider rival theories without prejudice.

The back-to-back reviews [53, 54] compared the strengths and weaknesses of two models and reached opposite conclusions. Ter Haar concluded that "the equilibrium theory offers a better solution how to account for the observed abundances of the chemical elements than the $\alpha - \beta - \gamma$ theory".[1] Not surprisingly, Alpher and Herman argued for "a number of indications that element formation probably took place in an early prestellar

[1]The $\alpha - \beta - \gamma$ notation refers to the seminal big-bang paper under the names Alpher, Bethe, and Gamow, *The origin of chemical elements*, published on 1 April 1948 [55].

state of the universe, in which it is difficult to conceive of the existence for a sufficiently long time of the physical conditions required for an equilibrium among nuclei." They added: "Difficulties of the kind mentioned and in particular the freezing-in problem have led to the development of non-equilibrium theories of element formation, unattractive as they may be in requiring the discussion of specific nuclear reactions."

Alpher's preference is clearly conditioned by cosmology rather than science, but in the event it led to the total elimination of the equilibrium model from subsequent enquiry. A complicating factor in the argument is the continuously changing evidence provided by astronomical observation. In both theories, the success is measured in terms of predicted nuclear abundances and invariably these predictions depend on the nature and characteristics of known types of star, assumed as the seat of nucleogenesis. As more powerful telescopes identify new types of heavenly body, new possibilities of nuclear synthesis open up and the model has to be reworked. This process continues for the $\alpha - \beta - \gamma$ model only. The equilibrium model was abandoned before the discovery of quasars and black holes, that obviously provide more attractive environments for nuclear synthesis. The only mechanism for the dispersal of freshly synthesized material is still assumed to be supernovae and this assumption could also stand reassessment.

The Abundance Criterion

The main premise of both models is that a successful theory should predict the correct cosmic abundances of all nuclides. This idea, that abundances hold the clue to nucleogenesis probably stems from a proposal by Harkins in 1915 that the abundance of an element depends on two factors:

1. The structure of the nuclei of its atoms
2. The relative abundance of the materials used in the formation of the element in question

The Harkins theory of nuclear structure [56] was formulated before the discovery of the neutron, but predicted the existence of this particle [57]. An atom was described in terms of a mass number, P, that specifies the number of protons in a nucleus, an atomic number Z, that specifies the number of extranuclear electrons and the number of nuclear electrons, N [58]. Another fundamental quantity was defined as the isotopic number

$$n = P - 2Z = 2N - P = N - Z$$

which today would be called the neutron excess of a nucleus, (4.4).

The essence of Harkins theory is that all atomic nuclei are built up from α-particles and clusters H_x of hydrogen atoms, with $x = 0, 1, 2, 3$ to yield mass numbers of $4n$, $4n - 2$, $4n \pm 1$, together with charge compensating electrons. The formula was shown to hold for the most abundant isotopes of the elements of atomic number less than 30.

In his early work Harkins tested his ideas against the prediction, made on the basis of the He-H nuclear model, that there should be a marked difference in the abundance of elements with even and odd atomic numbers, respectively. This prediction was tested against elemental abundances measured in meteorites, which showed that the even-numbered nuclei were more abundant by a factor of 70.

In his later work Harkins estimated the added stability of helium by comparison to the mass of four hydrogen atoms and introduced the concept of a packing fraction that later led to the theory of nuclear fusion and defines the source of stellar energy. On the basis of nuclear build-up by the addition of α-particles Harkins predicted the existence of seven unknown stable nuclides. The subsequent discovery of all seven rates with the predictions that gained acceptance of Mendeléeff's table in the scientific world. Harkins discovered [59] a classification of the stable nuclides in terms of the ratio N/P and showed that this ratio never exceeds the value 0.62 in atomic species. The same classification was rediscovered independently many years later [5] and the maximum was shown, more precisely, to be the golden ratio $\tau = 0.6180\ldots$.

A summary of the scheme of nuclear synthesis in terms of what was termed the principle of regularity and continuity of series was published in 1949 [60].

Cosmic Abundance

Although the models of Harkins are now all but forgotten, the central idea that any theory of nuclear synthesis should account for the observed abundancies of nuclides, remains. Analytical techniques have been refined to the point where *cosmic abundances* can be stated with confidence for all naturally occurring nuclides. Cosmic abundances are derived from a variety of measurements on systems under vastly different conditions, e.g. meteorites, cosmic rays, stellar spectra, interstellar clouds and nebulae, supernova remnants, the earth's crust, the surface of the sun, and others.

Abundance stipulates the fraction of a given sample that represents the species of interest. To define cosmic abundance it is therefore necessary to find an agreed reference standard accessible in all situations listed above. In modern work the mass density relative to 10^6 silicon atoms is used as the standard. The mass density is defined in terms of the molar atomic mass of species j,

$$\begin{aligned} W_j &= m_j/m_u \qquad (m_u = m(^{12}\mathrm{C})/12 = 1/N_A = 1.66 \times 10^{-27}\mathrm{kg}) \\ &= \left[Z_j m_H + (A_j - Z_j)m_n - B_j/c^2\right]/m_u \end{aligned}$$

For ^{12}C

$$W(^{12}\mathrm{C}) \cdot m_u = m(^{12}\mathrm{C}) = 6m_H + 6m_n - B(^{12}\mathrm{C})/c^2$$

i.e.

$$\begin{aligned} \frac{B(^{12}\mathrm{C})}{m(^{12}\mathrm{C})} &= \frac{6(m_H + m_n)N_A}{12} - 1 \\ &= \frac{W_H + W_n}{2} - 1 \end{aligned}$$

Using this relationship, the expression for W_j reduces to

$$\begin{aligned} W_j &= \frac{1}{m_u}\left[Z_j m_H + \frac{1}{2}A_j m_n - Z_j m_n + \frac{1}{2}A_j m_n - B_j/c^2\right] \\ &= \frac{1}{m_u}\left[Z_j m_H - Z_j m_n + \frac{A_j m_n}{2} - \frac{A_j m_H}{2} - \frac{B_j}{c^2}\right] + A_j\left(\frac{m_n + m_H}{2}\right)N_A \\ &= \frac{1}{m_u}\left[\left(Z_j - \frac{A_j}{2}\right)\left(m_H - m_n\right)\right] + A_j\left(\frac{W_n + W_H}{2}\right) \\ &= A_j + A_j\left[\frac{B(^{12}\mathrm{C})/12 - B_j/A_j}{m_u c^2}\right] + \left(Z_j - \frac{A_j}{2}\right)(W_H - W_N) \end{aligned}$$

Because of the small difference between proton and neutron mass the last term can be ignored. Let N_j be the number density of species per unit volume. The mass density becomes $\rho_m = N_j m_j$. For a mixture of species $\rho_m = \sum_j N_j W_j/N_A$, which becomes

$$\rho_m = \sum_j \left(\frac{N_j A_j}{N_A}\right)(1 + b_j)$$

where

$$b_j = \frac{B(^{12}\mathrm{C}) - 12B_j/A_j}{m(^{12}\mathrm{C})c^2}$$

To obtain a density independent of composition the small factor, b_j is set to zero. Hence

$$\rho = \frac{\sum_j (N_j A_j)}{N_A}$$

and the nuclear fraction for species j is defined as

$$X_j = \frac{N_j A_j}{\rho N_A}$$

Finally, the ratio of the number of nuclei of species j to the total number of nucleons in the system

$$Y_j = \frac{X_j}{A_j} = \frac{N_j}{\rho N_A}$$

which is a mole fraction.

Abundances on the Si6 scale are defined as

$$\log y_i = \log f_{\mathrm{Si}} + \log Y_i$$

For $y_{\mathrm{Si}} = 10^6$ and assuming $Y_{\mathrm{Si}} = 2.529 \times 10^{-5}$ as the mole fraction of silicon, $\log f_{\mathrm{Si}} = 10.5971$.

The Anders-Grevesse [61] table of cosmic abundances, used in Figure 4.5, is based on this scale. The cosmic abundances of all non-radioactive nuclides are plotted in Figure 4.5 according to the (mod 4) classification of Harkins. The bottom frame is a composite of the three upper frames. There are two unmistakable features:

1. The three series exhibit the same trends.

2. For nuclei of comparable mass, abundances decrease in series order $4n > 4n + 2 > 4n \pm 1$.

Nucleogenesis

The modern view of nucleogenesis [62] discounts the equilibrium model as a viable proposition, having been discarded on the strength of an earlier analysis [63], summarized as follows [64]:

> The general trend of these isobaric abundances clearly approximates an exponential decrease with increasing A, until $A \sim 100$, above which the relative abundance is roughly independent of A. There is no overwhelmong distinction between even-A and odd-A.

It was concluded that, in "theories which invoke a great event rather than a continuum of creation" equilibrium conditions of synthesis would lead to an exponential decrease in abundance, such that for heavy nuclei the predicted abundance "is $\sim 10^{50}$ smaller than the observed abundances".

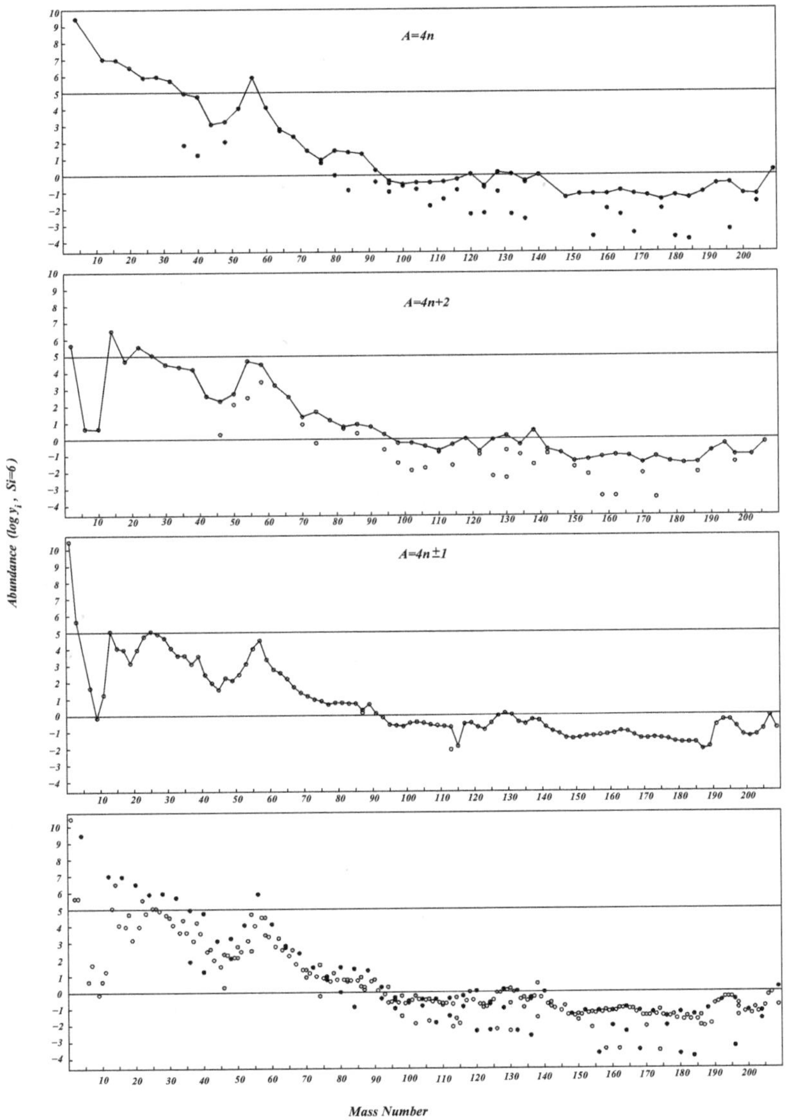

Figure 4.5: *Cosmic abundance of the stable nuclides.*

On re-examination of the abundances shown in Figure 4.5, there is little evidence to support these conclusions. There is a clear distinction between even-A and odd-A nuclei, already noted very early on [65], as well as between $A = 4n$ and $A = 4n + 2$ nuclei. These differences are most prominent at $A < 150$ and disappear altogether for the heavy nuclei. This is the exact behaviour to be expected for equilibrium systems that develop from different starting materials, but use a common building block.

It is instructive to examine the model that was used to simulate nuclear synthesis in a state of statistical equilibrium which was subsequently frozen down [53]. The probability of the system being in a given quantum state with N neutrons and Z protons with chemical potentials μ and λ respectively is given by the grand canonical ensemble

$$e^{[\mu N+\lambda Z-E(N,Z)]/kT}$$

with partition function

$$Q=\frac{(2\pi AkT)^{3/2}g}{h^3}$$

that treats the system as an ideal gas of protons and neutrons. Qualitatively, the concentration of nuclei in the ground state is given by

$$c_A=Qe^{\mu A/kT}=bA^{3/2}e^{-cA}$$

which represents the abundance as a function of mass number. The predicted exponential decrease of abundances is shown for the $4n$ series of nuclides in Figure 4.6. The curve of Figure 4.6 has $b=10^6$ and $c=1/6$. These parameters have been selected simply to demonstrate a measure of resemblance with observed abundances. Because of the scatter, it is no better and no worse than any other approximate smooth curve through the points. It shows that an exponential function need not be such a disastrously wrong predictor of abundances as to discount all equilibrium models, particularly if nuclear synthesis is driven by neither a great event nor continuous creation. The obvious resemblance between the abundance curves of the $4n$, $4n+2$ and $4n\pm1$ series, shown in Figure 4.5, points to an equilibrium model.

Detailed comparison of the models of nucleogenesis is complicated by the two assumptions that cosmic abundance is the same as solar system abundance [62] and that all nuclear synthesis after the big bang has happened in the familiar types of star that eventually disintegrate as supernovae. The second assumption is the most problematical. It provides the theoretical basis of the non-equilibrium or big-bang model of nuclear synthesis. It assumes that the big bang produced all matter in the form of light elements, mainly hydrogen and helium. Initially [66] it was assumed that all nuclides originated from the big bang, but the difficulty to bridge the gap due to the non-existence of nuclides of mass number 5 and 8 was interpreted to prevent any significant build-up to heavier elements by proton and neutron capture in the limited few minutes of reaction time.

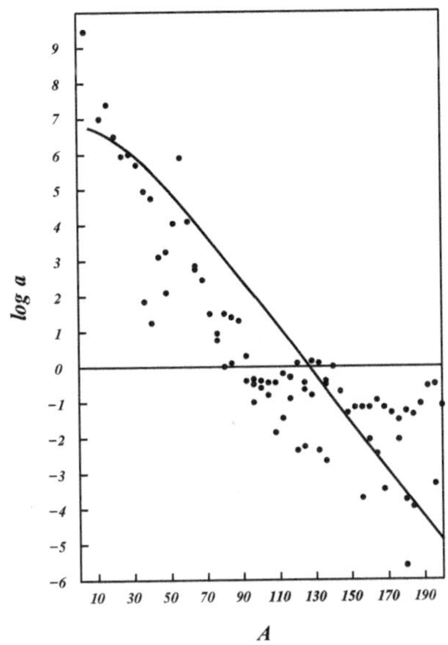

Figure 4.6: *Cosmic abundances of the non-radioactive nuclides with* $A = 4n$ *on a logarithmic scale. The superimposed curve represents the abundances* $a = bA^{3/2}\exp(-cA)$ *with* $b = 10^6$, $c = 1/6$.

The modern big-bang model postulates that all matter emerged from that event as hydrogen and helium, which at a later stage fused together in active stars to produce heavier nuclei. The reaction products would remain at the centre of the star so that the composition of the star gradually becomes richer in the heavy reaction products. An instability eventually sets in as the star runs out of fuel and a consequent gravitational collapse may initiate a catastrophic nuclear explosion that releases the heavy elements synthesised at the core, during a supernova. It is necessary to assume that material, enriched in heavy elements and ejected by many stars, accumulates in space, to become available for further condensation into new stars.

The role of the big bang in the non-equilibrium scenario is to account for the high abundance of H and He, relative to the heavier elements. The stellar synthesis of all other elements is assumed to proceed according to a proposal, originally made by Hoyle [67] to refute big-bang synthesis. It is therefore of interest to note that an alternative explanation of helium abundance exists. Only a small percentage of stars are massive enough to become supernovae. Many other aging stars simply blow off their outer layers while heavy elements remain trapped in their core remnants. The outer layers consist mainly of helium, produced by hydrogen-burning chain reactions, such as

$$4\mathrm{p} \rightarrow {}^{4}\mathrm{He} + 2e^{+} + 2\nu$$

Given sufficient time, helium should therefore accumulate in interstellar space. It appears that the helium abundance problem is largely self-generated by the theory that restricts the age of the universe to 10^{10} years.

The process of element building in stars is supposed to proceed by a variety of nuclear reaction sequences, such as

$$\begin{aligned} 3\,{}^{4}\mathrm{He} &\rightarrow {}^{12}\mathrm{C} + \gamma \\ {}^{12}\mathrm{C} + p &\rightarrow {}^{13}\mathrm{N} + \gamma \\ {}^{13}\mathrm{N} &\rightarrow {}^{13}\mathrm{C} + e^{+} + \nu \\ {}^{13}\mathrm{C} + p &\rightarrow {}^{14}\mathrm{N} + \gamma, \quad \text{etc.} \end{aligned}$$

to produce nuclei as heavy as Fe, but not beyond that.

The simplest sequence of nuclear reactions that occur in stars is known as *hydrogen burning*, with end products ^{4}He and ^{14}N. Towards the end of the hydrogen-burning cycle the temperature has increased sufficiently for helium burning to kick in. The main products, called *ashes*, of helium burning are ^{12}C, ^{16}O and ^{22}Ne. After helium, nuclear fuels that ignite in consecutive stages of increasing temperature, with their ashes, are:

$$\begin{aligned} {}^{12}\mathrm{C} &\rightarrow {}^{20}\mathrm{Ne}, {}^{24}\mathrm{Mg}, {}^{16}\mathrm{O}, {}^{23}\mathrm{Na}, {}^{25,26}\mathrm{Mg} \\ {}^{20}\mathrm{Ne} &\rightarrow {}^{16}\mathrm{O}, {}^{24}\mathrm{Mg}, {}^{28}\mathrm{Si}, \ldots \\ {}^{16}\mathrm{O} &\rightarrow {}^{28}\mathrm{Si}, {}^{32}\mathrm{S}, \ldots \\ {}^{28}\mathrm{Si} &\rightarrow {}^{56}\mathrm{Ni}, A \approx 56 \text{ nuclei} \\ {}^{56}\mathrm{Ni} &\rightarrow \mathrm{n}, {}^{4}\mathrm{He}, {}^{1}\mathrm{H} \end{aligned}$$

To account for the building up of elements above iron, it is necessary to resort to the idea of neutron-capture reactions. The neutrons may be supplied by (α, n) reactions, such as

$$\begin{aligned} {}^{13}\mathrm{C} + {}^{4}\mathrm{He} &\rightarrow {}^{16}\mathrm{O} + \mathrm{n} \\ {}^{17}\mathrm{O} + {}^{4}\mathrm{He} &\rightarrow {}^{20}\mathrm{Ne} + \mathrm{n} \\ {}^{21}\mathrm{Ne} + {}^{4}\mathrm{He} &\rightarrow {}^{24}\mathrm{Mg} + \mathrm{n} \end{aligned}$$

In the notation X(p,γ)Y that represents the nuclear reaction $\mathrm{X} + \mathrm{p} \rightarrow \mathrm{Y} + \gamma$, the more important types of reaction, responsible for nuclear synthesis in stars, include

$$(\mathrm{p}, \gamma), (\mathrm{p}, \mathrm{n}), (\mathrm{p}, \alpha), (\mathrm{n}, \gamma), (\mathrm{n}, \alpha), (\alpha, \gamma)$$

and their inverses, e.g.

$$^{16}\mathrm{O}\ (\alpha,\gamma)\ ^{20}\mathrm{Ne}$$

$$^{20}\mathrm{Ne}\ (\alpha,\gamma)\ ^{24}\mathrm{Mg}$$

$$^{18}\mathrm{O}\ (\mathrm{n},\gamma)\ ^{19}\mathrm{F}$$

$$^{22}\mathrm{Ne}\ (\alpha,\gamma)\ ^{26}\mathrm{Mg}$$

$$^{22}\mathrm{Ne}\ (\alpha,\mathrm{n})\ ^{25}\mathrm{Mg}$$

Also represented are heavy-ion reactions with emission of proton, α, neutron or deuteron, e.g.

$$^{12}\mathrm{C} + {}^{16}\mathrm{O} \rightarrow {}^{26}\mathrm{Al} + {}^{2}\mathrm{H}$$

There are myriads of possible nuclear reactions that might and probably do occur in the interior of stars and which could and probably do contribute to the formation of heavier, from lighter nuclei. The predominant type of reaction, responsible for the formation of elements beyond the iron group, is generally assumed to be neutron capture followed by β-decay to stable nuclides. One aspect common to all possible reactions, and especially to neutron capture, is the non-discrimination between nuclei of odd and even mass number that feature together in one complicated reaction network. There is nothing to suggest that a web of reactions, entangled in this fashion, should produce the separate series of products with comparable abundance distributions, clearly distinguishable in Figure 4.5.

The final word on nucleogenesis has obviously not been spoken. There are just too many loose ends and unwarranted assumptions to provide a consistent picture. Too many alternative mechanisms are ignored without mention or comment. The role of black holes, quasars, Seyfert galaxies and white holes, all of which could participate in a chain of nuclear synthesis, is not understood and therefore ignored. Even cosmic ray abundances, matched on the scale of solar abundances show up some important discrepancies [68]. Both H and He have low abundances in cosmic rays, whereas the elements Li, Be and B are five orders of magnitude more abundant. The relatively low abundance of the elements between Ca and Fe does not show up like that in cosmic radiation. The relative abundance of odd-Z elements is uniformly higher for cosmic rays. As a result there is much less variation in cosmic ray abundances than in solar and meteoric abundances. The general trend is a decrease in abundance from He to Fe with regular, though minor variation between even and odd-Z elements.

4.3 Theoretical Models

The Harkins model of nuclear structure antedates the discovery of neutrons, mesons and neutrinos, but it provides the correct atomic mass balance in terms of protons and electrons only. The equilibrium structure of an atom may therefore, to first approximation be described in terms of a single Hamiltonian pertaining to A protons and A electrons, of which Z are in extranuclear space.

4.3.1 The Shell Model

Schrödinger's equation for Harkins atoms has never been solved. The best alternative wave-mechanical model of an atomic nucleus is a semi-empirical shell model. Just like atoms with closed electron shells certain nuclei with "magic numbers" of nucleons (neutrons or protons) are more tightly bound than the average nucleus [69]. These nuclei have a lower mass and a greater abundance than would otherwise be expected. As an example the element Sn with $Z = 50$ has more stable isotopes (10) than any other element. Magic number nuclei also have very small quadrupole moments, indicating a nearly spherical shape. Such nuclei do not readily add an additional nucleon of the same type. An additional nucleon, when added, like valence electrons in an alkali metal, is loosely bound.

The shell structure suggested by these observations is difficult to visualize in a body as dense as a nucleus. A helpful suggestions due to Wigner can be understood in terms of a strong interaction between neighbouring nucleons that amounts to the exchange of a virtual particle and the interchange of nuclear identity.

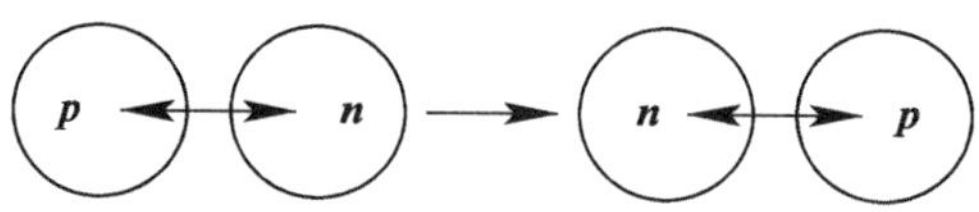

The same holds for interaction between neutron and proton pairs. In effect, two nucleons in contact appear to interpenetrate each other and hence move freely through the nucleus. Since the exact shape of the potential between nucleons is not known, the Wigner model suggests a shape between that of a square well and a harmonic oscillator. According to such a model [70], each neutron moves independently in a common potential well that is the spherical average of the nuclear potential produced by all other nucleons, and each proton moves independently in a common potential that is the average of the nuclear potential of all other nucleons, together with the coulombic potential of the other protons. This calculation produces groups of energy levels separated by gaps which indicate shell closure at the magic numbers 2, 8 and 20 states. Further gaps appear at non-magic numbers such as 40, but there is no gap at 50 or 82. The shell model, after further refinement gained general acceptance only when observed nuclear angular momenta were shown to agree with values predicted with the model.

Neutrons and protons are assumed to occur in potential wells that differ only in their potential-energy zeros. Actual potential wells are defined in terms of estimates for heavy nuclei, such as $^{208}_{82}$Pb. The nuclear states for a spherical well are defined by

$$-\frac{\hbar^2}{2m}\nabla^2\psi = E\psi \tag{4.5}$$

For infinitely high walls at $r = R$, it is assumed that $\psi = 0$ at $r \geq R$. As (4.5) is separable in spherical polar coordinates, $\psi(r,\theta,\phi) = u_l(r)Y_{lm}(\theta,\phi)$, and the radial function satisfies

$$-\frac{\hbar^2}{2m}\frac{1}{r}\frac{\mathrm{d}^2}{\mathrm{d}r^2}(ru_l) + \frac{\hbar^2}{2m}\frac{l(l+1)}{r^2}u_l = Eu_l \tag{4.6}$$

with boundary conditions that $u(r)$ is finite at $r = 0$ and zero at $r = R$.

The solutions to (4.6) are spherical Bessel functions $j_l(kr)$, with energies of the type

$$E(n,l) = \frac{\hbar^2}{2m}\left(\frac{x_{nl}}{R}\right)^2 \quad , \quad x = kR \tag{4.7}$$

The labelling of energy levels for varied l differs from the convention for electronic levels. Their sequence in terms of increasing energy is

$$1s < 1p < 1d < 2s < 1f < 2p < 1g...$$

As each nucleon has two possible spin states, $m_s = \pm\frac{1}{2}$, there are $(4l + 2)$ degenerate states for given (n, l).

A crucial step in establishing the nuclear shell model was the recognition [71] that there must be a term for spin-orbit coupling in the self-consistent potential felt by the nucleons, of the form $U_{so}(r)\boldsymbol{L}.\boldsymbol{s}$. Angular momentum is conserved only in the form of $\boldsymbol{L^2}$, $\boldsymbol{s^2}$ and $\boldsymbol{J} = \boldsymbol{L} + \boldsymbol{s}$. It has been shown that for given l and $s = \frac{1}{2}$, the allowed values of j are: $l + \frac{1}{2}$, with $(2l + 2)$ allowed values of j_z, and $l - \frac{1}{2}$, with $2l$ allowed values of j_z. On introduction of spin-orbit coupling, the $(4l + 2)$-fold levels therefore split into two levels each, which may be labeled $nl_{l+\frac{1}{2}}$, $nl_{l-\frac{1}{2}}$. Experiment shows that the state with $j = l + \frac{1}{2}$ invariably has less energy than the state with $j = l - \frac{1}{2}$.

The sequence of energy shells (with occupation numbers) inferred from experiment for both neutrons and protons is

$1s(1/2)$	$<$	$1p(3/2)$	$<$	$1p(1/2)$	$<$	$1d(5/2)$	$<$	$2s(1/2)$	$<$
2		6		8		14		16	
$1d(3/2)$	$<$	$1f(7/2)$	$<$	$2p(3/2)$	$<$	$1f(5/2)$	$<$	$2p(1/2)$	$<$
20		28		32		38		40	
$1g(9/2)$	$<$	$1g(7/2)$	etc.						
50		58							

The magic numbers of protons and neutrons for which the nucleus is particularly stable, as compared with neighbouring nuclei:

$$N, Z : 2, 8, 20, 28, 50, 82, 126 \text{ and } N = 184$$

coincide with completion of the subshells $1s(1/2)$, $1p(1/2)$, $1d(3/2)$, $1f(7/2)$ and $1g(9/2)$. Doubly-magic nuclei for which both neutron and proton number are magic are particularly stable and have relatively high abundances compared to their neighbours, e.g. ${}^4_2\mathrm{He}^2$, ${}^{16}_8\mathrm{O}^8$, ${}^{40}_{20}\mathrm{Ca}^{20}$, ${}^{208}_{82}\mathrm{Pb}^{126}$. Beyond $N, Z = 50$ the proton and neutron energy spectra are no longer identical.

It has been claimed [72] that in an atomic nucleus the close packing of spherons, consisting of two or four nucleons, in successive layers with packing sequences dictated by spin-orbit levels, reaches level completion at the magic numbers 2, 8, 20, 50, 82 and 126. An unconvincing aspect of the model is that inner layers may be of different composition in different nuclei. The completion of layers at selected magic numbers therefore depends on a unique and unpredictable choice of so-called inner-core, outer-core and mantle layers.

4.3.2 Strong Interaction

The presence of both protons and neutrons, tightly packed together in an atomic nucleus shows that the interaction that keeps these particles together cannot be of simple electromagnetic origin. The situation is somewhat akin to the formation of molecules by the interaction between positively charged atomic nuclei. However, in the latter case the mediation by oppositely charged electrons is readily understood as providing the necessary force of attraction. Quantum theory provides a convincing model of chemical bonding in terms of electron exchange between atomic nuclei [73]. In simple terms, a chemical bond may be viewed as the exchange of an electron between two positively charged ions, with the electron passing back and forth between them. The simultaneous exchange of two electrons between the ions ensures electric neutrality. Where Pauli's principle demands opposite spins for the bonding electrons, an electron-pair bond is formed.

It is not unreasonable to expect the interaction between nucleons in an atom to be of the same type, but possibly involving an exchange particle of higher mass than the electron. In this instance bond formation requires the exchange of positive and negatively charged, as well as neutral particles, according to the scheme, first proposed by Yukawa (1935):

$$\begin{aligned} p + \pi^- &\rightarrow n \\ n + \pi^+ &\rightarrow p \\ n + \pi^o &\rightarrow n \\ p + \pi^o &\rightarrow p \end{aligned}$$

Electrons and positrons are clearly not suitable to mediate this type of interaction, although their role in chemical bonding is parallel to the function of $\pi^\pm$ in the nucleus. The neutral particle π^o has a function, equivalent to that of a photon during electronic transitions in an atom.

The photon that mediates the electronic interaction is only observed when there is a transition to another energy level stimulated by energy absorbed from an external source. In the equilibrium state of an atom *virtual* photons are thought to be continually emitted and reabsorbed by electrons, only causing energy fluctuations within the limits of the uncertainty principle of quantum mechanics,

$$\Delta t \sim \hbar/\Delta E$$

The greater the distance between the charges the longer the time required for exchange and the smaller the energy of the exchanged photon. Although photons are not limited in range their momentum diminishes at long range and the force of interaction becomes weaker, as observed. The strong interaction within the nucleus was observed to operate at very short range only, as shown in Figure 3.10. These observations led Yukawa [74] to postulate the presence of three particles, called *mesons*, of equal mass, μ and different charges, within atomic nuclei.

In terms of Yukawa's model virtual mesons are exchanged between protons and neutrons in the nucleus. To explain why no fluctuation in proton or neutron mass is observed as a result, it is necessary to assume that the time elapsed between absorption and emission involving a given nucleon is within the uncertainty limit. As for photons the interaction is assumed to propagate at a velocity $v \sim c$, close to the speed of light. The range of interaction, unlike electromagnetic interaction, is limited and estimated to be of the same order as the classical radius of the electron $e^2/4\pi\epsilon_0 mc^2 = 2.8 \times 10^{-15}$ m (3.2).

The meson is a relativistic particle and in order to find its appropriate wave function it is necessary to start from the relativistic classical Hamiltonian for a free particle

$$H = \left(p^2c^2 + \mu^2c^4\right)^{\frac{1}{2}}$$

by introducing the momentum and energy operators (3.26 and 3.27)

$$p \rightarrow -i\hbar\frac{\partial}{\partial x}$$

$$H \rightarrow i\hbar\frac{\partial}{\partial t}$$

This substitution leads to the Klein-Gordon equation for a zero-spin particle:

$$\left(\nabla^2 - \frac{1}{c^2}\frac{\partial^2}{\partial t^2}\right)\psi = \left(\frac{\mu c}{\hbar}\right)^2 \tag{4.8}$$

For $\mu = 0$ the familiar equation for the electromagnetic field is obtained. The mass term

$$\frac{\mu c}{\hbar} = \frac{2\pi}{\lambda_C}$$

relates to the Compton wavelength of the meson and was interpreted by Yukawa to characterize the known range of the strong interaction, r_0, shown in Figure 3.10. To first approximation $r_0 \simeq r_{el}/2$ (3.2) and the ratio of mesonic to electronic mass follows as

$$\frac{\mu}{m} = \frac{8\pi\epsilon_0\hbar c}{e^2} = \frac{2}{\alpha} = 274$$

where α is the electronic fine-structure constant.

Assuming $\psi = \psi(r)$, (4.8) becomes

$$\frac{1}{r^2}\frac{\mathrm{d}}{\mathrm{d}r}\left(r^2\frac{\mathrm{d}\psi}{\mathrm{d}r}\right) = \left(\frac{\mu c}{\hbar}\right)^2\psi$$

which readily integrates to the Yukawa potential

$$\psi(r) = \frac{A}{r}\exp[(-\mu c/\hbar)r]$$

Soon after Yukawa's proposal a particle, called mesotron or *muon*, (μ) of mass $220m_e$, that decays into an electron and a neutrino

$$\mu^- \rightarrow e^- + \nu$$

was discovered in cosmic radiation. Whereas the Yukawa particle had to interact strongly with nuclear matter, the experimental particle was found to have little, if any, interaction, apart from electrical, with matter. In fact, the cosmic-ray meson was found capable of penetrating kilometers of solid rock without interacting with atomic nuclei.

The real Yukawa particle, called a π-meson or *pion* (π) of mass $273m_e$ was discovered in cosmic radiation in 1947. It escaped detection because it is extremely short-lived and therefore rarely occurs outside of nuclei. Charged pions decay into muons and neutrinos

$$\pi^{\pm} \longrightarrow \mu^{\pm} + \nu_{\mu}$$

and the neutral pion decays into either two photons

$$\pi^{o} \longrightarrow 2\gamma$$

or

$$\pi^{o} \longrightarrow e^{+} + e^{-} + \gamma$$

Before 1935 and the discovery of mesons that started the proliferation of nuclear particles, there were only protons, neutrons and electrons, together with their antiparticles. The photon, although well known to exhibit particle-like behaviour was considered, more conveniently as an excitation of the electromagnetic field because it lacked the attributes of mass, charge and half-integer spin, characteristic of fermionic particles. The electrical neutrality of the neutron could be ascribed to its composite nature as made up of proton and electron. The niggling observation that neutron decay did not preserve angular momentum and energy was addressed by invention of the neutrino, with zero mass and charge and half-integer spin, i.e.

$$n \longrightarrow p + e + \bar{\nu}$$

4.4 Particle Physics

For a brief period nuclear structure was well understood in terms of the five sub-nuclear particles, proton, neutron, electron, neutrino and photon. The discovery of positrons and mesons did not immediately complicate the issue. The Yukawa particles, or pions, postulated to mediate the strong interaction were demonstrated to be extremely short-lived, decaying within about 10^{-16}s, first to muons and eventually to electrons and positrons:

$$\pi^{\pm} \longrightarrow \mu^{\pm} + \nu \longrightarrow e^{\pm} + \nu + \bar{\nu}$$
$$\pi^{o} \longrightarrow 2\gamma \longrightarrow 2e^{+} + 2e^{-}$$

This simple picture was complicated with the discovery of new, more massive particles, in cloud chamber photographs of cosmic rays. These *strange* new particles included *hyperons*, that decay to protons, and *kaons* whose decay products consist of mesons. These observations were soon followed by the discovery of many *resonance* particles with widely varying properties. Classification of these new particles required the introduction of new quantum numbers to specify properties such as *isospin, strangeness, charm* and *baryon number*. The question remained whether or not to regard the hundreds of new particles as truly elementary. The theoretical consensus to resolve the issue is based on the symmetry properties of matter and particles. This symmetry includes *charge conjugation* that links matter to antimatter.

4.4.1 Antimatter

The idea of a shadowy parallel antiworld most certainly did not originate with Dirac's equation, nor with discovery of the positron. An argument based on annihilation of a positive and negative pair of electrons to create a quantum of energy, was used in 1904 by Jeans to explain radioactivity. Later on he gave credit to Newton for the idea. To state that negative-energy solutions of Dirac's equation provided the first indication of antimatter is to ignore the fact that, like Dirac's equation, the Klein–Gordon equation (4.8) also has negative-energy solutions. However, for spin half particles Dirac's equation is the appropriate form.

To obtain a relativistic wave equation for fermions, Dirac started with a Hamiltonian based on the operators (3.26) and (3.27), i.e.

$$H\Psi = \hbar i\frac{\partial\Psi}{\partial t} = E\Psi$$

Since the theory of relativity treats space and time variables on an equivalent basis the Hamiltonian operator

$$H = \left(p^2c^2 + m^2c^4\right)^{\frac{1}{2}}$$

should, like the time variable, be expressed in terms of linear momentum operators only,

$$\hat{p}_k \rightarrow \hbar i\frac{\partial}{\partial x_k} \quad (k = 1, 2, 3)$$

such that

$$E^2 = \left(\boldsymbol{\alpha}\cdot\boldsymbol{p}c + \beta mc^2\right)^2$$

This formulation leads to the Dirac equation

$$\left(E - \boldsymbol{\alpha}\cdot\boldsymbol{p}c - \beta mc^2\right)\Psi = 0\,, \quad \boldsymbol{p} = i\hbar\boldsymbol{\nabla}$$

or

$$i\hbar\frac{\partial\Psi}{\partial t} = -i\hbar c\boldsymbol{\alpha}\cdot\boldsymbol{\nabla}\Psi + \beta mc^2\Psi$$

which requires that the wave function has the form of a column vector $\Psi = \{\Psi_1, \Psi_2, \ldots, \Psi_N\}$, while $\boldsymbol{\alpha}$ and β are non-commuting $N \times N$ matrices [73].

In two-dimensional notation the four-dimensional energy and momentum eigenfunctions of $\Psi(z,t)$ are obtained as solutions of the secular equation

$$\begin{bmatrix} u_1 \\ u_2 \end{bmatrix}\begin{bmatrix} E + mc^2 & cp_z \\ cp_z & E - mc^2 \end{bmatrix} = 0$$

i.e.

$$E^2 - m^2c^4 - (cp_z)^2 = 0$$

whereby

$$E = \pm c\sqrt{p_z^2 + m^2c^2}$$

Dirac's equation describes particles with the correct relativistic relation between energy and momentum eigenvectors, but as for the KG equation, both positive and negative energy eigenstates occur.

The appearance of negative energy states was initially considered to be a fatal flaw in the Dirac theory, because it renders all positive energy states unstable. There is nothing to prevent an electron with positive energy from cascading down the negative levels *ad infinitum.* A solution to this dilemma was proposed by Dirac in terms of a many-particle theory. Observing that spin-$\frac{1}{2}$ particles, such as electrons, obey the Pauli exclusion principle, it was suggested that in the normal ground state (vacuum state) of Nature all possible negative energy levels are already occupied. This principle prevents any electron from falling into negative energy states and so ensures the stability of positive-energy physical states. For an electron of rest mass m_0 the energy gap separating the negative and positive energy solutions is exactly $2m_0c^2$, and it will be possible to excite an electron from the negative energy sea into a positive energy state. It then leaves a hole in the sea. This hole in the negative energy, negatively charged states, behaves exactly as a particle with the same mass as that of an electron, but having a positive electric charge (Figure 4.7). It may therefore be identified with a positron. If the excitation is effected by absorption of a γ-ray photon the result will be transformation of radiation energy into the creation of an electron–positron pair. Positrons created in this way will disappear rather fast due to the fact that the existence of a positron is equivalent to a hole in the negative energy sea. An ordinary electron must soon fall down into the negative energy state causing both electron and positron to disappear and the equivalent energy to be regained in the form of electromagnetic radiation. The four-component

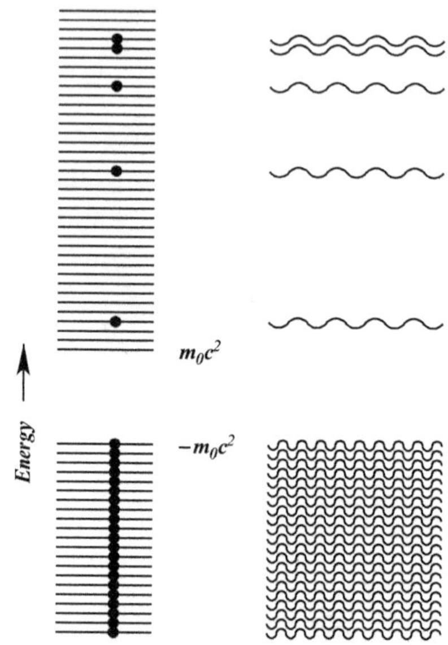

Figure 4.7: *Schematic drawing of the Dirac sea of negative-energy electrons, relative to real electrons. Particle and wave pictures are shown on the left and right respectively.*

spinors obtained as wave functions from the Dirac equation may now be interpreted as describing the spin-up and spin-down states of both electron and positron.

The arguments outlined here apply to any relativistic fermion. For each fermionic particle there must be a sea of negative-energy fermions, such as antiprotons, antineutrons and antineutrinos. Although mutual annihilation has not been observed experimentally, there is good reason to assume that $\pi^{\pm}$ and also $\mu^{\pm}$ mesons are antiparticles to one another. The two-quantum decay of π^o may be taken as experimental proof that this particle is identical with its antiparticle. Two π^o mesons could mutually annihilate each other to leave only energy quanta.

Antimatter in bulk is more than a theoretical curiosity. Several atoms of antihydrogen have recently been produced by the combination of antiprotons and positrons. Because of the preponderance of matter on the planet, larger quantities of more complex forms of antimatter become increasingly more difficult to handle. The possibility that large cosmic domains consist exclusively of antimatter is real, although the existence of such structures has not been confirmed experimentally. The generally accepted large-scale symmetry of the universe demands that half of all stuff should be in the form of antimatter. The debate on this issue is still open.

Awkward questions about the electromagnetic and gravitational fields of infinitely many particles in the vacuum remain unanswered. Also, the Dirac theory, amended by the hole proposition is certainly not a one-particle theory, and hence not a relativistic generalization of Schrödinger's equation. However, it is at the root of quantum field theory that provides the basis to modern symmetry theories of sub-nuclear structure. Most important of the fundamental symmetry laws is the charge–parity–time (CPT) relationship that differentiates between matter and antimatter.

4.4.2 The CPT Theorem

Reflection and inversion symmetry operations differ from common symmetry operations such as rotation and translation in being discrete rather than continuous. A law of Nature is said to be reflection (or inversion) invariant if the probability for any process equals the probability for the mirror image of that process. An example of such a symmetrical process is provided by energy emission when an excited atom returns to its ground state. If the laws that govern the structure of the atoms, and the emission of photons, are reflection invariant, the angular distribution must remain unchanged if reflected in a mirror that leaves the excited state unchanged. The process is described mathematically by means of an operator P which, when acting on any state, produces another state wherein all spatial coordinates are inverted. Since two reflections restore the original state, $P^2 = 1$. The operator has the two eigenvalues $\pi = \pm 1$, called *parity* states. Even and odd parity states are related in the same way as the enantiomers of a chiral molecule.

Another discrete symmetry, known as charge conjugation C implies the equivalence of a particle and its antiparticle. Like parity the wavefunction of a system may be even or odd under this operation:

$$C\psi = \pm\psi$$

The last of the three discreet symmetries, T or time reversal is visualized as a result of reversing the direction of all motion in a system. Symmetry under time reversal implies that if a system can evolve from a given initial state to some final state, then it is possible to start from the final state and proceed to the initial state by reversing all momenta.

Although each of the separate discreet symmetries may be approximate only, the CPT theorem is universally accepted to hold. The consequences of CPT are that particles and antiparticles should have the same masses and lifetimes. Furthermore, if any number of discreet symmetries are inexact, there must be a compensating symmetry of the remaining operations to cancel the first effect and ensure exact CPT symmetry. The only known laws of Nature that do not display reflection or, in some cases CP symmetry are weak interactions.

Weak Interaction

Neutrons are kept fixed in stable atomic nuclei by the strong interaction. Free neutrons however, decay by β-emission with a half-life of about 10 min. The interaction that keeps a free neutron intact is obviously much weaker than the strong nuclear interaction. It is for the same reason that unstable nuclei may decay by β-emission. The strong force that operates between protons and neutrons clearly has no effect on electrons. Particles with half-integral spin and sensitive to strong interaction are called *baryons* and those not affected are called *leptons*. The weak force, like its strong counterpart acts at short distance only.

An early puzzle about β-decay was the observation that the electrons emerge with a range of energies up to a maximum that corresponds to the energy equivalent of the mass loss. Most electrons seem to lose some of their energy during decay. To account for this energy loss and to ensure conservation of angular momentum, the existence of a neutrino, another lepton, was postulated. Neutron decay, formulated as $n \rightarrow p + e^- + \bar{\nu}$ requires release of an antineutrino to conserve spin angular momentum. This event corresponds to four fermions reacting at a point:

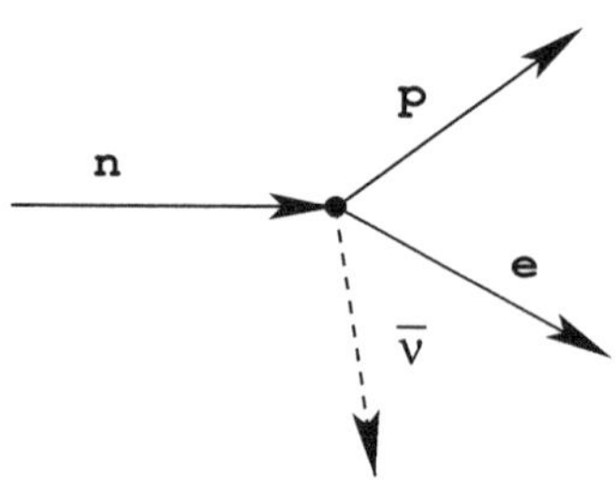

The wave function of the neutron is transformed into that of the proton and the wave function of an incoming neutrino (equivalent to an outgoing antineutrino) is transformed into that of the electron. The purely leptonic decay of mesons fits the same description, $\mu^{\pm} \rightarrow e^{\pm} + \nu + \bar{\nu}$:

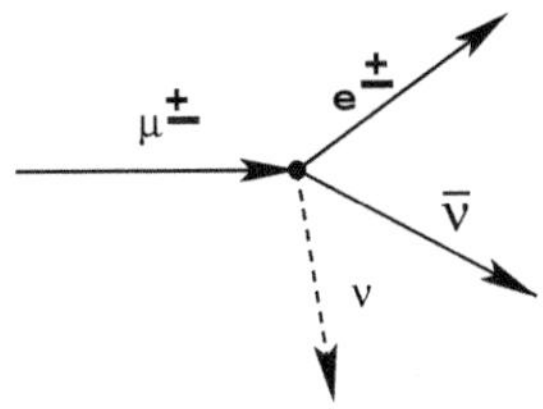

What seems to be an adequate description of the weak interaction cannot explain why a decay process such as

$$\mu^- \longrightarrow e^- + \gamma$$

is never observed. To account for this it was necessary to postulate that neutrinos are of two types, electron neutrinos ν_e and muon neutrinos ν_μ. With this distinction the β-decay of the neutron involves only electron neutrinos:

$$n \longrightarrow p + e^- + \bar{\nu}_e$$

and the muon decay of the pion involves only muon neutrinos:

$$\pi^- \longrightarrow \mu^- + \bar{\nu}_\mu$$

The muon decay into an electron therefore involves two types of neutrino:

$$\mu^- \longrightarrow e^- + \bar{\nu}_e + \nu_\mu$$

The distinction between neutrino and antineutrino allows formulation of a conservation law for lepton number, N_l. The law requires $N_l = 0$ for nucleons, $N_l = 1$ for e^- and ν, and $N_l = -1$ for e^+ and $\bar{\nu}$. Both electron-type and muon-type lepton number is conserved separately.

The most surprising finding about weak interactions was the observation that parity is not conserved in weak interaction. Parity violation occurs in the β-decay of ^{60}Co, spin-polarized in a magnetic field,

$$^{60}\text{Co} \longrightarrow {}^{60}\text{Ni} + e^- + \bar{\nu}$$

The heavy arrow in Figure 4.8 represents the nuclear spin. It generates rotation in the $x - y$ plane and is therefore not affected by reflection in this plane. The frame on the left shows the expected momentum vectors of decay electrons in a symmetrical situation. The emitted β-radiation has the same angular distribution around positive and negative field directions. In actual experiments the distribution shown on the right is observed. Spatial reflection is therefore not a symmetry of the β-decay interaction. Parity is violated.

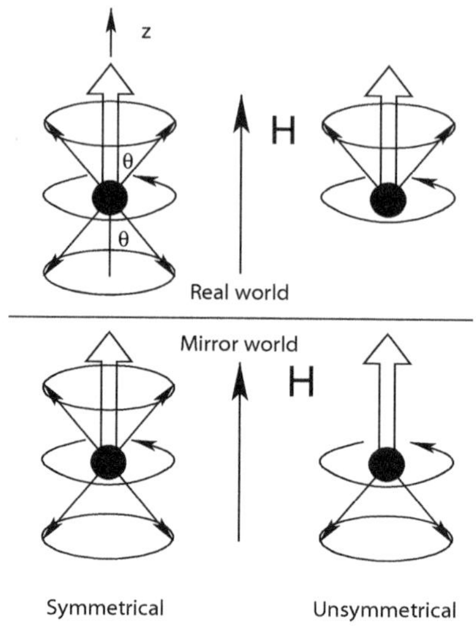

Figure 4.8: *The effect of mirror symmetry on the β-emission from ^{60}Co. The frame on the left shows the situation for emission that preserves parity. The frame on the right represents observed behaviour. The large arrow represents the nuclear spin in the direction of the applied magnetic field.*

The asymmetry in direction also has an effect on the spin of the β-electrons. The electrons show a net preference to spin in one of the two possible ways, as reflected in the polarization of the electrons, defined as

$$p = \frac{N_R - N_L}{N_R + N_L} = -\frac{v}{c}$$

where $N_L(N_R)$ is the number of left (right) spinning electrons in a measured sample; v and c are the velocities of electrons and of light. For $p = 1$ all electrons have the same (right) spin. The relationship between polarization and velocity can be measured experimentally. Scattering experiments have shown that electrons moving at relativistic speeds have a net negative polarization, $P \sim -1$ and for slow electrons $P \sim 0$.

In a decay process such as

$$^{152}\text{Eu} \rightarrow {}^{152}\text{Sm}^* + \nu$$
$$^{152}\text{Sm}^* \rightarrow \gamma$$

the spins of ν and γ must be correlated. From the fact that the emitted photon is polarized it has been deduced that neutrinos always have the same spin, opposite to that of the antineutrino.

From spin measurements on electrons and positrons in muon decay it was concluded that charge-conjugation symmetry C was also violated during weak interaction. Still, the combined operation CP could be obeyed. This possibility may be thought of as analogous to the conservation of baryon number $n = n(B) - n(\bar{B})$. In spite of the fact that the number of baryons can change with time, so can the number of antibaryons and their difference is conserved. However, the decay of neutral kaons into $\pi^+ + \pi^-$ has been interpreted to violate CP symmetry, and hence the weak interaction has none of the operations P, C or CP as an exact symmetry. The only discrete symmetry that could be exact is CP with time reversal – CPT.

4.4.3 The Quark Model

Since about 1950 the study of cosmic ray events in cloud chambers led to the discovery of many new particles with unexpected properties of mass and lifetime. To make sense of these new characteristics it was often necessary to invent new conservation laws and quantum numbers. Apart from parity, quantum numbers to specify *isospin (I)*, *strangeness (S)* and baryon number (B) came into general use.

Isospin The strong interaction between protons and neutrons is not affected by electric charge and the two types of particle may be considered as symmetry-related forms of the same particle, the nucleon. The nucleon, like particle spin, is a two-level system and may be described in the same formalism. Proton and neutron states are related by the *isospin* operator in abstract two-dimensional space. The total isospin $I = \frac{1}{2}$ and by analogy with directed spin, the projection $I_3 = \pm\frac{1}{2}$ distinguishes between proton and neutron states of the nucleon. A nucleon has both ordinary spin and isospin. The charge of a nucleon may be defined as $Q = I_3 + \frac{1}{2}$.

Baryon Number $B = 1, -1, 0$ for nucleons, antinucleons and for mesons and leptons respectively. Baryons are a subset of *hadrons*, or particles affected by the strong nuclear force. Hadrons, more massive than protons by factors up to 1,000, including kaons and hyperons which became known as strange particles, are known.

Strangeness What was strange about these particles is the relatively long life ($\sim 10^{-10}$ s) for particles, produced by strong interaction between pions and protons, normally expected to yield short-lived products with decay times of

$\sim 10^{-23}$ s. The particles appear to be formed by strong interaction between pion and proton

$$\pi^- + p \rightarrow \Lambda$$

However, their rate of decay is much longer than the rate of formation, and more appropriate to the decay of a particle held together by weak interaction. The problem was overcome by the suggestion that such particles are formed in pairs, much like the pair production that leads to the formation of electron and positron. Once formed, these particles are stable against further decay because of charge conservation. The e^+ and e^- pair are the lightest possible charged particles. By analogy it was assumed that hyperon and kaon must possess some internal property which is conserved. This property became known as *strangeness*. If K^0 is the lightest strange particle it should be stable once separated from Λ^0. The analogy is not perfect since the strange particles decay eventually. Strangeness appears not to be rigorously conserved and, in particular, is not conserved in weak interaction.

Strange particles always emerge in pairs, such as a hyperon and neutral kaon,

$$\begin{array}{cccc} \pi^- & + p & \rightarrow \Lambda^0 & + K^0 \\ (0) & (0) & (-1) & (+1) \end{array}$$

The quantum numbers for strangeness are shown in brackets. Strangeness is assumed to be a conserved quantity of the strong interaction. Decay of a single strange particle into non-strange components can therefore not proceed by strong interaction. Instead, particles such as hyperons are assumed held together by weak interaction that does not preserve strangeness and the decay

$$\Lambda^0 \rightarrow \pi^- + p \qquad \Delta S = 1$$

therefore is a slow process. Neutron decay shows that isospin is another example of a quantity which is not conserved in weak interaction:

$$n \rightarrow p + e^- + \bar{\nu}$$

Hadrons

With the proliferation of resonance particles efforts were made to classify these particles according to some scheme comparable to the periodic table of the elements. On the other hand, many of these entities appeared to be composite rather than elementary particles, judging by their high mass, spin and internal quantum numbers. Even the nucleon appears to be composite. The strong force, like a chemical interaction, seems to consist of an attractive and a repulsive part (Figure 3.10), suggesting some sub-structure. More evidence to this effect comes from the discovery of a short-lived excited state of the nucleon, called Δ and from lepton scattering with an angular distribution related to the spatial charge distribution of a target nucleon. Measurements of this type revealed that nucleons have charge radii of $\sim$0.8 fm, which was interpreted to mean that the nucleon has internal structure [75].

These observations inspired a search for a small number of fundamental units, dubbed *partons*, that may be combined in a variety of ways to generate all of the observed hadrons. Typical families of related hadrons can be identified amongst the excited states of the nucleon.

The Baryon Spectrum The lowest lying multiplets of the baryon spectrum is shown in Figure 4.9. Each of these baryon states has baryon number $B = 1$ and multiplicity $2I + 1$. The term nucleon refers only to the ground-state doublet that represents proton and neutron, $I_3 = \pm\frac{1}{2}$. Among the excited states there are other degenerate doublets with $Q = 1$ and 0, $I = \frac{1}{2}$.

There are also quartets of degenerate states, such as Δ, with isospin $I = \frac{3}{2}$, $I_3 = -\frac{3}{2}, -\frac{1}{2}, +\frac{1}{2}, +\frac{3}{2}$, $Q = -1, 0, +1, +2$. Not all multiplets have charges

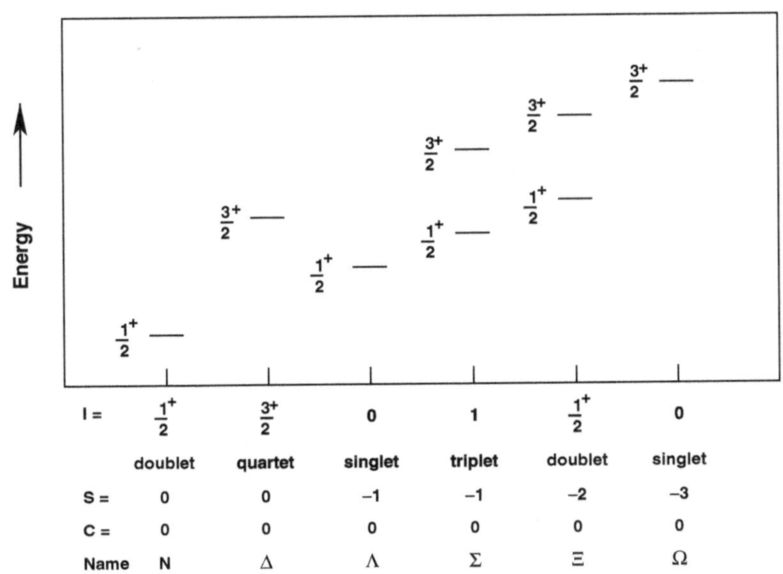

Figure 4.9: *The lowest lying multiplets that define an octet and a decuplet, of the baryon spectrum, after [75]. Numbers on the diagram specify ordinary spin and parity.*

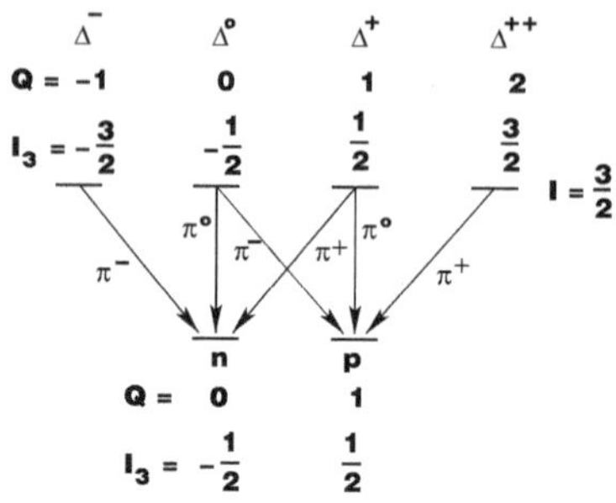

Figure 4.10: *Decay of the excited Δ baryon state.*

that agree with the simple charge formula. A more appropriate formula for all multiplets is

$$Q = I_3 + \frac{1}{2}Y, \quad Y = B + S + C \tag{4.9}$$

where S and C are the new quantum numbers for *strangeness* and *charm* assigned to the baryon states. The quantity Y is referred to as *hypercharge.* The quantum numbers Q, B, I_3, S and C are all additive.

The relaxation of an excited baryon quartet state is illustrated in Figure 4.10 and seen to be effected by the emission of charged pions. The π triplet must be assigned total isospin $I = 1$, so that $I_3 = 1, 0, -1$ for π^+, π^o and π^- respectively. For the charge formula to hold $Y = 0$ and since $B = 0$ for pions, it further follows that $S = C = 0$ for all π. Charge conjugation has the effect of replacing I_3 by $-I_3$ and hence π^+ and π^- are mutual antiparticles, whereas π^o, like a photon, is its own antiparticle. Hadrons have baryon number 1, and antihadrons have baryon number -1. Each stable hadron, like those in Figure 4.9 eventually becomes a proton, after a series of decays.

The attributes I, S and C are conserved in strong, but not in weak processes. Baryon number is conserved absolutely[2] and this prevents decay of the proton in processes such as $p \rightarrow e^+\pi^o$, or $p \rightarrow e^+\nu\nu$. The lower limit on the proton lifetime is $\sim 10^{31}$ years.

Quarks

For the postulate of a parton as a sub-nucleonic particle, to have any credibility, it should be defined in such a way as to account for all details of

[2]The massive preponderance of baryons over antibaryons in the observable universe provides a popular argument in favour of proton decay as a mechanism to destroy antimatter.

baryonic spectra and the behaviour of mesons, outlined in the preceding sections. The quark model, proposed 40 years ago by Gell-Mann and others, represents such a scheme, based on the following assumptions:

1. Quarks, denoted by q, are fermions, since they have to combine into objects of half-integer spin (baryons), and others of integer spin (mesons).

2. The baryon is composed of three quarks qqq and the baryon spectrum corresponds to the various quantum states of the qqq system. Since baryon number is additive it has the value $B = \frac{1}{3}$ for each quark. Antiquarks $\bar{q}$ have $B = -\frac{1}{3}$.

3. A meson consists of a quark and an antiquark, and therefore $B = 0$. As the meson's constituents can annihilate, there are no absolutely stable meson states.

4. To account for the additive quantum numbers I_3, S and C of hadrons, defined as either qqq or $q\bar{q}$ composites, at least five distinct types (*flavours*) of quark are required.

   (i) A pair of ordinary quarks, u(p) and d(own), that carry only isospin, but no other quantum numbers. Thus (u, d) is an $I = \frac{1}{2}$ doublet with $I_3 = \frac{1}{2}$ or $-\frac{1}{2}$ respectively. The antidoublet is $(\bar{d}, \bar{u})$.

   (ii) The strange quark s, that carries strangeness, but not charm or isospin. It is convenient to assign $s = -1$.

   (iii) The charmed quark, c, that carries charm, having $c = 1$.

   (iv) The fifth quark, b, that only carries the new attribute characteristic of some meson states.

5. A sixth t(op) quark is required by theoretical analysis [75].

Charge and Mass Using the additive charge formula (9) and assuming $B = \frac{1}{3}$ the following are obtained for the common quarks:

$$Q_u = \frac{1}{2} + \frac{1}{2} \cdot \frac{1}{3} = \frac{2}{3}$$

$$Q_d = -\frac{1}{2} + \frac{1}{2} \cdot \frac{1}{3} = -\frac{1}{3}$$

$$Q_c = 0 + \frac{1}{2}\left(\frac{1}{3} + 1\right) = \frac{2}{3}$$

$$Q_s = 0 + \frac{1}{2}\left(\frac{1}{3} - 1\right) = -\frac{1}{3}$$

Since only the ordinary quarks u and d carry isospin, all isospin properties of hadrons must come from these quarks. Any property of a hadron that depends on strangeness or charm must, by the same reasoning, be due to the presence of strange or charmed quarks among its constituents.

Since isolated quarks have never been observed an estimate of their mass remains ambiguous. It is however clear from hadron spectra that the mass of the c-quark, m_c is larger than that of u, d or s, and that the s-quark is heavier than u and d.

Interaction The strong force that keeps quarks together and which result in the strong interaction between hadrons is independent of quark flavour. In particular, the (u, d) isodoublet have the same strong interaction, despite the small $u - d$ mass difference, which is poorly understood in quark theory.

Classification of Hadrons To demonstrate how hadrons may be classified in terms of their quark structure, only the baryon octet shown in Figure 4.9 will be considered. The simplest structures are obtained from a combination of the ordinary u and d quarks only. There are two possible flavour combinations, $uud(Q = 1)$ and $udd(Q = 0)$, and since both have spin of $\frac{1}{2}$, these are the logical arrangements to represent p and n. The corresponding spin structure is shown in Figure 4.11. Symmetrical combinations uuu and ddd are ruled out for $\frac{1}{2}$ spin.

The state with two s-quarks follows the same pattern. These two quarks must have parallel spins and the u or d quark must have opposite spin, to yield a resultant $\frac{1}{2}$. As a result there are two isodoublets, one with $S = 0$, and another with $S = -2$:

$$\psi(u_\uparrow u_\uparrow d_\downarrow) = N^+ = p \qquad \psi(u_\downarrow s_\uparrow s_\uparrow) = \Xi^0$$

$$\psi(d_\uparrow d_\uparrow u_\downarrow) = N^0 = n \qquad \psi(d_\downarrow s_\uparrow s_\uparrow) = \Xi^-$$

The remaining spin $\frac{1}{2}$ combinations are uus, uds, dds. There are two possibilities: either the two ordinary quarks provide the total isospin, $I = 1$, or else $I = 0$. Because of the overall symmetry the spins of the ordinary quarks

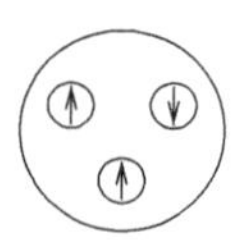

Figure 4.11: *Spin structure of a nucleon.*

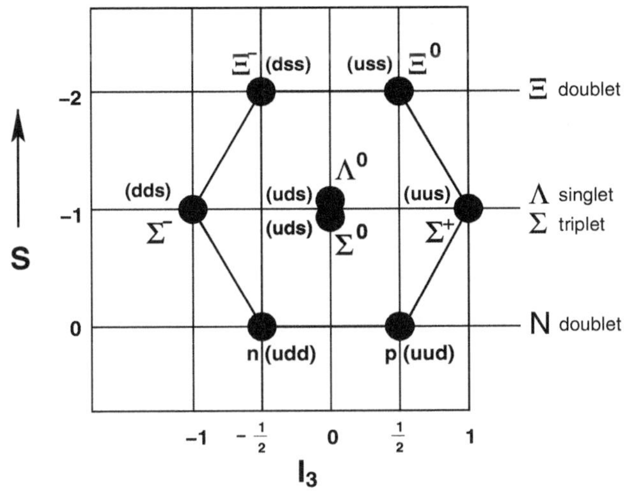

Figure 4.12: *The lowest spin $\frac{1}{2}$ baryon octet.*

must add up to 1 if $I = 1$, or 0 if $I = 0$. The spin of s must add appropriately to give spin $\frac{1}{2}$. $I = 1$ is a triplet state and $I = 0$ a singlet:

$$\begin{aligned}
\psi(u_\uparrow u_\uparrow s_\downarrow) &= \Sigma^+ \\
\psi(u_\uparrow d_\uparrow s_\downarrow) &= \Sigma^0 \qquad \psi(u_\uparrow d_\downarrow s_\uparrow) = \Lambda^0 \\
\psi(d_\uparrow d_\uparrow s_\downarrow) &= \Sigma^-
\end{aligned}$$

Altogether there are eight states in the spin $\frac{1}{2}$ baryon octet, shown in Figure 4.12 as an (I_3, S) diagram. The lowest lying states of the baryon spectrum falls neatly into the octet of Figure 4.12. The non-charmed baryons of spin $\frac{3}{2}$ forms a decuplet that fits the second multiplet of Figure 4.9.

4.4.4 Deep Inelastic Scattering

The elegance and utility of the mathematical scheme that led to the development of the quark model is not contingent on the existence of quarks as physical entities. To probe for the experimental evidence that confirms the existence of quarks the structure of nucleons continues to be explored. Although free quarks have never been observed, experiments referred to as deep inelastic scattering have provided circumstantial evidence of their existence. In these experiments the known interactions of leptons are used to probe the structure of nucleons.

A proton is about 10^{-15}m in diameter and to resolve any structure below that requires a probe of even smaller wavelength. The associated momentum, calculated as $p = h/\lambda$, exceeds 1 GeV/c and is sufficiently high to disintegrate any target on impact. The main measurement of the experiments is the variation of cross-section (effective target area) of the nucleon as a function of energy loss and scattering angle of the incident lepton. At high momenta the complicated scattering off a nucleon of finite spatial extent, is replaced by scattering off a point-like parton. The photon ceases to scatter off the nucleon as a coherent object and, instead, scatters incoherently off individual point-like partons.

Reversion to the simplicity of point-like scattering after a relatively more complicated transition phase is interpreted [76] to indicate that a more basic level within the nucleon has been reached.

The result of highly inelastic scattering of leptons by nucleons is the production of jet-like structures, consisting of hadrons. The interpretation is that the large momentum transfer ejects an individual parton from the parent hadron. However, free partons have never been observed. Instead, it is thought that before the struck parton can escape from the nucleon some confining mechanism comes into play by pulling parton–antiparton pairs from the vacuum. This sequence is illustrated in Figure 4.13.

The formation of jet-like structures is also observed during high-energy annihilation of e^+e^- pairs and hadron–hadron collisions. During e^+e^- annihilation it is assumed that a $q\bar{q}$ pair is formed:

$$e^+e^- \rightarrow \gamma \rightarrow q\bar{q}$$

Because of the strong interaction between the quarks the primary pair is prevented from separating. Instead, it is energetically favourable for the strong field to produce a secondary quark pair $q_1\bar{q}_1$ and to form singlet mesons, $m = q\bar{q}_1$ and $\bar{m} = q_1\bar{q}$, which are not sensitive to long-range confining forces.

The important assumption is that the cross-section for deep inelastic scattering can be calculated by addition over individual parton–probe interactions and that the complicated final-state interactions become important only over long space–time distances. This argument implies a force that varies with distance. At the short distances of initial interaction (10^{-17}m) the interquark force is weak and the quarks are essentially free. As the distance between two quarks increases beyond the nucleon diameter (10^{-15}m), the force increases indefinitely to confine the quarks, perhaps permanently, within the hadron. Any energy expended in the effort to separate the quarks will be sufficient to extract a new quark–antiquark pair from the vacuum and create a new hadron.

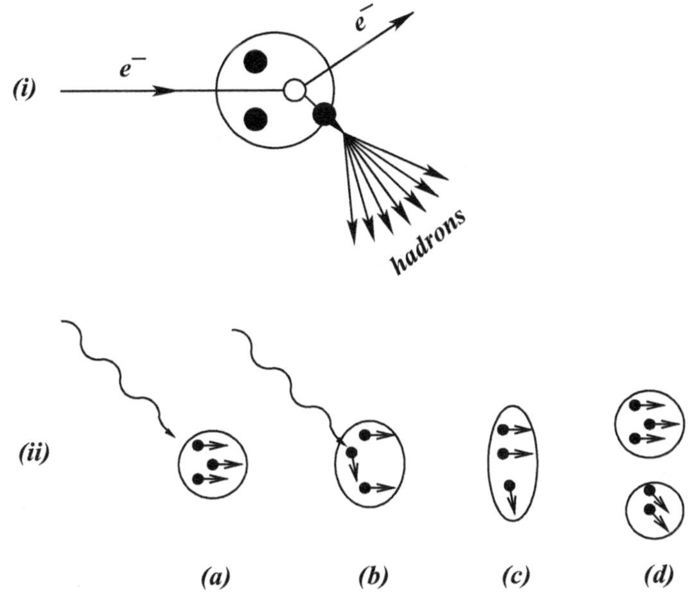

Figure 4.13: *(i) A high-energy electron hits a parton inside a proton. The parton is ejected, at the speed of light, and fragments into hadrons, forming a jet. (ii) A deep inelastic probe strikes a parton (a) and (b), which flies off with relativistic momentum until it is restrained by some confining mechanism (c) that pulls parton-antiparton pairs from the vacuum to create new particles (d).*

The tendency of the interquark forces to become weaker at small distances is known as *asymptotic freedom.* The other tendency of interquark forces to become increasingly stronger as the quarks are separated, is known as *confinement.* A picturesque description [77] of confinement is infrared slavery. In a way, the quarks may be thought of as bound like prisoners on a chain gang. If they stay close together nothing happens and they can move around freely. The problem starts when one prisoner tries to get away from the others. If that happens, all the prisoners are again reminded that they are in chains.

Theory of Quarks

An understanding of all interactions in Nature, including those in the subnuclear region seems to require four different forces with different characteristics. Most familiar is the gravitational force that pervades the entire universe and obeys an inverse square law:

$$F = G\frac{m_1 m_2}{r^2}$$

Best understood is the electromagnetic interaction. It is of infinite range and the strength of the force is characterized by the fine-structure constant

$$\alpha = \frac{e^2}{\hbar c} = \frac{1}{137}$$

The source of the force is electric charge which can be either positive or negative and appears as an integral multiple of the electronic charge, i.e. $Q = Ne$. The force between two separated charges is

$$F = K\frac{Q_1 Q_2}{r^2}$$

The resemblance between the formulae for electromagnetic and gravitational forces suggests that a common theory to embrace both phenomena could be formulated.

Electromagnetic phenomena are governed by Maxwell's laws, that introduce the photon as a particle that mediates the interaction, in the same way that the strong interaction between nucleons is mediated by pions. It is therefore tempting to use the theory of electromagnetism as a model to develop a single unified theory to describe the four forces of Nature in terms of a single formalism.

The most successful theory of the electromagnetic field is quantum electrodynamics (QED), the first example of what became known as gauge theories. A detailed, but simple introduction to the concept of gauge invariance has been published before [73] and only a brief summary is provided here.

Gauge Theory

QED is based on the conservation of electric charge. This means that the Lagrangian[3] density which describes the electron wave function must be invariant under symmetry transformation through time, space and rotation. Symbolically

$$G\mathcal{L}(\psi_e) \rightarrow \mathcal{L}(\psi_e')$$

[3]The Lagrangian is the dynamic function that links the energy, momentum and angular momentum of a system.

The symmetry group G, denoted $U(1)$, corresponds to a simple shift in the phase of the electron wave function. The transformation of phase is said to be *global* since it represents an identical operation at all points in space–time. Although the actual phase of the wave function is unobservable, differences in phase can be detected.

Global gauge invariance accounts for conservation of charge without allowing for local changes of phase, i.e. between neighbouring space–time positions. Such a local gauge transformation amounts to

$$G(x)\mathcal{L}(\psi_e) \rightarrow \mathcal{L}'(\psi_e')$$

where $x = (\boldsymbol{x}, t)$ denotes a space–time vector. Since the Lagrangian itself is changed by this transformation the theory is no longer invariant. To restore the gauge invariance it is necessary to introduce a second field to compensate for the local change in wave function. The required field must have infinite range, since there is no limit to the distances over which the phase conventions must be reconciled. The new field must therefore be massless. In fact, the field required for local gauge invariance can be shown to be the Maxwell electromagnetic field whose quantum is the photon. The way in which two electrons interact by exchanging a photon is shown diagrammatically in Figure 4.14. The changes in the photon wave function cancel out the changes in the Lagrangian resulting from a local phase transformation. So, the introduction of the photon leads to local gauge invariance. Symbolically

$$G(x)\mathcal{L}(\psi_e, A) \rightarrow \mathcal{L}(\psi_e', A')$$

where A denotes a four-vector that describes the electromagnetic field, viz. the photon's wave functions.

Electroweak Gauge Field

The formulation of a gauge theory for the weak interaction led to the unexpected result that electromagnetic and weak interactions are different manifestations of a single electroweak gauge field with broken symmetry.

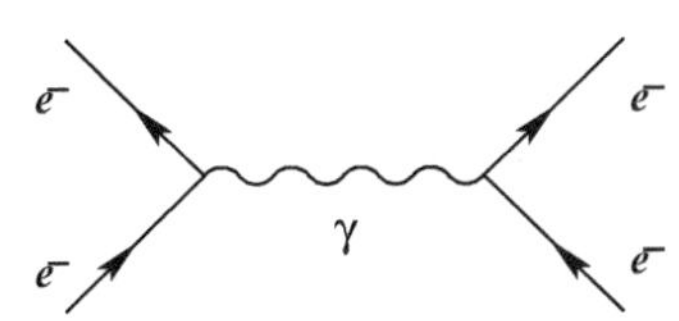

Figure 4.14: *The interaction between two electrons by the exchange of an electromagnetic gauge boson, or virtual photon.*

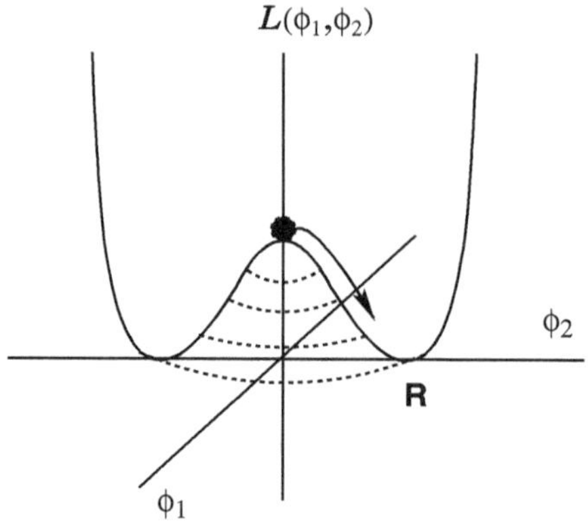

Figure 4.15: *Symmetry breaking as the marble rolls down from the centrally raised dome at the bottom of a wine bottle.*

The existence of asymmetric solutions to a symmetric theory comes about when the symmetric state is not the state of minimum energy and on transition to this ground state the intrinsic symmetry of the system is spontaneously broken. A spectacular example of this phenomenon is provided by a marble, precariously balanced at the raised centre at the bottom of a wine bottle (Figure 4.15). A small perturbation will cause the marble to fall into the trough, to a position of lower energy, which is one of an infinite set of rotationally symmetrical sites. Although the Lagrangian is symmetric, the actual ground state will be asymmetric.

On replacing the marble by a spinless particle with two components ϕ_1 and ϕ_2, the axes labeled ϕ_1 and ϕ_2 represent average values of the associated quantum fields. The energy is not a minimum at the zero values of the field (below the elevated central point), but along the circle defined by

$$\phi_1^2 + \phi_2^2 = R^2$$

This equation defines the vacuum states of the theory, characterized by non-zero average values for ϕ_1 and ϕ_2. For G, the rotation group in the plane, the Lagrangian is still symmetric under the global transformation

$$G\mathcal{L}(\phi_1, \phi_2) \to \mathcal{L}(\phi_1', \phi_2')$$

However, if in addition the Lagrangian is required to be invariant under local gauge transformation, it is necessary to introduce a compensating gauge particle, in order to maintain the invariance, i.e.

$$G(x)\mathcal{L}(\phi_1, \phi_2, A) \to \mathcal{L}(\phi_1', \phi_2', A')$$

If the fields are now redefined by shifting the origin of coordinates such that the axes pass through the point of minimum energy, the new field components

$$\phi_1'' = \phi_1, \quad \phi_2'' = \phi_2 - R$$

and the Lagrangian

$$\mathcal{L}(\phi_1, \phi_2, A) \equiv \mathcal{L}(\phi_1'', \phi_2'', A'')$$

In the new Lagrangian, instead of the two-component particle (ϕ_1, ϕ_2) and a massless gauge particle, there now appears a massive spinless particle ϕ_2 (Higgs boson) together with a massive vector gauge particle A'', that contains three polarization states. In summary

$$\left.\begin{array}{c} A \\ \phi_1 \end{array}\right\} \rightarrow \text{massive } A''$$
$$\phi_2 \rightarrow \text{massive } \phi_2''$$

Based on these ideas a unified theory for the weak and electromagnetic interactions was developed by Glashow, Weinberg and Salam. The theory describes the interactions of leptons by the exchange of W bosons and photons, and invokes the Higgs mechanism to generate the masses for the W bosons.

To ensure that the interactions between leptons conserve weak isospin[4] and weak hypercharge, the total Lagrangian was defined to be invariant under the groups $SU(2)$[5] and $U(1)$ at the same time. This procedure defines the gauge particle $W = (W^+, W^0, W^-)$.

Following the procedure for symmetry breaking outlined above, it was found that the $U(1)$ symmetry of QED remains unaffected. The photon remains massless and W gauge bosons become massive. The theory was confirmed by the discovery at CERN of the $W^\pm$ and W^0 bosons with the exact predicted masses.

---

[4]Weak isospin is the property that relates electrons to neutrinos in the same way that protons are related to neutrons by isospin.

[5]The rotation

$$G^{SU(2)} \begin{pmatrix} \nu_e \\ e^- \end{pmatrix} \rightarrow \begin{pmatrix} \nu_e^* \\ (e^-)^* \end{pmatrix}$$

in weak isospin space turns the leptons into equivalent mixtures of the electron–neutrino.

4.4.5 Quantum Chromodynamics

Soon after the proposal of the quark model it was realized that the suggested quark content of some particles violated the quantum-mechanical exclusion principle that demands wave functions to be anti-symmetrical under particle exchange.

The wave function that describes a hadron made up of three quarks is the product of three factors

$$\psi = \psi_{space} \times \psi_{spin} \times \psi_{flavour}$$

In some hadrons three quarks have the same flavour and hence the flavour factor must be symmetrical under the exchange of any two quarks. The same is true for the spin factors since the three quarks in many hadrons have the same spin. Since the total particle spin represents the sum of the three quark spins it means that there is no orbital angular momentum and that the three quarks must be positioned symmetrically. Consequently, the space factor is also symmetrical under the exchange of any two quarks. As all three factors are symmetric, the total wave function must be symmetric and the combination of three quarks seems to violate Pauli's principle.

Coloured Quarks

The way out of the symmetry problem was to assume that quarks carried another quantum number to distinguish between identical quarks and produce an antisymmetric wave function. The additional quantum number could, at the same time, be chosen to specify a new type of charge that represents the source of the strong field. By analogy with the electric charge, the new, so-called *colour charge* must be assigned to individual quarks in such a way that the combination is colourless for neutral particles. This statement is equivalent to saying that all hadrons are colour singlets, which also is the simplest symmetric multiplet, e.g. red, blue, green for qqq. In such combinations the colour of an individual quark, like the phase of an electron wave function remains invisible. The confinement of the quarks within hadrons can accordingly be restated as the confinement of the colour quantum number.

Colour Gauge Theory

The fundamental idea of QCD is that the colour charges of the quarks act as the sources of the strong chromodynamic force between quarks, just as electric charge acts as the source of the electromagnetic force between electrically charged particles. As the quarks carry both colour and electric charge, they experience both the strong and electromagnetic forces. However, the chromodynamic force is so much stronger than the electromagnetic, that the latter may effectively be ignored.

To incorporate these ideas into a gauge theory it is noted that the quark colour triplet may serve as a fundamental representation of the symmetry group $SU(3)$. The fundamental symmetry of the colour force may be taken as invariance under the redefinition (transformation) of quark colours. Like the phase of the electron wave function in QED, colour is unobservable in QCD and no real phenomena should depend on the convention that defines it.

Let the initial quark colours be defined by the multiplet

$$q = \begin{pmatrix} r \\ b \\ g \end{pmatrix}$$

The colour transformation

$$G^{SU(3)}q \rightarrow \begin{pmatrix} c_1(rbg) \\ c_2(rbg) \\ c_3(rbg) \end{pmatrix} \equiv \begin{pmatrix} v \\ y \\ o \end{pmatrix}$$

mixes the colours into three different proportions of r, g and b, say violet, yellow and orange. Since the colour coding may be different at different points in space, the Lagrangian is required to be locally gauge invariant under $SU(3)$. For the Lagrangian to remain invariant a new gauge field must be introduced:

$$G^{SU(3)}(x)\mathcal{L}(\psi_1, \psi_2, \tilde{A}) \rightarrow \mathcal{L}(\psi_1', \psi_2', \tilde{A}')$$

As in other gauge theories local invariance requires the introduction of a new gauge field to communicate the local colour convention between the quarks. The quanta of this gauge field are massless spin-1 particles, called *gluons*. They mediate the chromodynamic forces between quarks: whereas photons cannot interact directly among themselves, gluons carry colour charge and so give rise to their own colour fields (Figure 4.16). It means that they can interact amongst themselves directly. By analogy with Van der Waals forces between neutral atoms, the residual chromodynamic effects that remain between colour singlet states after internal colour neutralization, are responsible for the strong forces between hadrons.

Grand Unification

It appears feasible in principle to formulate a gauge theory based on a super group that incorporates $SU(3) \times SU(2) \times U(1)$ in a unified description of

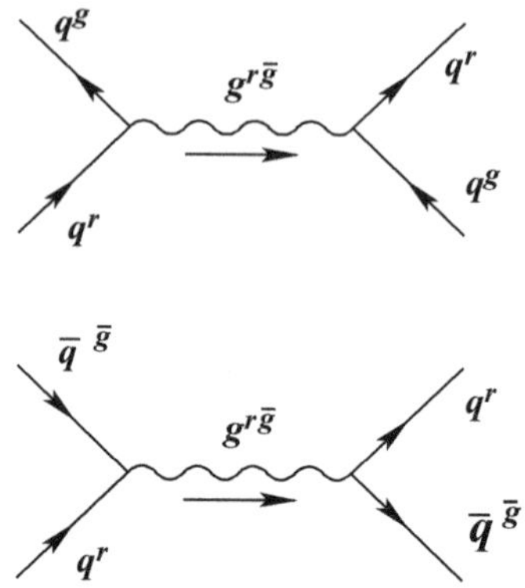

Figure 4.16: *Quarks interact by gluon exchange. The colour quantum numbers flowing into the gluons are equivalent to those of a $q\bar{q}$ pair.*

strong and electroweak fields. That would certainly solve the unexplained matching of charges carried by colour singlet leptons (-1) and colour triplet quarks ($3 \times \frac{1}{3}$).

Following the discovery of τ mesons and their neutrinos there appears to be a symmetric relationship between quarks and leptons:

$$\begin{pmatrix} u & c & t \\ d & s & b \end{pmatrix} \leftrightarrow \begin{pmatrix} \nu_e & \nu_\mu & \nu_\tau \\ e^- & \mu^- & \tau^- \end{pmatrix}$$

Both quarks and leptons appear to be point-like objects.

A grand unified theory should incorporate a hierarchy of spontaneous symmetry breaking which reflects transition from very high energy simplicity to the observed low-energy complexity. One expectation of grand unification is the elimination of baryon-number conservation, by allowing transformation of a ($B = \frac{1}{3}$) quark into a lepton ($B = 0$) or an antiquark ($B = -\frac{1}{3}$). The consequence is possible decay of the proton that could account for the cosmic imbalance between matter and antimatter.

4.4.6 Primary Structure

The theory of quarks, leptons and gauge fields defines an elegant logical construct, but not without its drawbacks as a theory of matter. The quest has been the ultimate elementary zero-dimensional point particle. At present 19 empirical parameters are required for such identification, without any certainty that the ultimate fermions from which all matter is constructed have been identified. Grand unified theories all suggest that the grand symmetry in itself must be badly broken, which indicates at least one deeper level of structure. The argument seems to be trapped in an endless cascade.

Even more worrisome is the concept of a point particle as the building block of higher dimensional structures. The current theory will most likely have to give way to a more abstract one without fundamental particles, but which describes phenomena that share the attributes now ascribed to quarks and leptons. A search for a physically acceptable description of the electron that accounts for its behaviour in macroscopic environments [36], identified a wave description as the most appropriate. That avoids the necessity of dealing with point objects and the infinities that it generates.

Chapter 5

Elements of Cosmography

There are three recognized disciplines designed to provide a description of the world, or cosmos, and these are often badly mixed up and even used interchangeably. Cosmogony attempts to document the origins and history of the cosmos, mainly in terms of myths and theology. Cosmology is that part of metaphysics that views the world as the totality of all phenomena in space and time. Plato distinguished between opinion or belief and knowledge by identifying objects of opinion as appearances and the objects of knowledge as realities. The role of metaphysics is to identify the ultimate realities at the basis of knowing the whole world. It is necessary to challenge all assumptions before finally arriving at an account of the nature of all things that is fully coherent and fully thought out. Modern cosmology has failed this objective by embracing the concept of an expanding universe as first-principle reality, without question. The third discipline, known as cosmography, is the science that describes and maps the general features of the universe. It is, by definition based on observation, mainly astronomical, from which it derives its theories.

Most authors fail to distinguish between the three disciplines and present a hodge-podge of the three under the term *cosmology*. No further effort will be made to correct the accepted terminology here and the remainder of this chapter will be presented in the same modern style that amounts to an unstructured mix of science, mathematics, speculation and dogma.

Because of the massive appeal of big-bang cosmology to religious and scientifically lay communities, and the uncritical support of the mass media, an unbiased opinion on the cosmological issue is rare and can only be expressed with caution. Two exceptions that strive to document the extensive volume of data, and the slippery ground trodden by theories based on extrapolation from local observation to the edges of the universe, are the works of

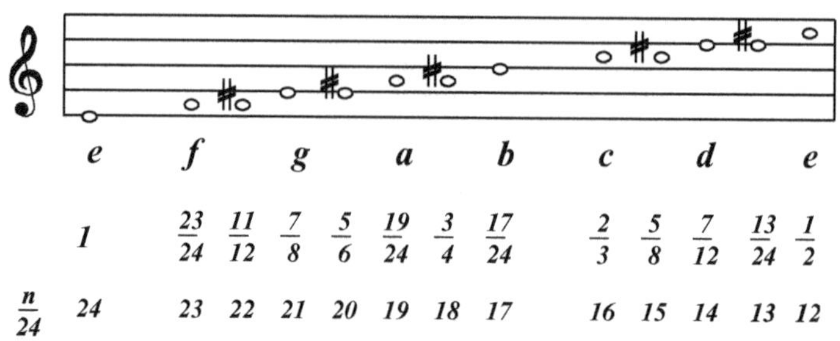

Figure 5.1: *The notes in one octave and their relationship to the fractional lengths of vibrating segments of the string.*

Satterthwaite [78] and of Narlikar [79]. The discussion that follows is based on a review by one of the authors in the 1993 Year Book to an encyclopedia [80].

5.1 Historical

The earliest known metaphysical cosmology was that of the Pythagorean cult, who based their model on the assumption that numbers provided the only intelligible structure of what is out there. Their dictum, *all is number*, probably derived from noting the functional significance of numbers in the objective world and, particularly in music.

The string of a lyre when plucked, vibrates over its entire length and produces a sound of specific pitch. When pinched at half length the pitch becomes one octave higher and sounds in harmony with the fundamental note. Further harmonious overtones are produced by pinching the string so as to restrict its vibration to segments of length corresponding to rational fractions of the total length.[1] When pinching the string at a point which is not a simple fraction, the generated tone is not in harmony with the others.

[1]Strictly Pythagorean tuning on 12 uniform strings per octave, like a piano keyboard, would result from the fractional lenghts, starting from any key, as shown in Figure 5.1. However, to ensure maximum harmony of chords the uniform spacing, a between adjacent notes should be such that

$$a^{12} = \frac{1}{2} \text{ i.e. } a = \sqrt[12]{\frac{1}{2}}$$

an irrational number. The problem of tuning exists in finding a compromise between the two scales.

This observation probably represents the first demonstration of a fundamental relationship between science and mathematics.

The Pythagoreans concluded that a harmonious relationship of similar type pervaded the whole world and reflects the ultimate nature of reality. These relations, like the multiplication table are timeless and changeless, and define a static world. The cult even denied the existence of incommensurable fractions, considered to be irrational, and only looked for perfect structures in Nature.

The cosmology that views the world as a static whole and considers motion and change as an illusion, is linked to the name of Parmenides of Elea. His pupil Zeno gave several convincing proofs of the doctrine, such as the analysis of an arrow in flight. At any instant the tip of the arrow occupies a definite position. At that instant the arrow cannot move, for an instant has no duration. Hence at each instant the arrow is at rest, which means that it never moves.[2]

Although the static model of the universe is rarely taken seriously in modern times, the link forged between number theory and cosmology remains in vogue. The ancient conviction that the truth about the world is locked up in numbers is still alive. The prevailing cosmology at any given period, cannot be more advanced than the current number system. With the acceptance of irrationals into the number system, to enable general root extraction, the static world order lost its appeal. Although the restriction on roots of negative numbers remained, a less eccentric world view, based on real numbers, became possible. With motion accepted as scientific fact, the static universe was replaced by a mechanical system with plausible theories on cosmogony, mainly due to Laplace. He maintained that, given the initial position and velocity of every particle in the universe at any particular instant, and given all the forces at work in Nature, a super-intelligence could calculate with precision the entire past and future history of the cosmos.

The most advanced modern cosmologies have the freedom of complex numbers, quantum theory and non-Euclidean geometry. However, the available scope is even wider. Hypercomplex numbers are the equivalents of complex numbers in n dimensions. The algebra to handle these numbers, initiated by Grassmann (*c.*1840), has never been worked out in detail and, except for the occasional use of n-dimensional vectors, still awaits proper implimentation in science.

---

[2]Conversely, since an instant has no duration the arrow tip passes through each point without coming to rest.

For centuries after Pythagoras, numbers that could be used to express incommensurable quantities were banned for being irrational. At a later stage extension of the number system to incorporate complex numbers was also resisted by mathematicians, branding these as imaginary, until they found application in science. The n-dimensional hypercomplex numbers, and with them cosmologies that require multidimensional curved space, are still on the black list. Until all available scientific evidence has been re-examined in terms of this more extensive number system, any claim of a metaphysically sound cosmology is therefore premature and probably wrong.

The modern number system is still incomplete in one last respect, namely the infinite. Infinity paradoxes have been around since the time of Zeno and, despite impressive progress over the last century, they still abound. From the work of Cantor there emerged the idea of transfinites, that transcend the infinite. However, the new paradox of an ultimate transfinite re-introduced all of the problems formerly associated with infinity, in a new guise.

Another anomalous feature of the infinite is demonstrated by the tangent function, as shown in Figure 5.2. From the definition, $\tan\theta = a/b$, it follows that $\tan 0 = 0$ and $\tan \pi/2 = \pm\infty$. As $\theta \to \pi/2$ from below, $\tan\theta \to \infty$, and as $\theta \to \pi/2$ from above, $\tan\theta \to -\infty$. The only logical conclusion to be drawn is that $\infty = -\infty$ and that the real line intersects itself at infinity. If an infinite universe is interpreted in the same sense it therefore means that space must close onto itself, not in one, but in n dimensions. Unless the implications of this requirement are understood, it makes no sense to entertain notions like the beginning of time and the universe.

The idea of closure at infinity is not new in mathematics, being the principle that underlies perspective geometry, that assigns a single extra point to the real line at infinity. Since there are no parallels in perspective space, cosmography that recognizes this principle cannot be based on affine geometry. Most current cosmologies are crippled by this requirement.

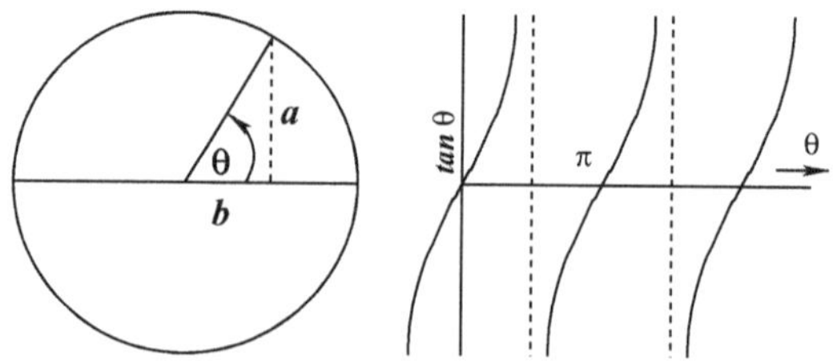

Figure 5.2: *The trigonometric tan function.*

5.2 Cosmological Paradoxes

Although the principle is not always respected, there is general agreement that an acceptable cosmological model should provide a plausible explanation to a number of scientific facts based on observation. Three of the major issues are usually formulated as paradoxes, representing experimental observations at variance with intuition. These paradoxes are:

- The Olbers or dark-night paradox
- The Zwicky or dark-matter paradox
- The chirality paradox

In addition, factors such as the cosmic microwave background and the large-number coincidences, are also considered important.

Common ground for discussion of these issues is found in the theory of general relativity that provides a geometric model to account for gravity, the dominant interaction on a cosmic scale. It is formulated in a four-dimensional manifold of space and time. This manifold assumes its simplest form in the absence of matter. In this hypothetical state the manifold is euclidean and represents the four-dimensional equivalent of a flat plane of infinite extent. Since the vast intergalactic spaces are considered to be empty, the large-scale structure of space–time is considered to be essentially euclidean, with local distortions in the vicinity of large accumulations of mass. The gravitational field is defined as the gradient of the manifold that slopes towards massive bodies. An experimental test of the model is the shift in apparent positions of stars during a solar eclipse, shown in Figure 5.3. It shows that light rays do not move along euclidean straight paths, but along the geodesics of the non-euclidean geometry of space–time. The assumed euclidean nature of space–time is an axiom of special relativity and is not dictated by the general theory. The possibility that space–time is curved on a large scale can therefore not be discounted, especially not, in view of the earlier observation about infinity and perspective geometry.

5.2.1 Olbers Paradox

It is fair to assume that the heavenly bodies are distributed evenly through space and considering their vast number, it is very likely that an observer who focuses in any randomly selected direction, should be looking directly at some star or galaxy. Since all galaxies are not at the same distance from earth they will not appear equally bright. However, if the density of galaxies

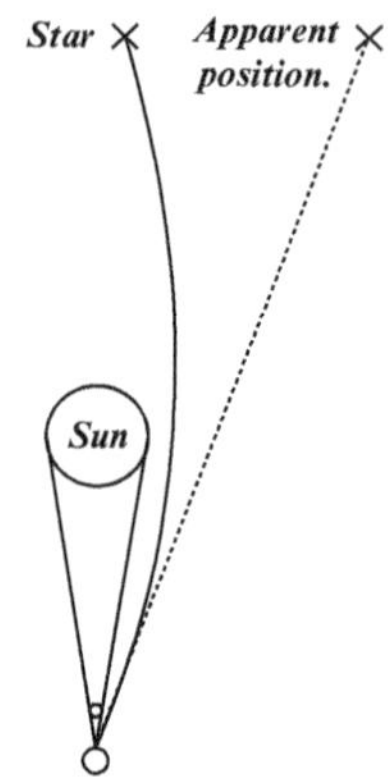

Figure 5.3: *Apparent displacement of stellar position due to gravitational interaction with the sun observed during an eclipse.*

through space is constant the number of galaxies at a given distance must increase with a factor which is proportional to the square of the distance. This conclusion follows immediately from the simple formula that specifies the surface area of a sphere, $A = 4\pi r^2$.

Since light from a source is radiated equally in all directions the intensity beamed in a certain direction must decrease by the same quadratic factor with distance. It follows that galactic light that originates in a spherical shell centred on the earth, reaches the planet with the same constant intensity, irrespective of distance,

$$I = \frac{I_0}{r^2} \cdot 4\pi r^2 = \text{constant},$$

for all r. On this basis it was argued by Olbers that the night sky cannot be dark.

So convincing is this argument that any cosmology without an explanation of the effect, fails directly. The simplest explanation in plain English is that light which travels over large distances becomes tired. Although such light is supposed to travel through empty space, during a journey over billions of light years a photon must encounter many dust particles, gas molecules, ionized material, atoms, plasma and other cosmic debris, despite the low density of these materials. Each encounter affects the energy of the photon, which is inversely proportional to the wavelength of the photon, $E = h\nu = hc/\lambda$. As the photon looses energy, the wavelength increases. The longer the path length, the larger the shift in wavelength would be. This explanation works well for a universe which is infinite in space and time. Other cosmologies require alternative explanations, such as an expanding universe or curved space–time.

5.2.2 Zwicky Paradox

This paradox refers to an inconsistency in the mass of galaxies as estimated by two different, equally valid methods. Galactic mass is estimated either from brightness or from rate of rotation. Since both methods are based on sound physical principles it means that some aspect is overlooked and the necessary assumptions required to convert either luminosity or angular-velocity data into a measure of mass, need to be re-examined.

The relationship between luminosity (L) and mass (M) for a star of radius R was first estimated by Eddington from Stefan's law (3.11) that relates energy density in a radiation field to the mean temperature, $\varrho \propto T^4$. Assuming $T \propto M/R$ and $\mathrm{d}T/\mathrm{d}r \simeq T/R$, the energy flux per unit area follows as

$$\frac{1}{\rho}\frac{\mathrm{d}\varrho}{\mathrm{d}r} \propto \frac{M^3}{R^2}$$

where $\rho = M/R^3$, is the mass density. The luminosity over the whole surface area, $L \propto M^3$, was thus found to be independent of the radius and proportional to the third power of the mass of a star.

The second method to estimate galactic mass assumes that all galaxies rotate around a central attractor in the same style as the milky way and the solar system. This means that galaxies do not rotate like a rigid wheel, where all particles rotate with the same period. Their rotation is differential and the period of each object depends on the distance from the galactic centre. If the period of revolution can be measured at a known distance from the galactic centre, the mass of the galaxy can therefore be estimated by the same formula (Kepler's harmonic law) that calculates the solar mass from the earth's orbital period (P) and its mean distance (R) from the sun, i.e. [81]

$$M = \frac{4\pi^2 R^3}{GP^2}$$

G is the gravitational constant.

If the galactic plane of rotation is directed towards the earth the orbital period can be measured by observing the relative Doppler shifts (see Section 5.3.1) of light sources at opposite edges of the galaxy, as shown in Figure 5.4. From the difference in wavelength the period of revolution and hence the mass of the galaxy can be calculated. As for the solar system, the total mass may be assumed to good approximation to be concentrated in the centre.

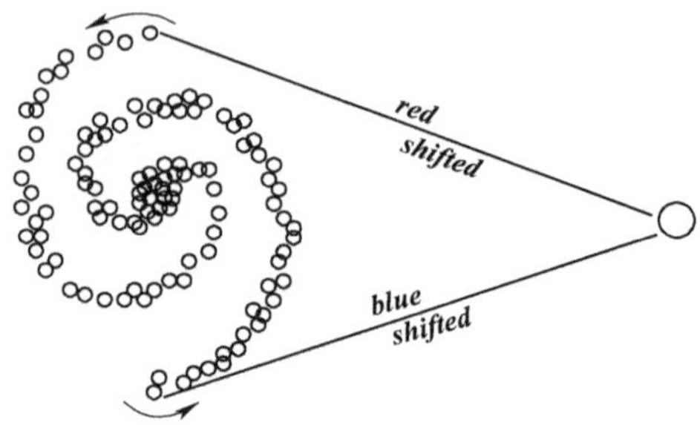

Figure 5.4: *Schematic diagram to illustrate the astronomical observation of the period of galactic rotation.*

The paradox consists therein that the second method consistently yields masses significantly larger than the luminosity method. Two explanations that immediately spring to mind are that the galactic mode of rotation resembles that of a rigid wheel, or that a large percentage of invisible mass resides in the galactic halo. For the second reason the difference in mass is colloquially known as dark matter.

5.2.3 Antimatter Paradox

The highly respected CPT theorem (Section 4.4.2) dictates a universal balance between matter and antimatter, which is nowhere observed. Although this is a real paradox it features less prominently in cosmological discourse. It is commonly stated that a slight imbalance, which occured without cause at the time of the early universe, initiated the annihilation of most matter and all of antimatter. This argument is too weak to resolve the paradox and the challenge remains for any serious cosmology to locate the missing antimatter and explain the chirality of space.

5.3 Cosmological Models

The leading cosmological models and some others with potential strength can now be submitted to the test of resolving the paradoxes. Although there are too many rival models to list, they all belong to one of three classes: expansion models, plasma models and curvature models. Each of these will be discussed separately.

5.3.1 The Expanding Universe

The most important factor in modern cosmology is Olbers' paradox. Because of subsequent spectroscopic analysis of galactic light that highlighted the same problem, the paradox has become linked to the name of Hubble.

Many experiments have shown that, in spectral composition, the light from remote galaxies differs from terrestrial light. Each chemical element radiates and absorbs light at characteristic frequencies and leaves a finger-print when it interacts with a light beam. In this way elements in the solar corona absorb their characteristic frequencies, with the effect that sunlight shows dark lines at these frequencies when analyzed on earth. Although the finger-prints of the elements also occur in galactic light they are no longer at the exact same frequencies as in the solar spectrum. In most cases the entire pattern is shifted towards longer wavelength, in a process referred to as a red shift.

In static models of the universe the tired-light hypothesis accounts for these red shifts, especially since the magnitude of the shifts correlates well with distance from the galactic source. Since however, cosmological models based on the theory of general relativity and large-scale euclidean geometry, appear to be dynamic rather than static, an alternative explanation of red shifts, offered by Hubble, gained general acceptance. Hubble associated the red shifts with the Doppler effect of receding sources.

The Doppler Effect

If a radiating body is in motion with respect to an observer, the apparent wavelength of the radiation is shortened if the motion is towards the observer and lengthened if the motion is away from the observer. This phenomenon, known as the Doppler effect may be observed as the change in pitch of its whistle when a train passes an observer. The same effect also occurs in all forms of electromagnetic radiation. The mechanism of the phenomenon is shown in Figure 5.5, [78]. A stationary body is assumed to radiate at a constant wavelength (λ) and the wavefronts will be concentric spheres as shown on the left. For a body in motion the wavefronts are compressed in the direction of propagation and the apparent wavelength for an observer in the direction of approach, λ_B, will be shortened, or blue shifted. In the opposite direction the wavefronts appear further apart and the wavelength, λ_R appears red shifted.

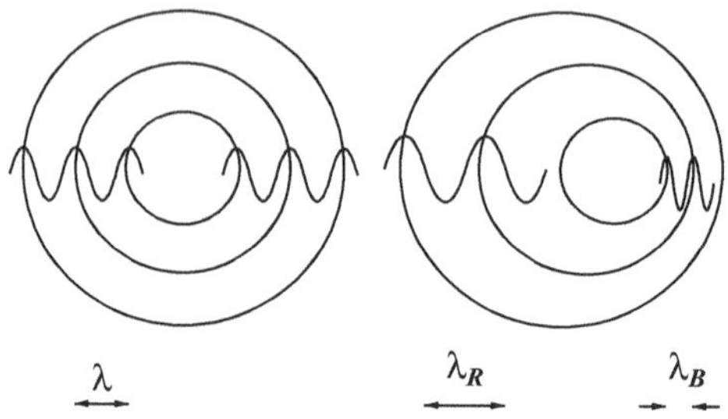

Figure 5.5: *Waves radiated by stationary (left) and moving (right) sources respectively.*

Big-Bang Theory

Hubble, who discovered the relationship between red shift and distance, proposed a new model of the universe on the assumption that galactic red shifts had their origin in the Doppler effect. With the exception of a few blue shifts observed for not too distant galaxies, galactic light in general is red shifted by an amount proportional to their distance from earth. All of these galaxies therefore appear to be receding from the milky way at velocities that depend on their distance from the observer.

At first glance this condition seems to imply a privileged position of planet earth at the centre of the universe. In fact, the interpretation stays the same if all galaxies are assumed to move apart at a rate that depends on the pairwise distance between individual galaxies. An analogy often used to simulate the situation is that of a spotted balloon that is being inflated. All the spots move apart at a rate that depends on the distance between the spots. A raising plum pudding is a better three-dimensional analogon.

An immediate implication of an expanding universe is an unusual situation at both beginning and end of the process. Extrapolation into the future suggests that an increasing number of galaxies would approach the horizon at which they recede at the velocity of light and out of the observable universe. At some stage the expansion must progress to the point where only the milky way remains in the locally observed universe and eventually the dispersion must approach a zero-density vacuum state.

Extrapolation into the past suggests the inverse scenario of increasing density, until the singular state of infinite density is approached at the very beginning, or the moment of the big bang. At this instant of infinitely high temperature there is no distinction between matter and energy and all forces of Nature converge into a single totally symmetric interaction. Even space and time are assumed to have their beginning at this fateful moment.

The big-bang theory has had a chequered career. The idea was first mooted by the novelist Edgar Allan Poe in his essay *Eureka* in 1849. Based on the ideas of Olbers he postulated a closed universe and an explosive beginning to prevent immediate gravitational re-collapse. The same idea was put forward as serious science by Lemaître in 1931, arguing that the increasing entropy of the universe implies that everything started from a single atom. The idea was grabbed by Millikan to explain cosmic radiation. This early version of the big-bang theory is now all but forgotten.

The concept was revived by Gamow in a paper published with co-authors Alpher and Bethe (α, β, γ) on 1 April 1948 [55]. It was argued that only some nuclear explosion of cosmic proportion could furnish the conditions under which heavy elements could be obtained by the fusion of hydrogen and helium. Although the synthesis argument has been superseded by less extravagent models, the grand concept survived. At that stage however, the big bang temporarily made way to steady-state theory, to be discussed next, but made a comeback when radio-astronomical observations suggested an evolutionary cosmic structure at variance with steady state.

The microwave background With both cosmological models under stress an alternative model of a big bang in a closed universe was proposed by Dicke. The objective was to calculate the helium content as a fraction of the total mass of the universe by theoretical variation of the photon:proton ratio. The correct percentage was found at a ratio that predicted a radiation density to persist in the closed universe at a temperature of 30 K.

Before the prediction could be tested experimentally microwave background radiation at an apparent temperature of 3 K was discovered and interpreted as final vindication of the big bang. To account for the difference between predicted and observed temperatures the theory was adapted by tracing back the origin of the radiation to a time at which matter and radiation first became decoupled and the early universe transparent. This happened when the temperature had dropped sufficiently for electrons and protons to combine into hydrogen atoms and photons were no longer confined by scattering off free electrons. The observed Planckian distribution (Figure 3.15) of the background radiation agrees with thermalization of the radiation prior to decoupling.

The observed microwave background turns out to be remarkably smooth and isotropic. So much so that the expansion since the big bang was found to be hopelessly inadequate to account for this observation. At that stage particle physics came to the rescue with the postulated inflationary stage of expansion.

Inflation Some of the most fundamental elementary particle theories apply at such high energies that experimental testing is not feasible. Grand unified theories (GUTs) are of this type (see Section 4.4.4). The operative assumption is that, at sufficiently high energy (temperature) the different forces of Nature become indistinguishable. As a testing ground for GUTs the conditions that presumably prevailed at the big bang are therefore highly appropriate. It is natural to argue that strong, weak, electromagnetic and gravitational fields were but one at the beginning. As the early universe cooled down through expansion the symmetry that equipoised the different forces broke down spontaneously (see Section 4.4.3).

Other processes of this type are well known as phase transformations. When a chemical system cools down, interactions that are too weak to manifest at high temperature, become dominant in producing additional structure which lowers the symmetry of a new emerging phase. From the theoretical models that are used to describe the different forces, the energies at which symmetry breaking should occur, can be calculated and correlated with the age of the early universe. One of the features that can be calculated is the time at which matter and radiation decoupled to initiate the spreading of background radiation. On the basis of these calculations the size of the universe at that stage was much too small to account for the isotropic spread of the background radiation as observed today.

To remedy the situation it may be recalled that in many chemical systems phase transitions do not always happen immediately the transition temperature is reached. The system is said to be supercooled and one effect thereof is that the latent heat of transition is not released. The inflationary scenario consists of postulating a supercooled universe and assuming that the latent heat is utilized to drive a hugely accelerated expansion before symmetry breaking sets in.

This suggestion has spawned several mathematical models of the inflationary period, none of which has been totally successful. What they have in common is to predict an expansion factor of about 10^{30} over a period of 10^{-32} seconds, probably driven by anti-gravity. This may sound like science fiction but it provides the only explanation in big-bang theory of the microwave isotropy and predicts sufficiently high density of the universe to eventually arrest the expansion. During the inflationary period all temperature differences are wiped out and the geometry of space is flattened. The theory now prescribes the mass of the universe and the 1% dark matter inferred from the Zwicky paradox is increased to 99%. According to particle theory all of this is non-baryonic mass or wimps.[3] A convenient feature of wimps is that they cannot be detected by any means. Had it not been for the explicit assurance of the experts, many a layman could mistake this scheme for mythology rather than science.

Measured against the stated criteria, the big bang accounts for the Olbers paradox, aggravates the Zwicky problem and does not address the chirality issue.

The steady-state Universe

An alternative to big-bang cosmology in an expanding universe was proposed, almost simultaneously with the Gamow theory, by Hoyle and his co-workers. Their theory became known as the steady-state model. The model is based on the idea of continuous creation of matter to compensate for diminishing density brought about by expansion of the universe. New matter is proposed to appear in the form of hydrogen which becomes incorporated into condensing clouds that contract to form new galaxies to replace those that disappear at the edge of the observable universe. This elegant idea gets around the unpalatable singularity which is the big bang, but like the latter is impossible to test experimentally.

The essence of a steady state, its homogeneity and isotropy, was the very reason why it fell into disfavour. The rise of radio astronomy led to the discovery of many unusual extragalactic sources associated with supernovae, galactic centres, pulsars and quasi-stellar objects (QSOs) or quasars. Many of these are associated with faintly visible objects with enormous red shifts which were therefore interpreted as objects that only occured in the distant past. This implied evolutionary history of the universe is incompatible with the homogeneity of a steady state. More recent observations suggest a non-Doppler origin of the intrinsic red shifts of quasars, which may therefore be much closer and of more recent vintage than thought before. Although the evolutionary argument then looses its force, the entire basis of the expanding universe is also called into question. A more devastating blow to the steady-state model was delivered by the discovery of the microwave background for which it had no reasonable explanation.

QSSC At the time, when both expanding-universe theories encountered serious problems, a hybrid model of big-bang and steady-state cosmologies was countenanced [82]. This *quasi steady-state cosmology* is formulated in

[3]Weakly interacting massive particles.

terms of a cosmic field that allows interparticle interaction at a distance, as required by Mach's principle.[4] Instead of continuous creation of matter it is here assumed that matter has limited life and that particle world lines are not endless. Such a world line may begin at a finite point in space-time and end at another. This means that a particle is created and, after a finite lifetime annihilated. It is further assumed that a typical particle to be created is a particle of Planck mass,

$$m_P = \left(\frac{\hbar c}{G}\right)^{\frac{1}{2}}$$

The Planck particle decays into a number $n \sim 6 \times 10^{18}$ of secondary baryons, mesons, leptons and photons. The sites of matter creation are potentially explosive and the explosions drive the expansion of space. Sites where X-ray and γ-ray activity occurs are the sites where creation of matter currently takes place, and include QSOs and active galactic nuclei. The reason why creation seems to happen in compact regions close to an event horizon is because the theory cannot be formulated in flat space-time. A strong gravitational disturbance is necessary to trigger a creation event through explosive acceleration of the Mach-field bosons.

From this point onwards all the useful arguments developed in big-bang cosmology are invoked as features of the mini-bang creation events, without the incommodious singularity of the former. The microwave background is ascribed to thermalization due to a postulated fog of metallic whiskers in intergalactic space. Dark matter is ascribed, not to wimps, but to the gravitational effects of invisible remnants of burnt-out stars. In big-bang cosmology there is not sufficient time to allow this interpretation.

5.3.2 Plasma Cosmology

One of the defects of expansion cosmologies is their exclusive reliance on gravity, while ignoring electromagnetic interaction. Cosmic radio sources attest to electromagnetic activity at an enormous scale. A variety of spectroscopic measurements suggest the presence of ionized plasma particles in intergalactic space.

The study of plasmas, especially in the presence of magnetic fields, is an important branch of physics. Since the sun, and possibly all stars, have magnetic fields it is reasonable to look for plasma effects on a cosmic scale.

[4]The principle implies that the laws of mechanics depend on interaction with all matter in the universe.

The leader of this quest is Hannes Alfvén, who has already succeeded to simulate a variety of observed cosmic structures and apparent interactions in the plasma laboratory. The simplest and smallest of these is the *aurora borealis* or northern lights. The largest includes vortex filaments and whirlpools with dimensions of billions of light years. The fact that such structures are continually observed around galactic centres and often between galaxies within a cluster, together with the fact that no other reasonable explanation of those phenomena exists, provide conclusive proof that the structure of the universe cannot be explained adequately without consideration of electromagnetic effects.

The difference in approach between Alfvén and big-bang cosmologists is instructive. The latter concentrate exclusively on the presumed first few seconds of creation, the one aspect not amenable to experimental test. Plasma cosmology, on the other hand, relies on astronomical observation and laboratory tests. The simulation of the most diverse galactic structures known, by computer modelling based on simple plasma models, is particularly impressive. It is clear that any future cosmology has to take plasma effects into account. At this stage it is already obvious that magnetic fields and electric currents are more effective than gravity in concentrating matter into plasma strings or filaments that could act as nucleation centres for the growth of solar systems, galaxies and larger clusters. It is therefore important to note how plasma arguments may serve to resolve the three paradoxes.

The most important of the three is Hubble expansion which the plasma model addresses together with the antimatter problem. It is argued that plasma processes not only lead to a large-scale separation of matter and antimatter, but also prevent their subsequent mixing. During chance encounters large explosions result. Maybe not as large as a big bang, but large enough to drive local expansion, observed as red shifts.

The dark-matter paradox is not addressed.

The microwave background is explained quite simply. High-energy electrons trapped in a magnetic field absorb and re-emit microwaves at the same wavelength. Over billions of years this process creates an isotropic background and represents a diffuse glow from the fog of plasma filaments that occur in any galaxy.

5.3.3 Curved-space Cosmology

Although the Doppler interpretation of galactic red shifts is generally accepted as the most likely, it is by no means a unique explanation. It is well known that red shifts may be explained equally well in terms of a static universe with curved space and time. In this case the galactic light source and the

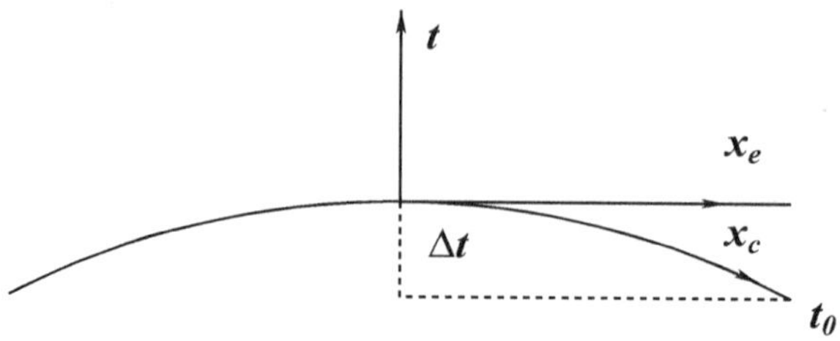

Figure 5.6: *Red shift as a function of space curvature.*

observer are no longer on the same time coordinate and if the curvature is elliptical the light beam has to catch up with the observer along the time coordinate. This situation is pictured in Figure 5.6. The apparent euclidean and actual curvilinear spatial axes do not coincide and galactic light emitted at time t_0 reaches the observer at time $x/c+\Delta t$. It is the time difference that creates the illusion of the galaxy receding at position x_e.

The rotation of a galaxy will also appear differently in flat space and in intrinsically curved space. The apparent and actual radii of rotation in curved space will be different and cause an interpretational error that leads to Zwicky's paradox. The amount of curvature needed to account for the mass difference can be used to calculate the radius of a closed universe and this value should correlate with the Hubble radius, derived from observed red shifts. Such a calculation therefore provides a falsifiable test of the model in the Popper sense of theoretical prediction against experimental observation.

Sporadic attempts are made to analyze cosmological effects in curved space, but no comprehensive reformulation of general relativity and its consequences in space–time with large-scale curvature, has been attempted. Speculative analysis of a closed universe in continuously curved space with an involution [9] shows that such a construct accounts for all three paradoxes. The Planckian microwave spectrum is a property of any closed universe. In such a universe, as on a Möbius strip, two sides of the same surface are separated by an interface, interpreted as the vacuum. Matter on opposite sides of the interface is of opposite chirality. Transplantation through the involution imperceptibly transforms matter into antimatter, to explain why no traces of substantial concentrations of antimatter have ever been detected.

5.3.4 The Anthropic Principle

No discussion of modern cosmology can be complete without mention of the large-number coincidences and their interpretation within the big-bang scenario. This issue concerns the values of three cosmic quantities in dimensionless units. The origin of the argument is obscure but it has been commented on by Eddington and by Dirac.

The three numbers (N_3, N_2, N_1) represent the baryonic mass number of the universe, the strength of gravitational interaction and the Hubble age of the universe, respectively. An apparent relationship between the three numbers is suggested by their generally accepted values of $N_1 = 1.5 \times 10^{40}$, $N_2 = 5 \times 10^{-39}$ and $N_3 = 1.5 \times 10^{78}$.

A supposed mystery lies therein that this simple relationship must disappear as the world ages. Maybe there is no mystery apart from a meaningless coincidence. However, prominent cosmologists maintain that a profound balance in the make-up of the universe is responsible for shaping a unique environment in which sentient beings may flourish.

As explained by Dicke, the link between N_3 and N_2 is provided by Mach's principle. The meaning of N_1 depends on the need for it to come into register during the epoch when conditions favour the appearance of intelligence. The assertion that initial values were set by design at the time of the big bang in order to prepare the ground for the advent of mankind on planet earth, is known as the anthropic principle. The appearance of humans at time 10^{40} corresponds to the time needed for a first generation of stars to burn out and release the elements required for biological life on a new planet. Without the balance between N_3 and N_2 this development would not be possible.

The anthropic argument completes the myth that universal structure has developed to accommodate mankind, the centre of the universe. This argument puts back the clock to pre-Copernican times. The flaw in the argument is that it only holds in an expanding universe. In curved space the meaning of the dimensionless quantities changes. The number N_1 no longer measures a Hubble age, but rather a Hubble radius of the universe, to yield a dimensionless volume, $V \sim 10^{120}$. If N_3 is the total mass, $N_2 = N_3/V = 10^{40}$ is the density of the universe, about which there is no mystery.

5.3.5 Elemental Synthesis

Elemental synthesis, the most speculative issue in cosmology tends to be of pivotal importance in the argument between rival theories. It could, at the same time, therefore serve to identify sufficient common ground for the formulation of some general model.

One point of probable consensus is that elemental synthesis occurs in the plasma state during violent events characterized by extreme electromagnetic activity. In some models these synthetic events happen at singular points such as the big bang, or perhaps black holes at galactic centres. In other models synthesis occurs in normal stars while quasars and other objects associated with x-ray and γ-ray emission are being implicated more often. Many such objects remain to be characterized more closely. Because of this uncertainty it is probably somewhat premature to ponder mechanisms of elemental synthesis too closely. The same is true about the mode of dispersal of the newly synthesized material through the cosmos. Having quasars and other active regions involved, it is no longer necessary to rely on supernovae as the only vehicle of dispersal.

Another common principle that seems to emerge is cyclic annihilation and creation of matter, for instance in a universe that pulsates between big bangs and crunches. In a like manner, matter disappears in black holes and re-emerges through gushers, also known as white holes. The advantage of this type of scenario is non-violation of the first law of thermodynamics. The disadvantage, to many minds, probably is the total silence about initial creation. However, should the observable universe be in equilibrium, such enquiry is idle speculation. One assumption that renders speculation redundant is the assumption that matter is no more than a special configuration of world space. This being the case the only approach towards a general cosmology is through specification of the nature of the vacuum substratum and the geometry of space–time.

5.4 Chirality of Space–time

Pasteur, who was first to isolate optical isomers from racemic mixtures, both by crystallization and by specific biological activity, is purported [84] to have made the grand conjecture

> L'universe est dissymmétrique

before the French Academy of Sciences, expressing his conviction that the chirality of molecules and of life processes was a function of the asymmetry of the universe. Today there is no further doubt that Pasteur was right, although most cosmological models simply ignore this important fact.

The electromagnetic right-hand rule is the most direct demonstration that space is chiral. In fact, it provides a simple way of communicating to a remote intelligence the local definition of left and right, assuming familiarity with physical phenomena at both ends. The observations that space, matter and biological activity are all of fixed chirality in this corner of the universe must mean more than mere coincidence. The known chirality of space, and in turn the interaction of matter with electroweak fields dictates the chirality of biological activity. To put all of this into perspective it will be necessary to clarify the notion of chiral space.

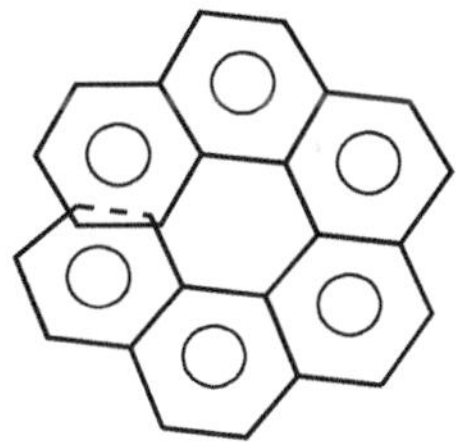

Figure 5.7: *One enantiomer of hexahelicene which is a fragment of a helical surface.*

This problem was addressed [83] by identification of infinite chiral structures of dimension $(n-1)$ that completely fill an isotropic continuum $\mathbb{R}^n$. That means chiralizing an isotropic straight line $\mathbb{R}^1$, plane $\mathbb{R}^2$, space $\mathbb{R}^3$ and hyperspace $\mathbb{R}^4$.

Existence of the chiral hexahelicene molecule, one enantiomer of which is shown in Figure 5.7, prompted efforts to fill the space $\mathbb{R}^3$ with a chiral two-dimensional helical figure.

5.4.1 The Helicoid

The construction starts with a right circular helix defined as the locus of a point moving in space under the action of a continuous twist [22]. The cylindrical coordinates of a point P on the helix are

$$x = a\cos\theta, \quad y = a\sin\theta, \quad z = a\theta\tan\alpha \equiv c\theta, \quad (c = a\tan\alpha)$$

The helix, shown in Figure 5.8, resembles the rail of a spiral staircase and has the equations

$$a^2 = x^2 + y^2, \quad \frac{y}{x} = \tan\theta = \tan\frac{z}{c}, \quad z = c\theta$$

which express it as the curve of intersection of two surfaces: the circular cylinder, $a^2 = x^2 + y^2$, and the helicoid,

$$\frac{y}{x} = \tan\frac{z}{c} = \tan\theta$$

The helicoid is defined as the conoid[5] generated by lines parallel to the plane XOY which intersects the Z-axis. The shape of the helicoid which is swept

[5]A conoid is the locus of a line that always intersects a fixed line (Z) and a given curve, and is parallel to a given plane [85].

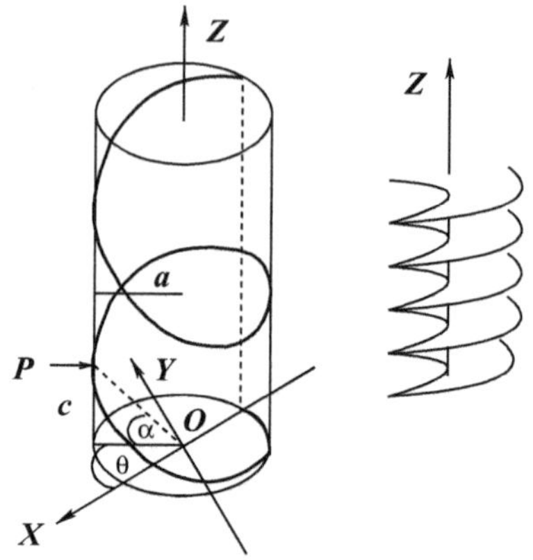

Figure 5.8: *Circular helix and helicoid, on the right.*

out by the parallel lines, is like the ceiling of the staircase, or a propellor blade. With decreasing c the space between successive planes of the helicoid also decreases and eventually as $c \to 0$ and $a \to \infty$ the helical surface of the helicoid completely fills $\mathbb{R}^3$. The chirality of this space, like the absolute configuration of the unit propellor depends on the direction in which the locus that defines the helix, is followed.

5.4.2 The Chiral Plane

Perhaps easier to envision is the way in which the plane $\mathbb{R}^2$ can be filled by a chiral curve, in this case an Archimedean spiral, $\rho = a\theta$, shown in Figure 5.9.

From $x = \rho\cos\theta$, $y = \rho\sin\theta$ follows the parametric equations

$$x = a\theta\cos\theta, \quad y = a\theta\sin\theta$$

It is evident that as $a \to 0$ the distance between adjacent coils of the spiral becomes infinitesimally small, until the space $\mathbb{R}^2$ is completely filled, as demonstrated in Figure 2.13.

5.4.3 General Theory

From the first example follows that $\mathbb{R}^3$ is filled by a two-dimensional chiral plane that screws along a special axis. Likewise the one-dimensional Archimedean spiral fills $\mathbb{R}^2$ starting from a special point at the spiral centre. These observations may be generalized into the conjecture that an isotropic continuum $\mathbb{R}^n$ is chirally filled by an $(n-1)$-dimensional chiral motif that coils

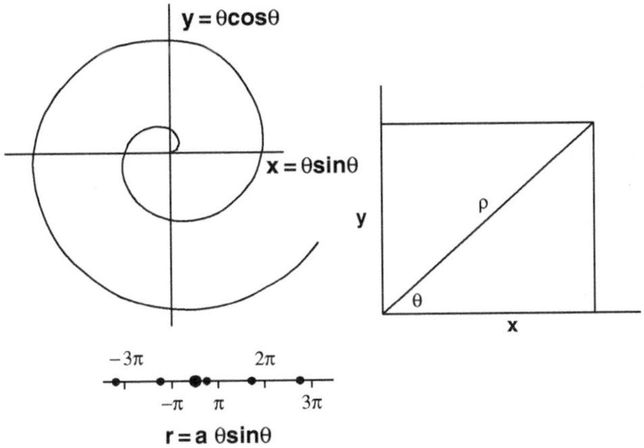

Figure 5.9: *Archimedean spiral to demonstrate the chirality of two-dimensional space.*

around a special $(n-2)$-dimensional site. To test this proposition it is noted that an infinite straight line $\mathbb{R}^1$ is filled by a sinusoidal chiral sequence of points that satisfy the law $r = a\theta \sin\theta$, as shown in Figure 5.9. Filling of $\mathbb{R}^1$ by points according to this law requires the presence of a central zero point. Compared to the central point at the origin of the Archimedean spiral, this point is of higher multiplicity and hence of dimension less than zero. The law $r = a\theta\cos\theta$ leads at $\mathbb{R}^1$ to an achiral sequence of points with an inversion centre at $r = \pi/2$. It may be inferred that the chiral filling of continua $\mathbb{R}^n$ by figures of $(n-1)$-dimensionality is also possible for $n > 3$ if a sinusoidal coordinate is retained.

It seems likely that hyperspace $\mathbb{R}^4$ may be chirally filled by helical three-dimensional space that requires a two-dimensional plane as a special element of the whole system. It may, in general, be conjectured that an n-dimensional isotropic space is filled by an $(n-1)$-dimensional chiral figure coiled around a special $(n-2)$-dimensional achiral site.

It is of special interest to note that a closed five-dimensional universe of the type $(\mathbb{R}^4, t)$ proposed by Thierrin [86] could be viewed as filled by a four-dimensional chiral hyperspace, wrapped around an achiral three-dimensional space, which is the vacuum as defined in Section 5.3.3. As already pointed out, the 4D enantiomers interconvert on displacement along the achiral vacuum. Since the fifth dimension in this model represents time, it means that the material 4D universe that completely fills five-dimensional space–time persists for all time. Time has no beginning or end in such a closed universe.

5.5 The Vacuum Substratum

Cosmology, which by definition provides the most comprehensive description of all physical phenomena, is required to be consistent with the most advanced theories of the physical world, at present recognized as quantum field theory (QFT) and the theory of relativity.

Relativity restricts the maximum velocity of matter to be less than that of a light signal, which is restricted to c. Should an object go faster than light, the electromagnetic fields that hold its atoms together are left behind,[6] causing the atoms to disperse and the matter to fall apart. Because of these restrictions it is not possible to obtain a relativistically consistent definition of an extended rigid body, because this would imply signals faster than light. To overcome the problem it was necessary to define particles such as electrons, as extensionless points, or singularities in a field, which in QFT generate infinite fields, that many scientists find objectionable. One way out is to accept that, like classical mechanics, quantum theory has its own limitations and breaks down outside of a given region of applicability. A reasonable assumption [87] is that the limits to valid theories of the physical world are related to the three fundamental constants of the observable world, $\hbar$, c and G.

Planck's constant $\hbar$ specifies the minimum amount of angular momentum, $c=(\epsilon_0\mu_0)^{-\frac{1}{2}}$ is the maximum rate of transmission of an electromagnetic signal as determined by the permeability and permittivity of the vacuum, and G, the gravitational constant specifies the inverse-square law of gravity. The force exerted on a mass m_1 in the gravitational field of another mass m_2, is given by

$$F = m_1\frac{\mathrm{d}^2r}{\mathrm{d}t^2} \propto \frac{m_1m_2}{r^2}$$

i.e.

$$\frac{\mathrm{d}^2r}{\mathrm{d}t^2} = \frac{Gm_2}{r^2}$$

For mass points in the gravitational field of the earth the acceleration due to gravity is

$$\frac{\mathrm{d}^2r}{\mathrm{d}t^2} = g = \frac{GM_E}{r^2}$$

It is noted that the three fundamental constants reflect the macroscopic physical properties of space–time, or the vacuum.

[6]Sound waves produced by an aircraft are left behind when it goes faster than sound.

Natural limits to the domain of quantum mechanics are next assumed to correspond to the elementary quantities of mass, length, time and energy, represented in Planck units, derived from the fundamental constants by:

$$m_P \equiv \left(\frac{\hbar c}{G}\right)^{\frac{1}{2}} \simeq 2.2 \times 10^{-8}\,\mathrm{Kg}$$

$$l_P \equiv \left(\frac{G\hbar}{c^3}\right)^{\frac{1}{2}} \simeq 1.6 \times 10^{-35}\,\mathrm{m}$$

$$t_P \equiv \left(\frac{G\hbar}{c^5}\right)^{\frac{1}{2}} \simeq 5.4 \times 10^{-44}\,\mathrm{s}$$

$$E_P = \left(\frac{c^5\hbar}{G}\right)^{\frac{1}{2}} \simeq 1.2 \times 10^{19}\,\mathrm{GeV}$$

The electromagnetic and gravitational fine-structure constants relate the fundamental constants to the charge of the electron and mass of the proton, respectively.

$$\alpha = \frac{e^2}{\hbar c} \simeq \frac{1}{137}$$

$$\frac{Gm_p^2}{\hbar c} = 5 \times 10^{-39}$$

the latter corresponding to the large number, N_2 of the anthropic argument.

The implications of restricting the use of quantum theory and relativity to lengths in excess of l_P and energies less than E_P only, are profound. When moving into the sub-quantum region the familiar notions of space and time fade away into something which, at present is unspecifiable. The infinite fields around singularities disappear automatically since Coulomb's law no longer applies, and the restrictions on propagation of signals become irrelevant.

The most general model of the vacuum substratum has been formulated by Bohm [87] in terms of an implicate order and holomovement. Substructure models in terms of particles and antiparticles of Planck mass, or standing waves formed by the superposition of advanced and retarded spherical waves have also been proposed [11].

5.5.1 Implicate Order and Holomovement

Bohm's model is based on the assumption that the undivided wholeness of the universe, recognized by both quantum theory and relativity, originates from a sub-quantum plenum. In the same way that a hologram enfolds all aspects of a three-dimensional structure in an apparently featureless pattern, the macroscopic world is enfolded or implicated at the substratum level. A hologram does not look at an object at all. It gives rise to an image when suitably illuminated. Whereas the laws of physics refer mainly to the explicate order, it is now proposed that primary relevance be given to the implicate order in the formulation of these laws. The motion of particles at the quantum level is also in explicate order, but not always at the manifest level because it is affected by the active information represented by the quantum potential.

In the production of a hologram or the modulation of a radio wave, a structure is enfolded in an electromagnetic wave. The universe, which includes the whole of existence, contains not only the electromagnetic and other fields that are known now, but also an indefinitely large set of further fields that are unknown and may never be known in totality. Recalling that the essential qualities of fields exist only in their movement, this total ground is called the *holomovement*. It is undefinable and immeasurable. Observable particles are the ripples on this sea of underlying activity. It is this activity which is responsible for zero-point energy. It can also be regarded as the source of the quantum potential, which has the form

$$V_q = -\frac{\hbar^2}{2m}\frac{\nabla^2 R}{R}$$

where R is the amplitude of the quantum field. The quantum potential is responsible for the non-local interaction in many-body quantum systems.

Non-locality is reflected therein that if an order is enfolded throughout all of space and time, it cannot coherently be regarded as constituting a signal that would propagate information from one place to another over a period of time. This means that where implicate order is involved, the descriptive language of relativity theory will no longer apply.

It is not clear how to interpret implicate order in an expanding universe, but not difficult to imagine holomovement to happen at the interface of the curved-space structure described in Section 5.3.3.

5.5.2 Information Theory

Another approach that may help to overcome some of the awkward conclusions of modern physics is based on information theory as developed by Stonier [88]. The basic postulate is that the universe is made up of three components : matter, energy and information. In a sense, what mass is to the manifestation of matter, and heat is to energy, organization is to information. Heat is the form of energy that lacks information. All other, more structured forms of energy contain an information component.

Like mass and energy, information may also be assumed quantized in the form of *infons*. In view of the interconvertibility of energy and information, a close relationship between photons and infons is implied. The relativistic energy equation

$$E = \frac{m_0 c^2}{\sqrt{1 - v^2/c^2}} \tag{5.1}$$

shows that a massless particle has zero energy unless it travels at the speed of light, in which case $E = \frac{0}{0}$ becomes indeterminate. Although the value of the energy cannot be calculated from (5.1), it is non-zero. The same applies to relativistic momentum,

$$p = \frac{m_0}{\sqrt{1 - v^2/c^2}}$$

A massless particle, at any speed but c however, has zero energy and momentum. Such a particle differs from a photon which has energy ($h\nu$) and momentum. An infon could be of this type, with wavelength

$$\lambda = \frac{h}{p} = \frac{h\sqrt{1 - v^2/c^2}}{m_0}$$

Only at velocity $v = c$ does the wavelength become indeterminate, but finite. For $v \neq c$, $\lambda \to \infty$. Thus, an infon is a photon whose wavelength has been stretched to infinity. Conversely, a photon is an infon that travels at the speed of light.[7] The fact that only quanta which travel at velocity c can be perceived as receptors is not unlike the way in which the absorption of radiation requires a receptor, attuned to the incoming frequency according to the Bohr frequency condition, $\Delta E = h\nu$. It seems that the receptor must in fact be attuned to both entrance velocity of the incoming particle, as well as its alternating magnetic field.

In the same way that different photons define an electromagnetic spectrum, there may be many different infons that together comprise an information spectrum. Some obvious examples are phonons, excitons, and the holes left in atomic shells by ejected electrons. The electron cloud that surrounds an atomic nucleus has the structure of organized energy. This structure, or *morphe* arises from interaction with the field of the nucleus that imposes a

[7] This result has an interesting corollary for the theory of energy transmission according to the absorber-theory model [11]. The supposed standing wave or virtual photon that connects emitter and receptor, is recognized as equivalent to the infon as defined here.

unique pattern on the surrounding intra-atomic space. Should an electron be ejected from the cloud, a hole, which continues to interact with the field, is left behind. Such a hole is equivalent to an infon. When the electron drops to a lower level the excess energy combines with the hole in the outer shell to form a photon that may escape from the atom, provided its velocity is modulated by the structured space to match c. The same applies to incoming energy approaching at a velocity near c. Before absorption can occur the incoming velocity, once more, needs modulation to c. That is the role of structured intra-atomic space. It appears that, whereas each electron in an atomic charge cloud can interact with a photon of characteristic frequency, all electrons only recognize photons with velocity c.

The notion that infons are interconvertible with photons and can be propagated at speeds different from c has important cosmological consequences. Consider a star that recedes at velocity v. Assume that emitted photons are transformed into pure information quanta, that would impinge on earth with velocity $c-v$. Unlike photons, these quanta have no momentum, frequency or wavelength and remain undetected by conventional receptors. In order to be detected it is necessary for those quanta to be accelerated to velocity c when they reach a receptor. Gravitational curvature could cause such acceleration.

Of all the quanta emitted by the star that recedes at velocity v, only those accelerated to velocity $c + v$ would reach earthly receptors at velocity c, to be converted to and detected as light quanta. Now, if the frequency of the quanta, $\nu = v/\lambda$, remains constant as set on original emission, acceleration to c must cause an increase in wavelength, observed as a red shift. Also, all other things being equal, the greater the velocity at which the star recedes, the smaller will be the fraction of quanta that can be accelerated to reach velocity c. The effect of this is that stars with large red shifts must appear faint, not by virtue of their distance from earth, but because of their high speed of recession. Many galaxies may therefore not be as remote as suggested by Hubble's law and the universe much smaller than commonly thought. Furthermore, infons that originate at stars that move towards earth, arrive with velocities $v + c$ and there is no device to slow them down to the observable velocity of c. Such stars therefore remain invisible and hence, all galactic objects appear to be red-shifted. This argument undercuts the entire rationale of an expanding universe inferred from observed red shifts.

Chapter 6

The Periodic Laws

> The properties of the elements are the properties of numbers. – Alexandre Émile Beguyer de Chancourtois, 1862.

6.1 Introduction

The periodic law of the elements is the most fundamental principle of chemistry. It became firmly established in about 1875 amidst one of the most bitter controversies since the collapse of the phlogiston theory. The principle at issue is best remembered as Prout's hypothesis. To account for the fact that most of the atomic weights, that provided the basis for periodic classification, were measured to be uniformly close to whole numbers, it postulated hydrogen as the basic building block of all matter. Of the common elements some typical (modern) relative atomic weights that illustrate the point, include H(1.008), He(4.003), Be(9.012), C(12.011), N(14.007), O(15.999), F(18.998), Na(22.990), and many more. The most conspicuous exceptions to the rule, the elements Cl(35.453), Cu(63.546) and Rb(85.468) became the subject of intense research aimed at the refutation of Prout's hypothesis, by the measurement of the most accurate atomic weights. The results refuted the hypothesis which eventually however, prevailed with the discovery of isotopes.

With hindsight Prout's hypothesis is seen to reflect the composition of atoms in terms of protons, neutrons and electrons. Each proton–electron pair in an atom may be thought of as either a hydrogen atom or a neutron. The identity of an atom is fixed by the number of protons in the nucleus. Isotopes are atoms in which different numbers of neutrons are associated with the same fixed number of protons. A mixture of different isotopes in an elemental sample leads to non-integral atomic mass and apparent break-down of Prout's conjecture. A surprising aspect of the periodic law, that applies to elements only, is that no effort has been made, until very recently [5], to obtain a generalized formulation of the law that applies to all stable nuclides in terms of neutron count N and/or mass number A.

In that study [5] it was found that the 264 stable (non-radioactive) nuclides are distinguished by two factors, a proton (atomic) number and a proton:neutron ratio. When examined in terms of these factors the nuclides neatly divide into two classes, of even and odd mass number ($A = Z + N$) respectively. The plot of Z/N vs Z for even mass numbers is shown in Figure 6.1.

Nuclides are represented by small circles on a set of easily discernable festoons. Consecutive nuclides on the same festoon differ by the equivalent of an α-particle, i.e. by two protons and two neutrons. Each festoon terminates in two radioactive nuclides, indicated by open circles. The terminal nuclide on the high-Z/N side decays by positron emission and/or electron capture. The first stable decay product from this process lies on the next lower festoon of the same family. The nuclides of even mass number separate into two 81-member families of mass $4n$ and $4n + 2$ respectively, distinguished by two differently filled type of circle in Figure 6.1. The terminal nuclide on the low Z/N side of any festoon decays by β emission. The straight lines AB and BC were found [5] to provide a simple prescription for predicting the range of stable isotopes of any given element. AB represents the relationship

$$N = \frac{Z}{1 - Z \tan \theta}$$

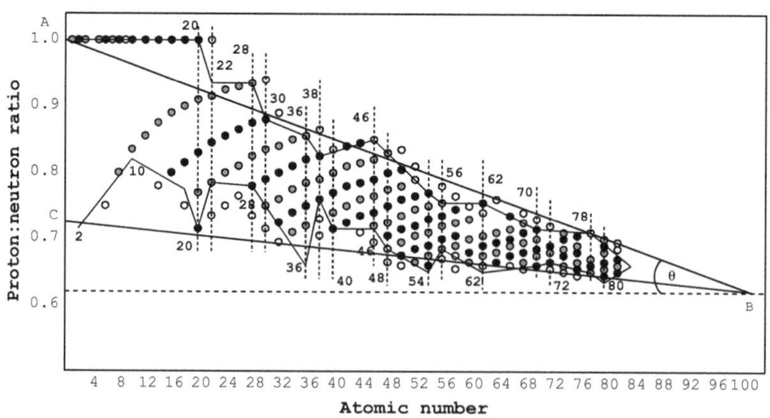

Figure 6.1: *Plot of the even mass number stable nuclides to show their build-up from α-particles. The limiting lines AB and BC predict the same region of stability as demarcated by the irregular profile drawn to emphasize periodicity.*

Like Z, the number N must be a discrete whole number. The integer, obtained from N as calculated, by discarding fractions, correctly describes the minimum number of neutrons for given Z, that can be associated with the lightest stable isotope of atomic number Z. The line BC identifies the heaviest stable isotope for given Z in like manner. These observations imply that each of the even-mass series can be generated by adding consecutive α-particles to a single starting entity, as shown in Figure 6.2 for some nuclides of the series $A = 4n$.

When a nuclide that lies outside the area of stability, as calculated from the line AB is reached in the course of this progression, positron emission and/or electron capture returns the sequence into the area of stability.

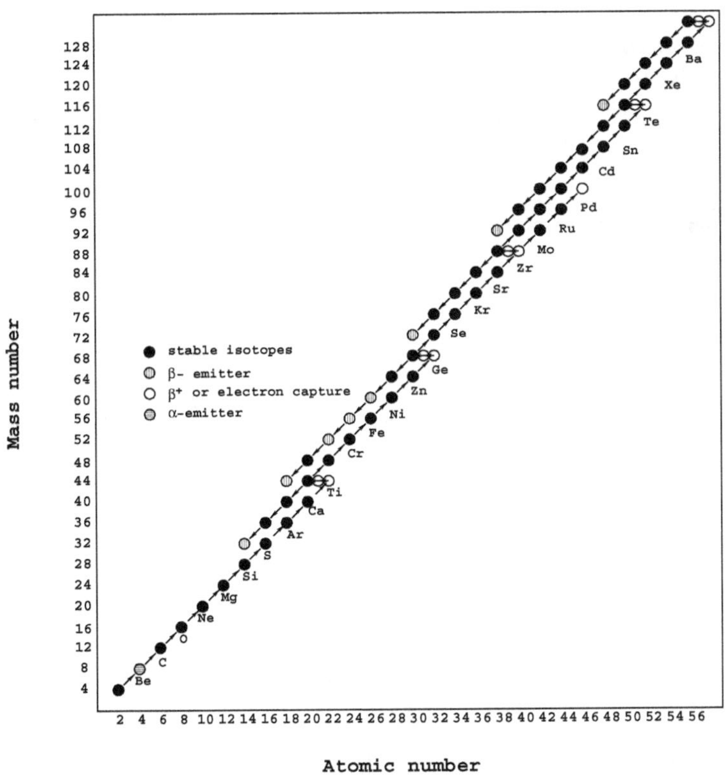

Figure 6.2: *The 81 nuclides of the series $A = 4n$ are defined by the addition of α-particles to ^{12}C. Whenever a radioactive species is generated the process is continued by restarting from the first stable decay product, as demonstrated in the diagram.*

No additional rules are required to correctly predict all stable nuclides of atomic number not exceeding 83. The only unexplained feature is a small number of α-unstable nuclides that occur within the triangle of stability. Their positions are left blank in Figure 6.1.

The odd-mass nuclides occur in analogous fashion as two 51-member sequences ($4n \pm 1$) in a triangle, completely enclosed within the triangle of Figure 6.1.

It is of interest to identify those nuclides that lie within the excluded triangle at the tip of the triangle of stability, as shown in Figure 6.1. The (radioactive, α-emitter) nuclides, predicted to lie in this region by extension of the four series that generate the stable (non-radioactive) nuclides, are shown in Table 6.1 and in Figure 6.3. The 36 additional nuclides within the tip of the triangle increase the number of possible nuclides to 300 and the number of elements to 100.

Another way of defining the area in Figure 6.1 to which stable nuclides are confined, is by drawing straight-line segments that separate filled and empty circles, respectively. The turning points of the discontinuous boundary lines are found to correspond surprisingly well with atomic numbers, known to

Table 6.1: *The α-unstable nuclides shown in Figure 6.3. The experimental evidence to support the selection is not too reliable.*

Z		$4n$	$4n+1$	$4n+2$	$4n+3$
84	Po	212	213		211
85	At		213		215
86	Rn	216	217	218	215
87	Fr		221		219
88	Ra	224	225	226	223
89	Ac				227
90	Th	228	229	230	
91	Pa		233		231
92	U	236	237	238	235
93	Np				239
94	Pu		241	242	243
95	Am		245		
96	Cm	248	249		
97	Bk				251
98	Cf				255
99	Es		257		
100	Fm		261		

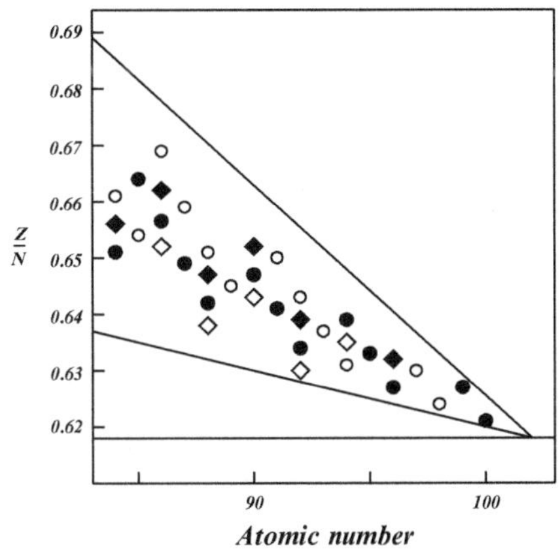

Figure 6.3: *Nuclides with* $83 < Z < 102$ *generated by the rules for* α*-addition derived for the stable nuclides.*

represent closed-shell electronic configurations in the periodic table of the elements. As shown in Figure 6.1 these line segments can be constructed so as to match an eight-group observed table, shown in Figure 6.4, almost perfectly. This unexpected result implies that the periodic table of the elements is a subset of a more general periodic law of the stable nuclides. A new feature of Figure 6.4, not noticed before in any periodic classification, is the spread of the 14-element f-series across two periods, such that the sixth member (^{62}Sm,^{94}Pu) occurs in the same group as the elements with filled p and d sub-shells. It will be shown that this property, rather than an accidental curiosity, introduces a fundamental feature of periodic classification.

Had the rare earth elements been known to Lothar Meyer, this extra element of periodicity would have been revealed in his atomic volume plot, Figure 3.3. A modern form of the LM curve, showing atomic volume as a function of atomic number is in Figure 6.5. Instead of connecting all points in sequence of Z, the curve, in Figure 3.3, is interrupted at those points where completion of s, p, d and f energy levels are known to occur. By this procedure the curve fragments into regions made up of either two or eight elements, except in the rare-earth domain. Taken together with the preceeding couple of elements, Cs and Ba, a break at Sm divides this domain into two eight-member segments at an obviously special point on the curve. All segmented regions now occur in strict alternation as negatively and positively sloping curves. Not only does the fragmented curve therefore fit the observed periodic table of Figure 6.4 in all respects, but furthermore, reflects

Observed

	1	2	3	4	5	6	7	8	
1s	1 H	2 He							
2s	3 Li	4 Be	5 B	6 C	7 N	8 O	9 F	10 Ne	*2p*
3s	11 Na	12 Mg	13 Al	14 Si	15 P	16 S	17 Cl	18 Ar	*3p*
4s	19 K	20 Ca							
	21 Sc	22 Ti	23 V	24 Cr	25 Mn	26 Fe	27 Co	28 Ni	*3d*
	29 Cu	30 Zn	31 Ga	32 Ge	33 As	34 Se	35 Br	36 Kr	*4p*
5s	37 Rb	38 Sr							
	39 Y	40 Zr	41 Nb	42 Mo	43 Tc	44 Ru	45 Rh	46 Pd	*4d*
	47 Ag	48 Cd	49 In	50 Sn	51 Sb	52 Te	53 I	54 Xe	*5p*
6s	55 Cs	56 Ba	57 La	58 Ce	59 Pr	60 Nd	61 Pm	62 Sm	
	63 Eu	64 Gd	65 Tb	66 Dy	67 Ho	68 Er	69 Tm	70 Yb	*4f*
	71 Lu	72 Hf	73 Ta	74 W	75 Re	76 Os	77 Ir	78 Pt	*5d*
	79 Au	80 Hg	81 Tl	82 Pb	83 Bi	84 Po	85 At	86 Rn	*6p*
7s	87 Fr	88 Ra	89 Ac	90 Th	91 Pa	92 U	93 Np	94 Pu	
	95 Am	96 Cm	97 Bk	98 Cf	99 Es	100 Fm	101 Md	102 No	*5f*

Idealized

	1	2	3	4	5	6	7	8	
1s	1 H	2 He							
2s	3 Li	4 Be	5 B	6 C	7 N	8 O	9 F	10 Ne	*2p*
3s	11 Na	12 Mg	13 Al	14 Si	15 P	16 S	17 Cl	18 Ar	*3p*
3d	19 K	20 Ca	21 Sc	22 Ti	23 V	24 Cr	25 Mn	26 Fe	
	27 Co	28 Ni							
4s	29 Cu	30 Zn	31 Ga	32 Ge	33 As	34 Se	35 Br	36 Kr	*4p*
4d	37 Rb	38 Sr	39 Y	40 Zr	41 Nb	42 Mo	43 Tc	44 Ru	
	45 Rh	46 Pd	47 Ag	48 Cd	49 In	50 Sn	51 Sb	52 Te	
	53 I	54 Xe	55 Cs	56 Ba	57 La	58 Ce	59 Pr	60 Nd	*4f*
5s	61 Pm	62 Sm	63 Eu	64 Gd	65 Tb	66 Dy	67 Ho	68 Er	*5p*
5d	69 Tm	70 Yb	71 Lu	72 Hf	73 Ta	74 W	75 Re	76 Os	
	77 Ir	78 Pt	79 Au	80 Hg	81 Tl	82 Pb	83 Bi	84 Po	
	85 At	86 Rn	87 Fr	88 Ra	89 Ac	90 Th	91 Pa	92 U	*5f*
6s	93 Np	94 Pu	95 Am	96 Cm	97 Bk	98 Cf	99 Es	100 Fm	*6p*

Inverted

	1	2	3	4	5	6	7	8	
			1 H	2 He	3 Li	4 Be	5 B	6 C	
	7 N	8 O	9 F	10 Ne	11 Na	12 Mg	13 Al	14 Si	*4f*
	15 P	16 S	17 Cl	18 Ar	19 K	20 Ca	21 Sc	22 Ti	
3d	23 V	24 Cr	25 Mn	26 Fe	27 Co	28 Ni	29 Cu	30 Zn	*2p*
1s	31 Ga	32 Ge	33 As	34 Se	35 Br	36 Kr	37 Rb	38 Sr	*3p*
2s	39 Y	40 Zr	41 Nb	42 Mo	43 Tc	44 Ru	45 Rh	46 Pd	
	47 Ag	48 Cd	49 In	50 Sn	51 Sb	52 Te	53 I	54 Xe	*5f*
			55 Cs	56 Ba	57 La	58 Ce	59 Pr	60 Nd	*4p*
	61 Pm	62 Sm	63 Eu	64 Gd	65 Tb	66 Dy	67 Ho	68 Er	
4d	69 Tm	70 Yb	71 Lu	72 Hf	73 Ta	74 W	75 Re	76 Os	*5p*
3s	77 Ir	78 Pt	79 Au	80 Hg	81 Tl	82 Pb	83 Bi	84 Po	*6p*
	85 At	86 Rn	87 Fr	88 Ra	89 Ac	90 Th	91 Pa	92 U	
5d	93 Np	94 Pu	95 Am	96 Cm	97 Bk	98 Cf	99 Es	100 Fm	*7p*

Idealized : Inverted

	1	2	3	4	5	6	7	8	
			1 H	2 He	3 Li	4 Be	5 B	6 C	
	7 N	8 O	9 F	10 Ne	11 Na	12 Mg	13 Al	14 Si	*4f*
	15 P	16 S	17 Cl	18 Ar	19 K	20 Ca	21 Sc	22 Ti	
3d	23 V	24 Cr	25 Mn	26 Fe	27 Co	28 Ni	29 Cu	30 Zn	*2p*
1s	31 Ga	32 Ge	33 As	34 Se	35 Br	36 Kr	37 Rb	38 Sr	
	39 Y	40 Zr	41 Nb	42 Mo	43 Tc	44 Ru	45 Rh	46 Pd	*5f*
	47 Ag	48 Cd	49 In	50 Sn	51 Sb	52 Te	53 I	54 Xe	
4d	55 Cs	56 Ba	57 La	58 Ce	59 Pr	60 Nd	61 Pm	62 Sm	*3p*
2s	63 Eu	64 Gd	65 Tb	66 Dy	67 Ho	68 Er	69 Tm	70 Yb	
	71 Lu	72 Hf	73 Ta	74 W	75 Re	76 Os	77 Ir	78 Pt	*6f*
5d	79 Au	80 Hg	81 Tl	82 Pb	83 Bi	84 Po	85 At	86 Rn	
	87 Fr	88 Ra	89 Ac	90 Th	91 Pa	92 U	93 Np	94 Pu	*4p*
3s	95 Am	96 Cm	97 Bk	98 Cf	99 Es	100 Fm	101 Md	102 No	

Figure 6.4: *Four periodic tables of the elements in compact form.*

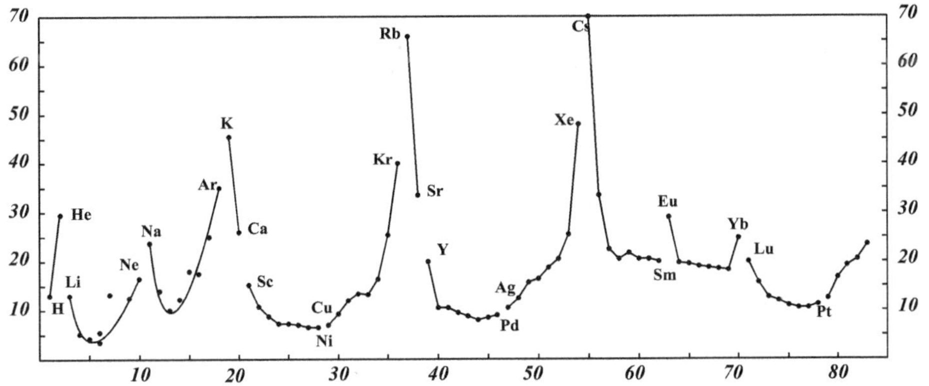

Figure 6.5: *Modern version of Lothar Meyer's atomic volume curve.*

the precise details on which electropositive and electronegative elements used to be distinguished traditionally [31]. The fundamental basis of the elusive *electronegativity* concept is here revealed for the first time as a periodic property of the elements. Calculated from ionization radii [6], as the electrochemical potential of the valence state [89], it displays the same periodicity as the LM curve.

Even more surprising was to find that nuclidic periodicity matches the distribution of prime numbers.

6.2 Number Spiral and Periodic Laws

A remarkable property of the eight-group table of the elements is that it has several features in common with the 24-group spiral of the natural numbers, shown in Figure 1.2.

Number theorists refer to prime numbers as the atoms of mathematics. In the same way that atoms are the building blocks of all matter (nuclides) all numbers are created by multiplication of combinations of prime numbers. Reminiscent of the 8-group table, prime numbers occur on eight arms at $6n \pm 1$ along the spiral.

To understand the compact form of the observed periodic table, periodic octets may be interpreted to be made up of either $(2s{+}6p)$, $(8{+}2)d$, or $(8{+}6)f$ electrons per subshell, arranged in such a way that closed-shell configurations occur at groups 2 and 8 only. In essence, the compact periodic table therefore defines the closure of electron pairs and octets in groups 2 and 8 respectively. An equivalent property is recognized in the number spiral by forming the sums

$$\sigma_{(j+1)} = \sum_{n=24j}^{24(j+1)} n = (2j+1)300\,,\; j = 0,1,2\ldots \tag{6.1}$$

over all natural numbers. The sums

$$\sigma_i = a, 3a, 5a, 7a\ldots\;, \;(a = 300)$$

have coefficients that match the numbers of s, p, d and f electron pairs required to complete the atomic sub-shells. Each of the sums σ_i may therefore be read to represent the total number of electron pairs of a given type, obtained by summing over a atoms. However, only one third of the numbers in each cycle are located on the prime-number arms. Any parallel between atomic structure and the prime-number distribution therefore implies:

(a) A total of 300 different nuclides

(b) A total of 100 different elements

(c) Nuclide periodicity of 24

(d) Elemental periodicity of 8

Each of these four conditions is now examined in more detail, starting with nuclide periodicity.

To explore the idea that stable nuclides have a periodicity of 24 the 264 nuclides are arranged in order of increasing mass and atomic numbers on a 24×11 matrix [5] shown as Figure 6.6. Plotted on the axes of Figure 6.1, each period is represented by a numbered block of 24 nuclides, separated by the set of bold hemlines in Figure 6.7. The α-emitters at atomic numbers beyond 83 could add further periods, as indicated by the dotted line in Figure 6.7.

Since the bold hem lines are not parallel vertical lines, the atomic numbers at which the eleven periods come to a close are functions of the ratio Z/N. As discussed before [5] an almost perfect match with the observed periodic table of Figure 6.4 is found at the ratio $\tau = Z/N = 0.6180\ldots$, also known as the golden ratio. At this ratio (line b in Figure 6.7) the eleven periods end at the atomic numbers 10, 18, 28, 36, 38, 46, 48, 56, 62, 70, 80. All of these numbers occur in either group 8 or 2 of the compact periodic table in Figure 6.4. This observation confirms the conclusion reached on the basis of Figure 6.1, to the effect that the stable nuclides obey a periodic law based on the same principle as the periodic law of the elements.

Periodic Table of the stable Isotopes

	1	2	3	4	5	6	7	8	9	10	11	12	13	14	15	16	17	18	19	20	21	22	23	24
1	2 H	3 He	6 Li	7 Li	9 Be	10 B	11 B	12 C	13 C	14 N	15 N	16 O	17 O	18 O	19 F	20 Ne	21 Ne	22 Ne	23 Na	24 Mg	25 Mg	26 Mg	27 Al	28 Si
2	29 Si	30 Si	31 P	32 S	33 S	34 S	35 Cl	36 S	36 Ar	37 Cl	38 Ar	39 K	40 Ar	40 Ca	41 K	42 Ca	43 Ca	44 Ca	45 Sc	46 Ca	46 Tl	47 Tl	48 Ca	48 Ti
3	49 Ti	50 Ti	50 Cr	51 V	52 Cr	53 Cr	54 Cr	54 Fe	55 Mn	56 Fe	57 Fe	58 Fe	58 Ni	59 Co	60 Ni	61 Ni	62 Ni	63 Cu	64 Ni	64 Zn	65 Cu	66 Zn	67 Zn	68 Zn
4	69 Ga	70 Zn	70 Ge	71 Ga	72 Ge	73 Ge	74 Ge	74 Se	75 As	76 Ge	76 Se	77 Se	78 Se	78 Kr	79 Br	80 Se	80 Kr	81 Br	82 Se	82 Kr	83 Kr	84 Kr	84 Sr	85 Rb
5	86 Kr	86 Sr	87 Sr	88 Sr	89 Y	90 Zr	91 Zr	92 Zr	92 Mo	93 Nb	94 Zr	94 Mo	95 Mo	96 Zr	96 Mo	96 Ru	97 Mo	98 Mo	98 Ru	99 Ru	100 Mo	100 Ru	101 Ru	102 Ru
6	102 Pd	103 Rh	104 Ru	104 Pd	105 Pd	106 Pd	106 Cd	107 Ag	108 Pd	108 Cd	109 Ag	110 Pd	110 Cd	111 Cd	112 Cd	112 Sn	113 Cd	113 In	114 Cd	114 Sn	115 Sn	116 Cd	116 Sn	117 Sn
7	118 Sn	119 Sn	120 Sn	120 Te	121 Sb	122 Sn	122 Te	123 Sb	124 Sn	124 Te	124 Xe	125 Te	126 Te	126 Xe	127 I	128 Te	128 Xe	129 Xe	130 Te	130 Xe	130 Ba	131 Xe	132 Xe	132 Ba
8	133 Cs	134 Xe	134 Ba	135 Ba	136 Xe	136 Ba	136 Ce	137 Ba	138 Ba	138 Ce	139 La	140 Ce	141 Pr	142 Ce	142 Nd	143 Nd	144 Sm	145 Nd	146 Nd	148 Nd	150 Nd	150 Sm	151 Eu	152 Sm
9	153 Eu	154 Sm	154 Gd	155 Gd	156 Gd	156 Dy	157 Gd	158 Gd	158 Dy	159 Tb	160 Gd	160 Dy	161 Dy	162 Dy	162 Er	163 Dy	164 Dy	164 Er	165 Ho	166 Er	167 Er	168 Er	168 Yb	169 Tm
10	170 Er	170 Yb	171 Yb	172 Yb	173 Yb	174 Yb	175 Lu	176 Yb	176 Hf	177 Hf	178 Hf	179 Hf	180 Hf	180 W	181 Ta	182 W	183 W	184 W	184 Os	185 Re	186 W	186 Os	187 Os	188 Os
11	189 Os	190 Os	191 Ir	192 Os	193 Ir	194 Pt	195 Pt	196 Pt	196 Hg	197 Au	198 Pt	198 Hg	199 Hg	200 Hg	201 Hg	202 Hg	203 Tl	204 Hg	204 Pb	205 Tl	206 Pb	207 Pb	208 Pb	209 Bi

Series ● 1 A=4n ◐ 3 A=4n+1
◍ 2 A=4n+2 ○ 4 A=4n+3

Figure 6.6: *The 264 stable nuclides arranged on a 24 × 11 grid.*

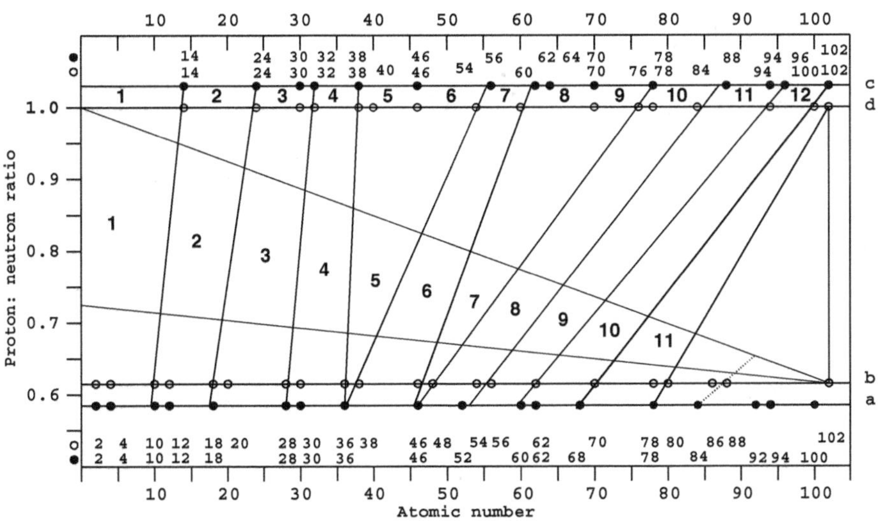

Figure 6.7: *Periodic blocks of the stable isotopes, demarcated by straight hem lines that predict four different forms of elemental periodicity along the lines a, b, c and d.*

The periodic principle is commonly taught to be defined by Schrödinger's equation of an electron in the coulombic field of a proton [37]. Strictly speaking however, the Schrödinger solution predicts the energy spectrum

1s	2s	2p	3s	3p	3d	4s	4p	4d	4f	...	etc.
2	4	10	12	18	28	30	36	46	60	...	etc.

This row of numbers, also marked along line a in Figure 6.7, corresponds with the closure of sub-shells in the Schrödinger spectrum. A hypothetical idealized eight-group table based on the Schrödinger spectrum is also shown in Figure 6.4. As before, closure of all sub-shells occurs in groups 2 and 8. All of the hemlines of Figure 6.7 extrapolate to closed sub-shell points along a and so predict the same periodicity as shown in Figure 6.4. Line a occurs at the ratio $Z/N = 0.6 - \delta$, where $\delta = \tau - 0.6$.

The differences between the observed periodic table and the table based on the wave-mechanical description of the hydrogen atom are largely due to interelectronic interactions that operate in all atoms except hydrogen. Stated differently, Schrödinger's analysis is based on a model that ignores all interaction except the coulombic attraction between an isolated proton and an electron. The model implies an essentially empty universe that only contains one hydrogen atom, whereas any real atom occurs in a universe which is densely populated by additional matter.

According to the theory of general relativity there exists a reciprocal relationship between matter and the curvature of space–time. An empty universe, by definition, therefore corresponds to euclidean flat space–time and any material universe exists in curved space–time [73].

Schrödinger's one-electron equation, in principle, predicts electronic configurations for atoms in flat space-time, different from those observed in the real world. It is inferred that the electronic structure of atoms may be sensitive to the environment, as determined by the local curvature of space–time. For this interpretation to hold it can be argued that extrapolation of the hem lines of Figure 6.7, towards higher values of Z/N, must reveal how the structure of the periodic table is affected by increased curvature of space–time in regions of high mass density, or pressure. In fact, a logical periodic structure occurs at $Z/N = 1$. At this ratio, line d, the first three periods close at the atomic numbers 14, 24 and 32 which seem to indicate filling of electronic levels in an inverted sequence of $f \rightarrow d \rightarrow p \rightarrow s$.

A periodic table constructed on the basis of the atomic numbers on line d and, once more closing all sub-levels at groups 2 and 8, shown in Figure 6.4, differs from the arrangement obtained by inversion of the Schrödinger table. Extrapolation to $Z/N = 1 + 2\delta$ however, produces the completely inverted table along line c. This table is also shown in Figure 6.4.

The observation [5] that all stable nuclides can be divided into four groups, in which neighbouring nuclides differ by the equivalent of a single α-particle, could indicate a build-up mechanism by α-addition, for the formation of all nuclides, starting from the four elementary units, neutron (n), proton (p), ^{2}H, and α. Since the Z/N ratio of an α-particle is unity, all four series approach this same ratio after a sufficient number of steps. Nuclide synthesis by this assumed mechanism may therefore occur in a state of curvature that limits nuclide stability by a Z/N ratio of one.

The conditions of curvature, envisioned for elemental synthesis, can only be realized in a massive stellar object. Since the form of the periodic table under these circumstances is assumed known, it is possible, in principle to calculate the nature of the stellar object. Compression of the hydrogen atom is known to shift electronic energy levels. In the highest state of compression, *i.e.* at a compression radius of 0.1 a_0, at which orbital energies have been calculated [90], the hydrogen sub-levels in order of increasing energy are

$1s$	$<$	$2p$	$<$	$3d$	$<$	$2s$	$<$	$4f$	$<$	$3p$	$<$	$5g$	$<$	$4d$
2		8		18		20		34		40		58		68

Some inversion is already apparent at this level of compression. The $2p$ and $3d$ levels are lower than $2s$, followed immediately by $4f$. Although more calculation is needed, it is therefore likely that total inversion of the levels could result from further compression. Since g-levels never occur in known electronic configurations, the appearance of $5g$ was disregarded. The energy spectrum defined by line d may then be assumed to prevail under conditions of synthesis.

Release of the synthesized heavy atoms is assumed to occur on disintegration of the stellar object. Many of these nuclides will be unstable against radioactive decay in the changed environment of decreased curvature and pressure. Within the solar system only 264 stable nuclides survive, while the energy spectrum inverts from d to b.

In terms of the prime-number model there should be 300 nuclides, spread over 100 elements. However, two atomic numbers, 43 and 61, are never reached during generation of the four series of nuclides, by α-addition. The maximum allowed atomic number for stable nuclides at $Z/N = 1$ therefore increases to 102. Speculations based on prime-number distribution support the conclusions drawn from Figure 6.7. There may well be sufficient α-unstable nuclides at the tip of the triangle of stability to increase the number of elements to 100 and their isotopes to 300.

The proposed parallel between the nuclear and extranuclear structures of an atom remains to be demonstrated. The inference that elemental periodicity is a subset of nuclide periodicity demands that the nuclear and electronic arrangements of an atom obey the same mathematical relationship. More precisely, that Schrödinger's equation for proton and electron describes both nuclear and electronic energy states. However, this equation has not been solved. The best alternative wave-mechanical model of the nucleus is the semi-empirical shell model (Section 4.3.1).

It is of interest to see whether any structure that reproduces the results of the shell model, or a related spectrum, can be recognized in the periodic function for stable nuclides. One position where to search for such a relationship is in Figure 6.7 at $Z/N = 0$, i.e. $Z = 0$, which probably describes the arrangement of nucleons in an atomic nucleus. Extrapolation of the hem lines to this zero ratio, shown in Figure 6.8, displays a prominent meeting of lines at proton numbers of 4, 10, 16, 24, 32 and 48.

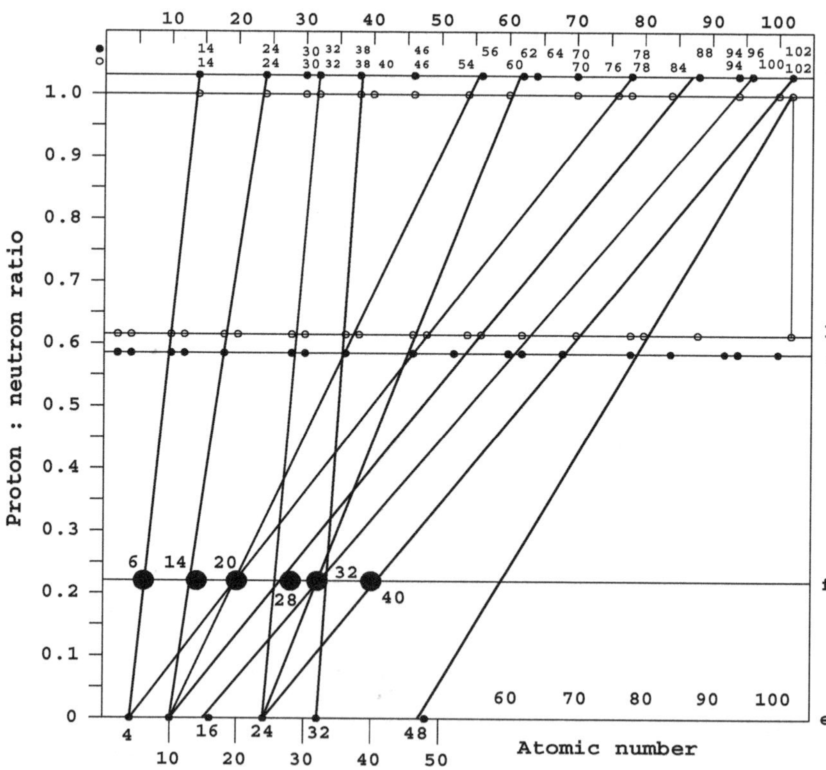

Figure 6.8: *Extrapolation of the hem lines that separate nuclidic periods to* $Z/N = 0$.

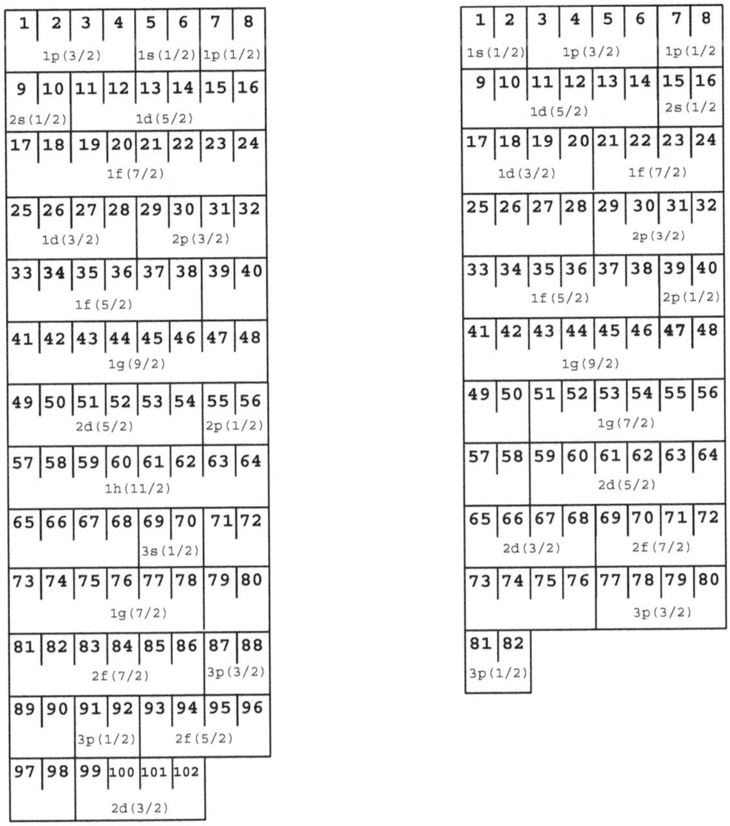

Figure 6.9: *Energy spectra of atomic nuclei according to a hypothetical one-particle model (left) and the shell model (right), presented in the form of compact periodic tables.*

This set of numbers allows construction of a reasonable spectrum that fits an eight-group table[1] to the energy sub-shells of the shell model, as shown in Figure 6.9. The spectrum defined by these points is related to, but not identical, with the experimental spectrum used in the definition of the

[1]The appearance of four- and sixfold degenerate subgroups requires that closed-shell arrangement could appear in groups 2, 4, 6 and 8. The sub-level notation used in the tables does not imply that spin-orbit coupling can be deduced from the extrapolation of hemlines.

shell model. At the ratio of $Z/N \simeq 0.22$ (line f) however, intersecting hem lines indicate a spectrum defined by the points 6, 14, 20, 28, 32 and 40, identical to an eight-group arrangement that satisfies the shell model, also shown in Figure 6.9, on the right hand side. The differences between the two sets are minor and not unlike those between the one-electron Schrödinger spectrum and the experimentally observed periodic table of the elements. It may therefore be argued that line e corresponds to a one-particle description of atomic nuclear structure, whereas line f also compensates for interparticle effects.

However, there is another interpretation of this set of numbers. Whereas the numbers at $Z/N = 0.22$ clearly refer to protons in the nucleus, those at $Z/N = 0$, i.e. $Z = 0$, can only refer to neutrons. Since the neutron energy spectrum is better defined in terms of neutron number, $N = A/Z$, it is necessary to convert the observed atomic numbers to the neutron scale. Assuming an increasing N/Z ratio, this transformation may be defined as $4 \rightarrow 6$, $10 \rightarrow 14$, $16 \rightarrow 20$, $24 \rightarrow 28$, $32 \rightarrow 38$, $48 \rightarrow 56$, $54 \rightarrow 68$, to yield essentially the same shell-model description for neutrons as for protons.

6.3 General Periodic Function

Four of the points along e (0, 4, 10, 24) (Figure 6.8) connect up with two sets of four points each on c, one set below 51 (0, 14, 24, 32) and the other higher than 51 (70, 78, 88, 102). The spacings between consecutive points of the lower set (14, 10, 8) are inverse to those between consecutive points of the higher set (8, 10, 14). Together with the pairs (38,64) and (46,56) the sequence

$$0, 14, 24, 32, 38, 46, 56, 64, 70, 78, 88, 102$$

along c, is symmetrical around 51. Because of this symmetry the entire diagram of Figure 6.8 may be rearranged to define a closed system.[2] The first step of the rearrangement consists of moving the set of points at $Z > 51$ to the opposite side of $Z/N = 0$, as shown in Figure 6.10.

[2]Some prominent points along c may appear to be arbitrarily deleted from the set that defines the symmetry. As shown in Figure 6.4, this happens because all of the atomic numbers in groups 2 and 8 do not represent closed sub-shell arrangements, although all of the latter are in groups 2 and 8. The exceptions in group 2 are numbers 8, 16, 40, 48, 72 and 80, which correspond to either f^8 or d^2 configurations. The three f^8 numbers are symmetrically transposed to p^6 arrangements in group 8 whilst the d^2 numbers correlate with d^8 arrangements in group 8, in line with the overall symmetry.

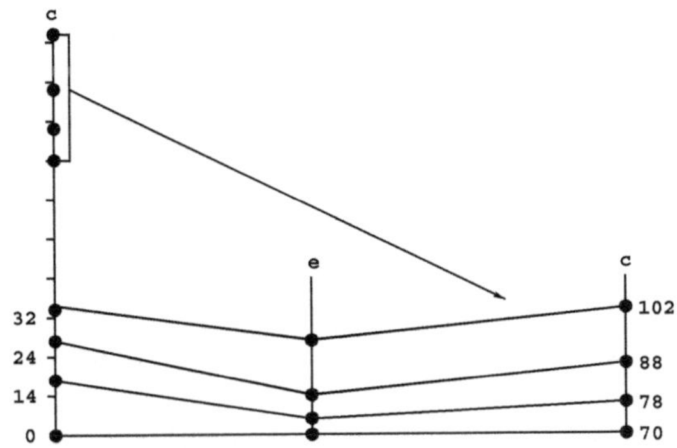

Figure 6.10: *Rearrangement of the periodic points along c of Figure 6.7 to emphasize the presence of a symmetry element at Z = 51.*

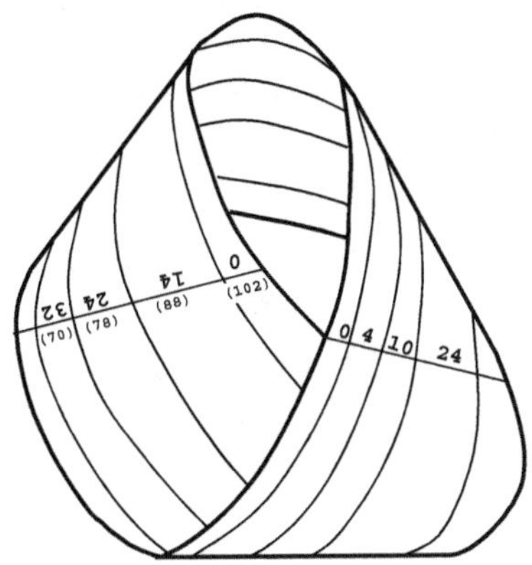

Figure 6.11: *The plot of Figure 6.10 projected on the surface of a Möbius band. Because of the double cover the area also doubles to make provision for both matter and anti-matter.*

The pairs of points $Z = 0, 102$ and $Z = 50, 52$ are next identified in an operation that requires the introduction of a topological twist. The result is a Möbius band, as shown in Figure 6.11. Only the first four lines are shown on the Möbius band. When plotted on the orientable double cover [91] of a Möbius band available room for the periodic function of elements and isotopes appears to be doubled. One possible interpretation is that the function describes atoms of matter and of anti-matter. Starting from zero space–time curvature along line a, the function follows the hem lines through the configurations represented by lines b and d to a situation of total inversion of electron configuration at infinite curvature along c. What appears to be a singularity at this point may therefore be interpreted as a transition from matter to antimatter instead. That may well be the fate of matter that enters a black hole.

The numbers shown in parentheses in Figure 6.11 represent points on the invisible underside of the Möbius double cover. The lines, shown extended beyond the points 0, 14, 24 and 32 are the mirror images of the lines joining these points to points greater than 51 on line c of Figure 6.11. These mirror lines describe the periodicity of anti-matter and where they pass behind line e(0,4,10,24), symmetry related points e(102,98,88,78) are inferred to occur. By adding these additional points and mirrored lines to Figure 6.8, the

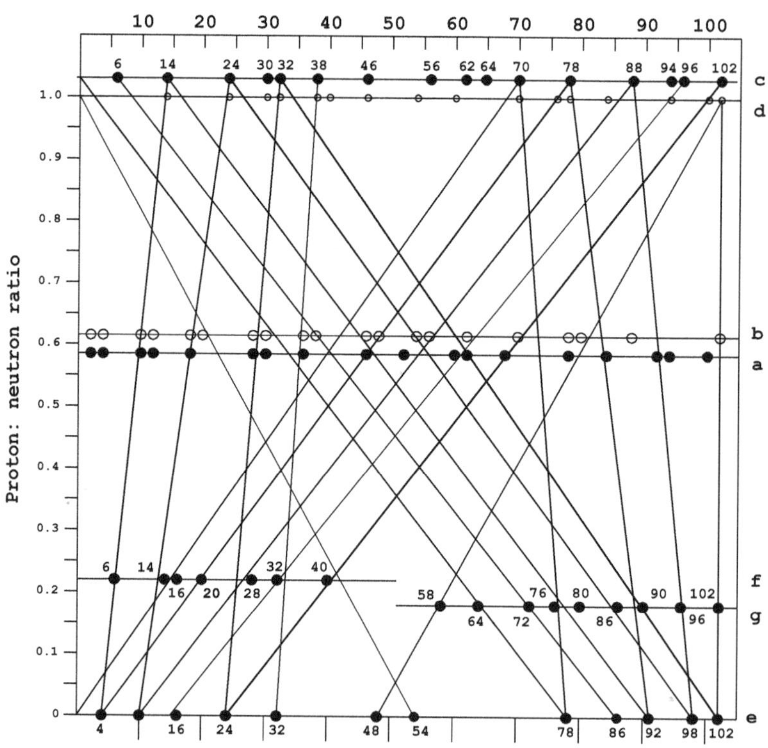

Figure 6.12: *Two-dimensional mapping of the Möbius double cover to show the appearance of the shell-model spectrum along f and g.*

symmetrical arrangement of Figure 6.12, which is a two-dimensional mapping of the Möbius band,[3] is obtained.

The symmetrical disposition around $Z = 51$ may now be used as a guide for connecting additional points along c to associated points on e. It is significant to note how the sixth position of the f-series ($f' = 6$) along c, once more appears with the filled sub-shell configurations.

By incorporating the symmetry-generated points on e with those (< 51) obtained before, the one-particle nuclear spectrum of Figure 6.9 (left) is readily extended to 102. However, the observed shell-model spectrum does not appear along f (at $0.2 + \delta$) as expected, but along g ($0.2 - \delta$), where it refers to anti-matter states. The equivalent lines for matter are not shown, but since these lines would not be displaced by the gap of 2δ between lines d and c, they would produce the expected periodic points for $Z > 51$ at the level f. The shell-model results are therefore extended by this construction to a value of 102, as shown in Figure 6.9 on the right.

The nuclear energy spectra derived from Figure 6.12 refer to periodic relationships based on atomic number Z, and strictly apply to proton states only. Nuclear states for neutrons are derived in 6.5 from periodicities based on $N = A - Z$.

6.4 Hidden Symmetry

Closure of the atomic number function over a period of 102 at level c provides the key to understanding elemental periodicity. It is remarkable how early workers like Reynolds [35] already likened the periodic table, then based on atomic weights, to vibrations on a stretched string. An improved representation was later proposed by Stewart [7] in terms of atomic numbers. Stewart's diagram comes tantalizingly close to summarising the observed periodicity of the elements in terms of a standing wave of period 8. It has nodes at 2, 10, 18, 27, 36, 45, 54, 62, 70, 78, 86, 95, almost identical to the periodicity defined by the compact periodic table (Figure 6.4), with nodes at 2, 10, 18, 20, 28, 36, 38, 46, 54, 62, 70, 78, 86, 94 and 102. The compact table avoids the periods of 9 introduced by Stewart, by the insertion of two additional short periods of 2. Probably because the analysis [7] was only partially successful

[3]Only those lines that also occur in Figure 6.7 refer to matter. The lines, generated by the mirror line at $p = 51$, and sloping in the opposite sense, strictly describe anti-matter only. The nuclear structures with $p > 51$, described in the next paragraph, are therefore anti-matter structures, assumed to mirror material nuclei.

it is now all but forgotten. However, the approximate symmetry observed by these workers may well reflect another, more fundamental hidden symmetry.

In pursuit of the proposed hidden symmetry it is found that the periodic table based on the antisymmetric arrangement at c (Figure 6.12) is seen to correspond well with a standing wave structure, obtained by arbitrary but equal displacement of consecutive pairs of periodic points in opposite direction, as shown in the upper frame of Figure 6.13. The nodal points between atomic numbers 3 and 99 are regularly spaced at intervals of 16, double the observed periodicity and identical to the periodicity of the Telluric helix proposed by De Chancourtois [92]. All of the nodal atoms appear in group 5 of the antisymmetric table (Figure 6.4) and on three arms of the number spiral, as shown in Figure 1.2.

To allow for the possibility that the displacement of periodic points need not be the same everywhere, the empirical wave of Figure 6.13 was redrawn as a sine wave,

$$z = \sin 2\pi \left(\frac{i-3}{32} \right), \quad i = 0, 102 \tag{6.2}$$

as shown in the second frame of Figure 6.13. The adjusted positions of the periodic points are identified. It is of significance that the periodic symmetry is represented by a sine, rather than a cosine curve. As discussed in 5.4.3 it may be inferred that elemental periodicity occurs in chiral space.

The symmetry observed along line c of Figure 6.12, is broken (hidden) at the other Z/N ratios where familiar forms of the periodic table occur, but it re-emerges along line e which represents the symmetrical state of nuclear structures. The construction is shown in the two lower frames of Figure 6.13. The upper frame shows the periodic wave obtained by arbitrary, but equal displacement of the pairs of points (10,14), (24,32), (38,48), (54,64), (70,78), (86,92) that appear along line e of Figure 6.12. In the lower frame is shown how this curve is modified into the same sine wave (6.2). The highlighted points are those that define the one-particle spectrum of Figure 6.9.

The sine curve (at both c and e) is closed by identification of points 0 and 102. In Figure 6.13 closure of the curve implies that the natural elements are restricted to atomic numbers $0 \rightarrow 101$, with the exception of numbers 43 and 61. The total of 100 natural elements therefore includes $Z = 0$, which is the neutron that appears to be stable under the conditions defined by the ratio $Z/N = 1$. The third frame of Figure 6.13 shows two forms of (6.2), with positive and negative arguments of the sine function, the latter suggested to represent atoms of anti-matter. The non-periodic region is blown up in Figure 6.14. It contains the elements ^{0}n, p(^{1}H), α(^{2}He), ^{100}Fm and ^{101}Md, and

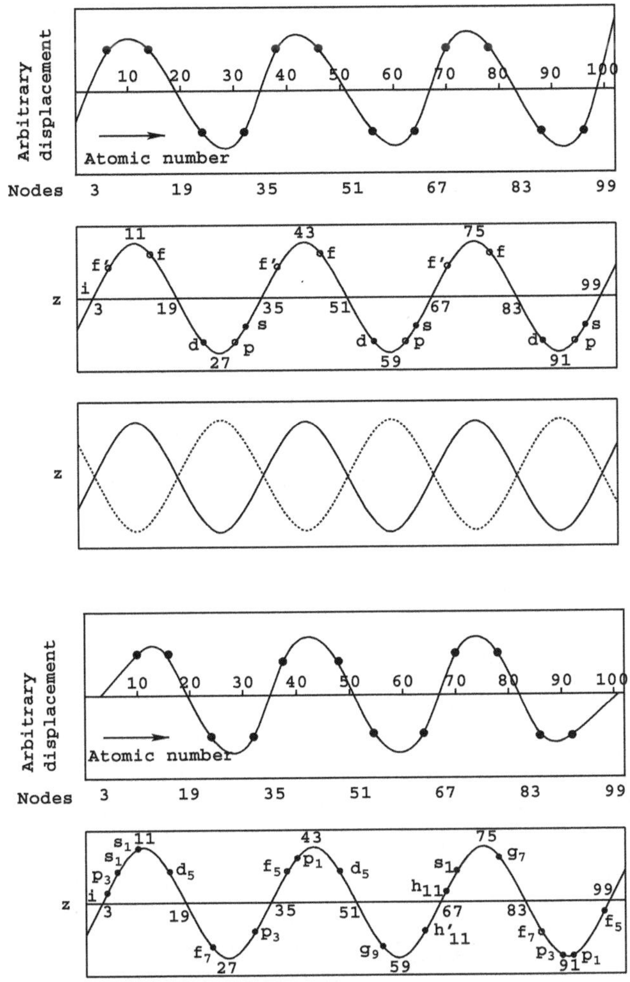

Figure 6.13: *The periodic arrangement at the ratio $Z/N = 1 + 2\delta$ (line c of Figure 6.7), represented as (i) an arbitrary regular wave (upper frame) and (ii) a sine wave (second frame). The third frame shows the superposition of positive and negative sine functions, to represent atoms of matter and of anti-matter. The two lower frames represent the identical operation applied to the one-particle nuclear spectrum at line e.*

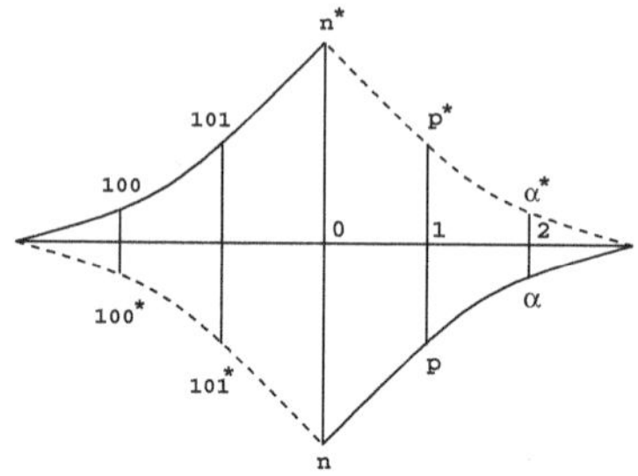

Figure 6.14: *Non-periodic part of the symmetric function that describes atomic periodicity. Stippled lines and asterisks indicate anti-matter. Atomic numbers 102 and 0 (neutron) coincide to yield a possible* $(101-2)+1$ *stable elements.*

their anti-matter equivalents, marked with asterisks. This region corresponds to the short period introduced by Reynolds and Stewart.

To understand how the periodic symmetry is hidden in periodic tables at different Z/N ratios, it is instructive to examine the familiar periodic table at $Z/N=\tau$. The energy spectrum is dictated by the local curvature and the only allowed states are those that produce closed sub-shell arrangements at groups 2 and 8, indicated by the set of short vertical lines in Figure 1.5.

The only way in which to meet this requirement is by introducing three blank regions, six numbers wide, between atomic numbers 2 and 3, 20 and 21, and 38 and 39, 18 blank spaces in total. It now becomes necessary to distinguish between atomic numbers that do not increase in the blank regions, and symmetry numbers that increase evenly from 0 to 102. The atomic number that coincides with the maximum symmetry number of $102 \equiv 0$, is 84. Only 84 (102-18) elements (atomic numbers 0 to 83) are therefore allowed at $Z/N=\tau$. However, under these conditions the neutron decays and since atomic numbers 43 and 61 cannot be realized, there are only 81 stable elements. The compact form of the periodic table of the elements in terms of both atomic and symmetry numbers is shown in Figure 6.15.

In Figure 6.13 the symmetry of the one-particle nuclear spectrum is mapped onto the sine curve. The non-periodic region now consists of structures with neutron and/or proton numbers of 0, 1 and 2, including n, p, ^{2}H and α, the starting materials of nuclear synthesis.

6.5 Neutron Periodicity

It can be demonstrated that the nuclear shell structure for neutrons is also a periodic relationship. The same procedure that was used to demonstrate

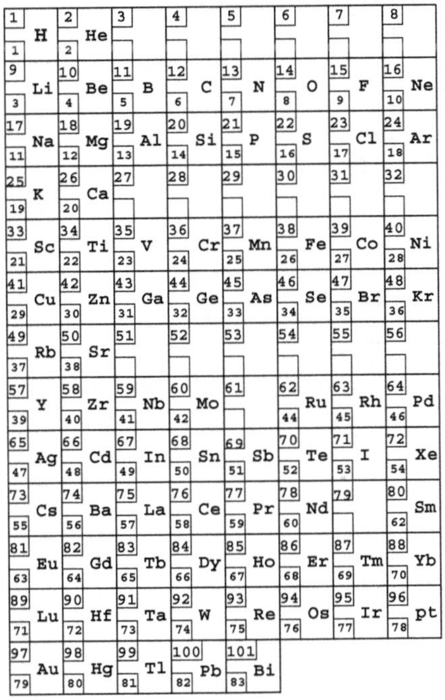

Figure 6.15: *Periodic table of the elements that incorporates all known features of hidden symmetry. Both symmetry numbers and atomic numbers are shown.*

the relationship between elemental and nuclidic periodicities may be used by plotting the distribution of nuclides as a function of $Z/(A-Z)$ vs $(A-Z)$, i.e. Z/N vs N. Such a plot is shown in Figure 6.16. The region of stability is readily shown to be enclosed by a profile with special points that define the neutron shell structure surprisingly well. Figure 6.17 shows how the nuclide periodicity of 24 generates the same periodic law in terms of neutron number at Z/N ratios of both τ and 1. A remarkable feature of the neutron energy spectrum derived from Figure 6.17 is that it remains invariant with respect to proton:neutron ratio. In terms of previous arguments this invariance must also hold with respect to environmental factors. The shell structure derives from a hypothetical potential somewhat between that of a square well and a three-dimensional isotropic oscillator. The predicted numbers of like nucleons that completely fill consecutive energy levels are shown along the top rows of Table 6.2. For nucleon number less than 50 these numbers coincide with the empirically established magic numbers that correspond to exceptionally stable arrangements. By assuming strong spin-orbit coupling a different order

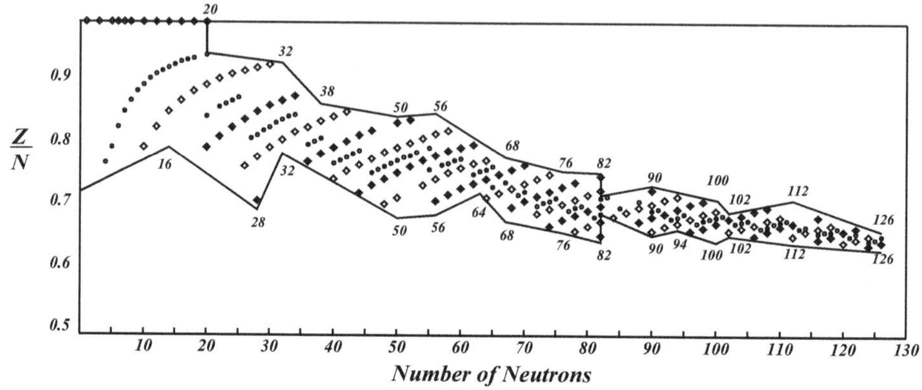

Figure 6.16: *Plot of neutron number, A−Z vs proton:neutron ratio, Z/(A−Z) for the four sets of stable nuclides. The proton numbers along the irregular profile correspond to closure of energy shells in the nucleus [70].*

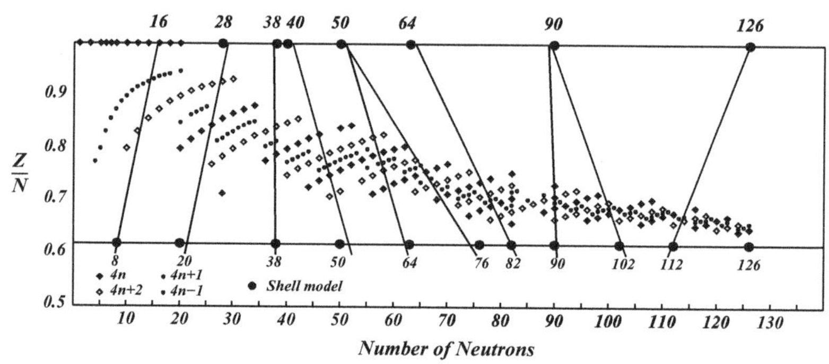

Figure 6.17: *Neutron energy levels derived from the assumed periodicity of 24 for the stable nuclides. Closed neutron levels of the shell model [70] are shown for comparison.*

of energy levels, shown in the middle rows is obtained. The filling of principal square-well levels now coincides everywhere with the magic numbers. However, the predicted spins of many nuclei do not agree with known experimental values. To improve the situation several pairs of spin sub-levels have been reversed in order, or superimposed where the energies are degenerate. This empirically improved scheme gives rise to the alternative closure of shells shown in the bottom rows of Table 6.2. Many important exceptions remain unresolved. In the present context it is important to note the fair match between the turning points of Figure 6.16 and the closed neutron levels of

Table 6.2: *Nucleon energy spectrum according to the nuclear shell model. Magic numbers are in the columns marked m.*

m		m			m	m				m
2		8			20					40
2	6	8	14	18	20	28	34	38	40	50
				16			32			
				m						m
				70						112
58	64	68	70	82	90	100	106	110	112	126
56		76	80		92		104	118	124	

the shell model, especially for $N < 82$. This match strongly suggests that the periodicity of the stable nuclides in terms of mass number, atomic number, or neutron number obeys related laws.

Figure 6.16 refers specifically to the neutron spectrum of the nucleus whereas Figure 6.12 reflects the nuclear proton spectrum.

Up to nucleon number 50 the two spectra are identical. The next level closes at $Z = 58$ in the proton spectrum and $N = 56$ in the neutron spectrum. This difference implies an inverse order of the levels $d_{5/2} < g_{7/2}$ for protons. From here on the neutron levels indicated by Figure 6.16 deviate appreciably from the accepted shell levels of Table 6.2. The two different schemes can be compared in terms of their predicted nuclear spins under spin-orbit coupling.

According to Figure 6.16 a reasonable level sequence at $N > 82$ is $2f_{7/2} < 3p_{3/2} < 2f_{5/2} < 3s1/2 < 1h_{9/2} < 1i_{13/2}$, compared to the scheme $1h_{9/2} < 2f_{7/2} < 3p_{3/2} < 1i_{13/2} < 2f_{5/2} < 3p_{1/2}$, still widely quoted, even in recent literature. Only the spins derived by fitting an energy spectrum to the quantum numbers obtained from Figure 6.16 are shown in Table 6.3.

The fit with measured spins is substantially better compared to the conventionally accepted scheme, but many of the heavy-nucleus spins had, admittedly, not been measured at the time of Goeppert-Mayers analysis [71]. Of the 102 nuclides with an odd number of nucleons only 62 spins[4] can be shown to be consistent with spin-orbit predictions. The high-multiplicity levels, designated g, h, and i show the poorest match. Of the 34 spins at these

[4]Numerical values of nuclear spin and other nuclear properties used in this work, such as thermal neutron cross section and coherent neutron scattering length, have been taken from the Rubber HCP [93].

Table 6.3: *Nuclear spins assigned according to the level structure of Figure 6.16 and assuming spin-orbit coupling.*

A	Z		2 × σ	Obs.
3		$1s_{1/2}$		
5				
		—— 2		
7	3		3	3
9		$1p_{3/2}$		
11	5		3	3
13				
		—— 6		
15	7	$1p_{1/2}$	1	1
17				
		—— 8		
19	9		5	1
21				
23	11	$1d_{5/2}$	5	3
25				
27	13		5	5
29				
		—— 14		
31	15	$2s_{1/2}$	1	1
33				
		—— 16		
35	17		3	3
37	17		3	3
39	19	$1d_{3/2}$	3	3
41	19		3	3
43				
		—— 20		
45	21		7	7
47				
49				
51	23		7	7
53		$1f_{7/2}$		
55	25		7	5
57				
59	27		7	7
61				
		—— 28		
63	29		3	3
65	29		3	3
67		$2p_{3/2}$		
69	31		3	3
71	31		3	3
73				
		—— 32		
75	33		5	5
77				
79	35		5	3
81	35	$1f_{5/2}$	5	3
83				
85	37		5	3
87				
		—— 38		
89	39		1	1
91				
		—— 40		
93	41		9	9
95				
97				
99				
101				
103	45	$1g_{9/2}$	9	1
105				
107	47		9	1
109	47		9	1
111				
113	49		9	9
		—— 50		

N		2 × σ	Obs.
1	$1s_{1/2}$	1	1
	—— 2		
5	$1p_{3/2}$	3	3
	—— 6		
7	$1p_{1/2}$	1	1
	—— 8		
9		5	5
11		5	3
	$1d_{5/2}$		
13		5	5
	—— 14		
15		1	1
	$2s_{1/2}$		
	—— 16		
17		3	3
	$1d_{3/2}$		
	—— 20		
23		7	7
25	$1f_{7/2}$	7	5
27		7	7
	—— 28		
29		3	1
	$2p_{3/2}$		
31		3	1
	—— 32		
33		5	3
	$1f_{5/2}$		
37		5	5
	—— 38		
	$2p_{1/2}$		
	—— 40		
41		9	9
43		9	1
	$1g_{9/2}$		
47		9	9
49		9	9
	—— 50		
51		5	5
53	$2d_{5/2}$	5	5
55		5	5
55		5	5
	—— 56		
57		7	5
59		7	5
	$1g_{7/2}$		
63		7	1
	—— 64		

A	Z		2 × σ	Obs.
113		—— 50		
115				
117				
119				
121	51		7	5
123	51		7	7
125				
127	53	$1g_{7/2}$	7	5
129				
131				
133	55		7	7
135				
137				
139	57		7	7
		—— 58		
141	59		5	5
143				
145		$2d_{5/2}$		
147				
149				
151	63		5	5
153	63		5	5
155		—— 64		
157				
159	65		3	3
161		$2d_{3/2}$		
163				
165	67		3	7
167		—— 68		
169	69		7	1
171				
173				
175	71		7	7
177		$2f_{7/2}$		
179				
181	73		7	7
183				
185	75		7	5
187				
		—— 76		
189				
191	77		3	3
193	77	$3p_{3/2}$	3	3
195				
197	79		3	3
199				
		—— 80		
201				
203	81	$3p_{1/2}$	1	1
205	81		1	1
207		—— 82		
209	83			9

N		2 × σ	Obs.
65			1
65			1
67			1
69	$1h_{11/2}$		1
71			
73			1
75			1
	—— 76		
77		3	3
	$2d_{3/2}$		
77		3	3
	—— 80		
81	$3s_{1/2}$	1	1
	—— 82		
83		7	7
85		7	7
	$2f_{7/2}$		
	—— 90		
91		3	3
93	$2p_{3/2}$	3	3
	—— 94		
95		5	5
97	$2f_{5/2}$	5	5
99		5	7
	—— 100		
101	$3p_{1/2}$	1	1
	—— 102		
103		9	5
105		9	7
107		9	9
	$1h_{9/2}$		
109		9	1
111		9	1
	—— 112		
113			3
117			1
119	$1i_{13/2}$		1
121			3
125			1
	—— 126		

levels only 8 are predicted correctly. It is clear that nuclear spins cannot be accounted for in terms of simple proton and neutron energy-level occupation according to Hund's rules and that an additional spin factor operates. However, it now appears doubtful that this additional factor is of the hypothetical type assumed by the shell model.

6.5.1 The Magic Diagram

Extrapolation of the hem lines of Figure 6.17 to $Z/N = 0$ should reveal beyond doubt any periodic relationship among neutrons in the nucleus. The extrapolation is shown in Figure 6.18, producing seven well-defined points. Only two of these are not among those already established at $Z/N = 1$ or τ. The two extra points fit in exactly with the magic-number pattern of Table 6.2. Altogether 16 distinct points are recognized on the diagram. Only three of these, 90, 102 and 112, are not members of the generally accepted most likely set of points that constitute the magic spectrum, according to Table 6.2. All the magic numbers, except for 2 which cannot be resolved, are however, there.

The distinction between neutrons and protons in the nucleus where they constitute a two-level system, is a convenience only. The magic numbers derived from Figure 6.18 must therefore refer equally to both types of nucleon and represent a periodic relationship, dictated by strong interaction. Superimposed on this strong periodicity there must be another function that minimizes coulombic repulsion and which is defined by the relative positions of protons and neutrons. The spectrum that appears along line e of Figure 6.8 will be interpreted as describing this distribution function. The same distribution is generated by inserting the line $Z/N = 0.22$ on the magic diagram, as was done in Figure 6.8 in order to identify the shell-model spectrum. However, as before, the numbers along $Z/N = 0.22$ on the N-scale need to be converted to atomic numbers when applied to protons. A reasonable transformation, $10 \rightarrow 8$, $38 \rightarrow 32$, $58 \rightarrow 48$, $80 \rightarrow 56$, $98 \rightarrow 68$, $104 \rightarrow 72$, $118 \rightarrow 80$, $126 \rightarrow 82$, reconfirms the conclusion that protons and neutrons are correctly

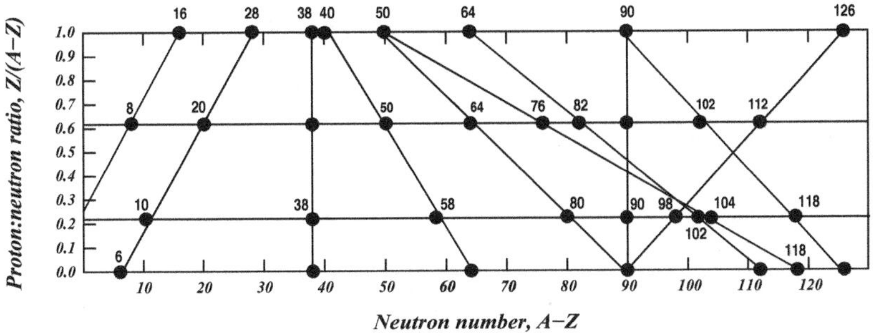

Figure 6.18: *Diagram to show the extrapolation of the hem lines of figure 6.17 to the proton:neutron ratio of zero. The three lines,* $Z/N = 1, \tau, 0$, *together define the neutron energy energy spectrum, widely known as magicity. The line at* $Z/N = 0.22$ *define points of the proton spectrum.*

ordered by virtually the same energy-level shell structure in the nucleus. A notable exception is the substitution of 48 for 50 in the proton spectrum and the appearance of 66, 72 and 80, not part of the pure neutron spectrum. With the full magic spectrum at hand there are sufficient data to explore the possibility of defining an energy spectrum, consistent with magicity and without invoking spin–orbit coupling.

A sub-level structure built around three basic sequences,

$$A(2,4,8)$$
$$B(4,6,10)$$
$$C(2,6,8,12)$$

is readily shown to account for the entire observed spectrum. Treating A, B, C as three types of principal energy level, the tentative sequence of sub-levels shown in Table 6.5 solves the problem directly. The implications are clear. There is no need to resort to the poorly understood model of spin-orbit coupling. The magic spectrum cannot be generated by a central-field potential and the shell model must be abandoned. The implications thereof will be examined more thoroughly in Chapter 8.

Apart from confirming the shell structure, which is directly revealed for protons at $Z/N = 0.22$ in Figure 6.8 and for neutrons at $Z/N = 0$ in Figure 6.18, the set of numbers that occurs at $Z/N = 0$ in Figure 6.8 and $Z/N = 0.22$ in Figure 6.18 contains information about the relative distribution of protons and neutrons in the nucleus. This distribution in the nucleus has an equally simple form, not of the central field type. As shown in Table 6.4 the most distinct feature is a periodicity of 16 in atomic number.

It is suggested that an acceptable model of the nuclear structure should reflect both of these periodicities related to neutrons and protons, as well as the periodicity in mass number.

Table 6.4: *Table to summarize the relative distribution of nucleons that minimizes coulombic repulsion, in terms of atomic number.*

A	*B*	*C*	*D*	*E*
4			*54*	*70*
10	*24*	*38*	*58*	*78*
16	*32*	*48*	*64*	*80*

Table 6.5: *The neutron spectrum, better known as magicity, based almost entirely on an interpretation of Figure 6.18 and the proposed sub-level structure.*

1A	2A	1B	1C	m	2B	2C	3A	3B	4A	m
2				2	4					80
4				6			2			82
	2			8		2				84
8				16		6				90
	4			20				4		94
	8			28	6					100
		4		32					2	102
		6		38	10					112
			2	40				6		118
		10		50		8				126
			6	56						
			8	64						
			12	76						

6.6 Nuclide Periodicity

The nuclidic periodicity of 24 which is fundamental to the scheme adopted here has been inferred from the rather tenuous link, assumed to exist between a number spiral and atomic structure. It is therefore gratifying to find that the same pattern appears naturally on arranging the stable nuclides by mass number as a function of Z/N. The resulting plot is shown in Figure 6.19. Linear-segment profiles drawn to enclose the area of stability are seen to correlate well with mass numbers at the closure of the periods of 24. The hem lines are all of infinite slope and the predicted periodicity is therefore independent of proton:neutron ratio. Nucleon numbers that close the observed nuclidic periods are shown in Table 6.6. Selected proton and neutron levels that provide a reasonable match with the observed numbers are also shown. This match, although convincing, is rarely perfect. Neutron (N) and proton (Z) levels reach saturation near the closure of the nucleon (A) levels. The slight mismatch may be expected to show up in the physical and symmetry properties for nuclides in these off closed-shell regions. An unexplained periodic feature, $Z, N, A = 8n, 11n, 19n$, is particularly striking for periods beyond 4, and although the match is no better than approximate, it suggests the operation of a simple number pattern that underlies the composition of

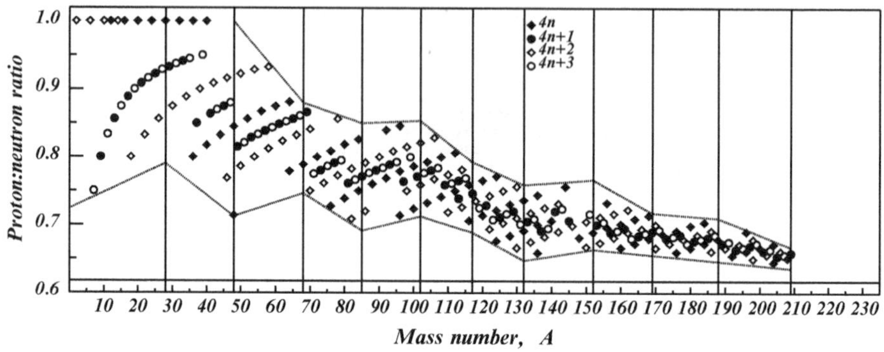

Figure 6.19: *Stable nuclides placed as a function of proton:neutron ratio and mass number.*

Table 6.6: *Comparison of the eleven nuclide periods of 24 with periodicity predictions and mean proton, neutron and mass number periods of 8, 11 and 19, respectively.*

				Observed			Derived		
n	8n	11n	19n	A	Z	N	A	Z	N
1	8	11	19	28	14	14	32	16	16
2	16	22	38	48	22	26	48	20	28
3	24	33	57	68	30	38	70	32	38
4	32	44	76	85	37	48	88	38	50
5	40	55	95	102	44	58	104	48	56
6	48	66	114	117	50	67	118	50	68
7	56	77	133	132	56	76	132	56	76
8	64	88	152	152	62	90	154	64	90
9	72	99	171	169	69	100	168	68	100
10	80	110	190	188	76	112	188	76	112
11	88	121	209	209	83	126	208	82	126

atomic nuclei. The most important conclusion to be drawn from Table 6.6 is that the periodicity, first inferred from the distribution of prime numbers, gets such overwhelming support from several unrelated constructions that the possibility of a lucky coincidence can be ruled out, in favour of finding a sound basis for the observations in number theory.

Chapter 7

Periodicity and Number Theory

The Farey sequence, as defined in number theory, provides a modular classification of entities made up of integral numbers of two types of particle, such as protons and neutrons. If the stability of such entities is assumed to depend on an increasing excess of one particle type (neutrons), compositions (nuclides) of constant excess are shown to be stabilized over limited regions, related in extent to the golden ratio, and leading to a periodic relationship that depends on relative stabilities. This stability trend is shown to be identical to the hypothetical periodicity amongst stable nuclides, postulated before on the basis of prime-number distribution on a spiral. Triangles of stability that limit the number of possible nuclides are shown to derive from limiting ratios, defined by fractions generated by the Farey procedure from Fibonacci numbers. The results correlate well with experimental stabilities inferred from measured mass defects and with solar abundances.

7.1 Introduction

It has been demonstrated [5] that an atomic model based on the distribution of prime numbers on a spiral may be used to analyze, not only the periodic law of the chemical elements, but also of the stable nuclides. The method relies on the assumption that all stable nuclides have been generated from four elementary units, by progressive addition of α-particles. Since there are no theoretical grounds for assuming such a mode of nuclear build-up it is of interest to establish if number theory provided an alternative to this assumption. The first objective is to find a generator from number theory to predict all possible combinations of neutrons and protons that may constitute a stable (non-radioactive) nucleus. The next step is to formulate appropriate bounds to distinguish between stable and radioactive nuclides. Finally, the known periodicity of the closed set of stable nuclides should be amenable to analysis in terms of modular arithmetic. Once the number theory is in place it should be possible to formulate a physical model in parallel with the number theory.

All of these aspects will be explored in this chapter, starting with a brief summary of the α-particle model.

7.2 Nuclear Synthesis by α-particle Addition

It is assumed that all stable nuclides can be obtained by the systematic addition of α-particles to four elementary units.

By assumption, the four elementary units are nucleons made up of neutrons and anti-neutrons (marked by an asterisk) only, i.e. n*n*, n*, nn*, n, with neutron numbers of −2, −1, 0, and 1. The first few steps of α-addition proceed as follows:

$$
\begin{aligned}
&\mathrm{n^*n^*} \xrightarrow{\alpha} {}^{2}\mathrm{He} \xrightarrow{-\beta+} {}^{2}\mathrm{H} \xrightarrow{\alpha} {}^{6}\mathrm{Li} \xrightarrow{\alpha} {}^{10}\mathrm{B} \longrightarrow (A = 4n-2)\\
&\mathrm{n^*} \xrightarrow{\alpha} {}^{3}\mathrm{He} \xrightarrow{\alpha} {}^{7}\mathrm{Be} \xrightarrow{-\beta+} {}^{7}\mathrm{Li} \xrightarrow{\alpha} {}^{11}\mathrm{B} \longrightarrow (4n-1)\\
&\mathrm{n^*n} \xrightarrow{\alpha} {}^{4}\mathrm{He} \xrightarrow{\alpha} {}^{8}\mathrm{Be(unstable)} \xrightarrow{\alpha} {}^{12}\mathrm{C} \longrightarrow (4n)\\
&\mathrm{n} \xrightarrow{\alpha} {}^{5}\mathrm{He(unstable)} \xrightarrow{\alpha} {}^{9}\mathrm{Be} \xrightarrow{\alpha} {}^{13}\mathrm{C} \longrightarrow (4n+1)
\end{aligned}
$$

After addition of three α-particles the series of mass number $(4n-2)$ and $(4n-1)$ have progressed to the boron isotopes ^{10}B and ^{11}B, whilst the $4n$ and $(4n+1)$ series have reached ^{12}C and ^{13}C respectively. The neutron excess, defined as $A-2Z$, for the four series at this stage is 0, 1, 0 and 1 respectively. It is only after 24 steps that the four are found in register again, all having reached isotopes of Ti with neutron excess of 2, 3, 4 and 5 respectively, an increase of four units each over the initial neutron numbers. The same happens again after 24, 36 and 47 steps. This manifestation of a periodic structure is at the basis of this entire analysis.

Empirical evidence is needed to establish its fate (i.e. mode of decay) once an unstable nuclide is generated in the course of α-addition. However, any α addition or β-decay changes neutron and proton counts by integral amounts. The proton:neutron ratio for any nuclide must therefore be a rational fraction. It will be examined how the fractional ratio $Z/(A-Z)$ changes with Z in the course of each of the four progressions that may be designated by $A(\text{mod}4) \equiv 3, 2, 0, 1$ respectively.

Table 7.1: *The four isotope sequences.*

0 *n*n** *n** *nn** *n*

n	4n−2	4n−1	4n	4n+1	n	4n−2	4n−1	4n	4n+1
1	H	He	He*	He*	27	Cd Pd	Ag	Cd Pd	Ag
2	Li	Li	Be*	Be	28	Cd Pd	Cd	Sn Cd	In Cd
3	B	B	C	C	29	Sn Cd	Sn	Sn Cd	Sn
4	N	N	O	O	30	Sn	Sn	Te Sn	Sb
5	O	F	Ne	Ne	31	Te Sn	Sb	Xe Te Sn	Te
6	Ne	Na	Mg	Mg	32	Xe Te	I	Xe Te	Xe
7	Mg	Al	Si	Si	33	Ba Xe Te	Xe	Ba Xe	Cs
8	Si	P	S	S	34	Ba Xe	Ba	Ce Ba Xe	Ba#
9	S	Cl	Ar S	Cl	35	Ce Ba	La#	Ce	Pr
10	Ar	K	Ca Ar	K	36	Nd Ce	Nd	Sm Nd*	Nd
11	Ca	Ca	Ca	Sc	37	Nd	Sm*	Sm* Nd	Sm*
12	Ti Ca	Ti	Ti Ca	Ti	38	Sm Nd	Eu	Gd* Sm	Eu
13	Cr Ti	V	Cr	Cr	39	Gd Sm	Gd	Dy Gd	Gd
14	Fe Cr	Mn	Fe	Fe	40	Dy Gd	Tb	Dy Gd	Dy
15	Ni Fe	Co	Ni	Ni	41	Er Dy	Dy	Er Dy	Ho
16	Ni	Cu	Zn Ni	Cu	42	Er	Er	Yb Er	Tm
17	Zn	Zn	Zn	Ga	43	Yb Er	Yb	Yb	Yb
18	Ge Zn	Ga	Ge	Ge	44	Hf* Yb	Lu	Hf Yb	Hf
19	Se Ge	As	Se Ge	Se	45	Hf	Hf	W Hf	Ta
20	Kr Se	Br	Kr Se	Br	46	W	W	Os W	Re
21	Kr Se	Kr	Sr Kr	Rb	47	Os W	Os	Os	Os
22	Sr Kr	Sr	(Zr) Sr	Y	48	Pt* Os	Ir	Pt* Os	Ir
23	Zr	Zr	Mo Zr	Nb	49	Pt	Pt	Hg Pt	Au
24	Mo Zr	Mo	Ru Mo Zr	Mo	50	Hg Pt	Hg	Hg	Hg
25	Ru Mo	Ru	Ru Mo	Ru	51	Hg	Tl	Pb Hg	Tl
26	Pd Ru	Rh	Pd Ru	Pd	52	Pb	Pb	Pb	Bi

The four-series table of stable isotopes is reproduced in Table 7.1 for easy reference. Underlined symbols indicate that the progression by α-addition is interrupted at the next step, which involves a nuclide that decays by positron emission and/or electron capture. The steps marked with a hash produce a product that decays by β-emission. Asterisks in the table identify α-emitters.

7.3 Nuclides in Farey Sequence

The allowed fractions for the $4n$, $A(\text{mod}4) \equiv 0$ series are identified graphically in Figure 7.1. The simplest row of fractions results from $Z = A - Z$, i.e. a uniform neutron excess of 0 and a proton:neutron ratio, $Z/(A-Z) = 1$. The next simplest sequence has a neutron excess of 4. After dividing out common factors, either 4 or 2, the fractions representing possible proton:neutron ratios are 1/3, 1/2, 3/5, 2/3, 5/7, 3/4, ..., with alternate differences, denominator-numerator, of 2 and 1, to be called the *rank of the fraction*. Further sequences, characterized by neutron excesses of $4k$, $k = 2, 3\ldots$, are shown in Figure 7.1. In each sequence, a number $k-1$ of additional fractions occur between each pair of ranks 1 and 2. These additional fractions have ranks equal to $2k$ and the factors of $2k$.

The simplest sequence of fractions in the $A(\text{mod}4) \equiv 2$ series, once more has neutron excess and rank zero, i.e. 1/1, 3/3, 5/5, etc., followed by a sequence with neutron excess 2 and rank 1, i.e. 2/4, 4/6, 6/8, 8/10, etc. The next sequence 2/8, 4/10, 6/12, 8/14, 10/16 has neutron excess of 6 and ranks in the order 1,3,3,1,3,3,1... In general, each sequence is characterized

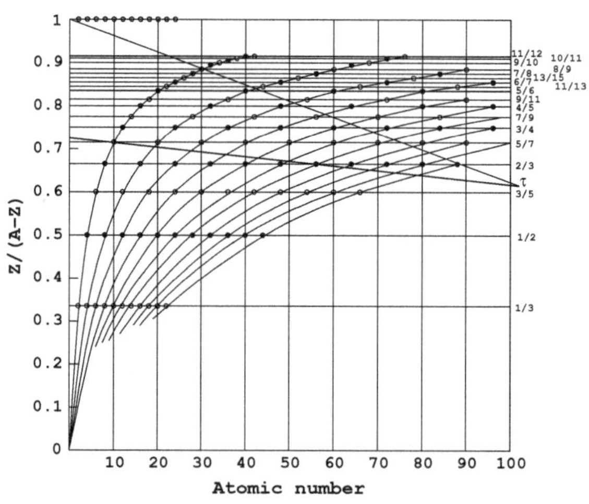

Figure 7.1: *Possible combinations of integers Z and $A-Z$ such that $A = 4n$, represented as rational fractions plotted against Z. Only fractions of ranks 1 and 2 are shown. Between each neighbouring pair of ranks 1 and 2 there appear $k-1$ ($k = 1, 2, \ldots$) additional fractions of rank equal to $2k$ or the factors of $2k$. The irrational number τ at $Z = 102$ represents a limiting point that will be shown to characterize the stability of possible nuclides.*

by a neutron excess of $4k + 2$, $k = 0, 1, 2...$, and $2k$ fractions of rank $2k + 1$, or the factors, if any, of $2k - 1$, between any neighbouring pair of unit-rank fractions.

The odd series $A(\text{mod}4) \equiv 1, 3$ $(4n \pm 1)$ add nothing new to the structure and together they define the same pattern of fractions as $A(\text{mod}4) \equiv 2$, but at neutron excess of $4k \pm 1$. The appearance of the only nuclide with negative neutron excess, ^{3}He, and first member of the series $2/1, 4/3, 6/5, \ldots$, of rank and neutron excess -1, shows that the pattern established for positive fractions extends unchanged into the negative region.

Table 7.2 shows the ranks of all fractions that define the unique motif that, when repeated, generates all allowed fractions consistent with the build-up of nuclides by successive α addition. A remarkable property of Table 7.2 is that it shows 1:1 correspondence with the well-known Farey sequence, first formulated to allow ordering and enumeration of all rational fractions [94]. One half of the Farey sequence is reproduced in Figure 7.2 to demonstrate this correspondence.

The rank, r of the simpler fractions is indicated by surrounding r-gons. The significance of the Farey sequence featuring in the α-addition process, is to confirm that all possible proton:neutron ratios are realized within this mechanism.

Table 7.2: *The ranks of fractions featuring in sequences with increasing complexity and representing possible proton:neutron ratios resulting from nuclear synthesis by α-particle build-up.*

-1 -2 -1

-1 -1

0

1 1

1 2 1

1 3 3 1

1 4 2 4 1

1 5 5 5 5 1

1 6 3 2 3 6 1

1 7 7 7 7 7 7 1

1 8 4 8 2 8 4 8 1

1 9 9 3 9 9 3 9 9 1

1 10 5 10 5 2 5 10 5 10 1

1 11 11 11 11 11 11 11 11 11 11 1

1 12 6 4 3 12 2 12 3 4 6 12 1

1 13 13 13 13 13 13 13 13 13 13 13 13 1

1 14 7 14 7 14 7 2 7 14 7 14 7 14 1

1 15 15 5 15 3 5 15 15 5 3 15 5 15 15 1

1 16 8 16 4 16 8 16 2 16 8 16 4 16 8 16 1

1 17 17 17 17 17 17 17 17 17 17 17 17 17 17 17 17 1

1 18 9 6 9 18 3 18 9 2 9 18 3 18 9 6 9 18 1

1 19 19 19 19 19 19 19 19 19 19 19 19 19 19 19 19 19 19 1

1 20 10 20 5 4 10 20 5 20 2 20 5 20 10 4 5 20 10 20 1

1 21 21 7 21 21 7 3 21 7 21 21 7 21 3 7 21 21 7 21 21 1

1 22 11 22 11 22 11 22 11 22 11 2 11 22 11 22 11 22 11 22 11 22 1

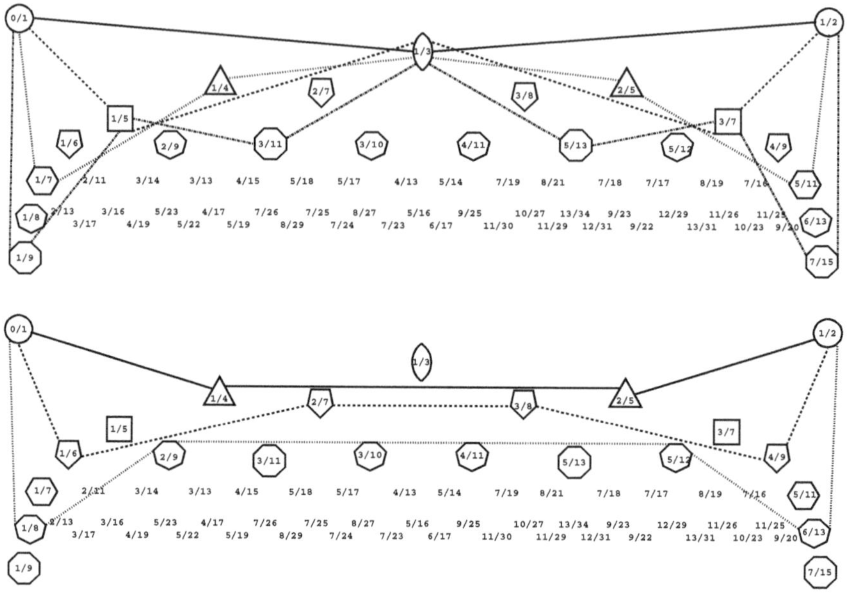

Figure 7.2: *One half of the Farey sequence reproduced twice to show the relationship with possible fractions generated during nuclear synthesis according to α-particle addition. The upper diagram highlights fractions that occur in sequences with $A = 4n$ and the lower one applies to all other sequences. The connecting lines identify the order in which rational fractions appear by rank. In this table each rational number is generated from the two that brace it from above by separate addition of numerators and denominators.*

7.4 Triangle of Stability

Having identified the Farey sequence as the generator of allowed proton: neutron ratios, it remains to establish the actual region over which the possible nuclides will be stable. To this effect it is noted that, as in Figure 6.2, each of the sequences of constant neutron excess plots on a straight line of slope 2 on axes A vs Z, shown in Figure 7.3 for actual nuclides.

The magnitude, x of each linear segment that represents the range of stability for given neutron excess is estimated by noting that invariably, $\Delta A = 2\Delta Z$, whereby

$$x^2 = (\Delta A)/2)^2 + (\Delta A)^2$$
$$x = (\sqrt{5}/2)\Delta A = (\Phi - 1/2)\Delta A = (2\Phi - 1)\Delta Z$$

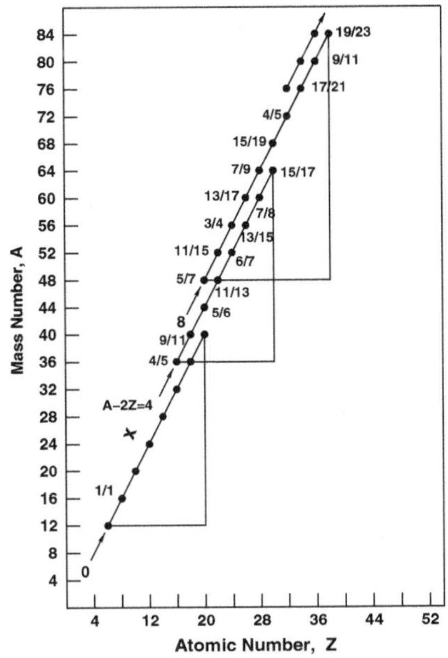

Figure 7.3: *The relationship between A and Z for some nuclides of the series* $A(\text{mod}4) \equiv 0$, *with ratios* $Z/(A-Z)$ *shown as fractions, for comparison with Figures 7.1 and 7.2.*

where $\Phi = 1.6180...$, known as the golden ratio or golden section.[1] The proton:neutron ratio at which each of the slope-2 line segments of Figure 7.3 terminates in an unstable nuclide is known from observation and relates to the triangle of Figure 7.1 that also involves the golden section.

Because of their implied relationship to τ the lines that define the stability triangle are most likely generated by Fibonacci fractions. The first few Fibonacci numbers are $0, 1, 1, 2, 3, 5, 8, 13, 21$, etc. The rational fractions that occur between the first non-zero Fibonacci fractions, i.e. $1/1$ and $2/3$ may be established by the Farey procedure, which is demonstrated in Figure 7.2. The resulting fractions are shown in Figure 7.4. These fractions converge to $1/1$ at the one end and $2/3$ at the other end. By choosing $5/8$ as the second limit the convergence proceeds in exactly the same way through $2/3$ towards

[1]The symbols Φ and τ are often used interchangeably for both the irrational number 1.6180.. and its reciprocal 0.6180.. To avoid confusion the symbol τ will be used here for the latter and Φ for the former.

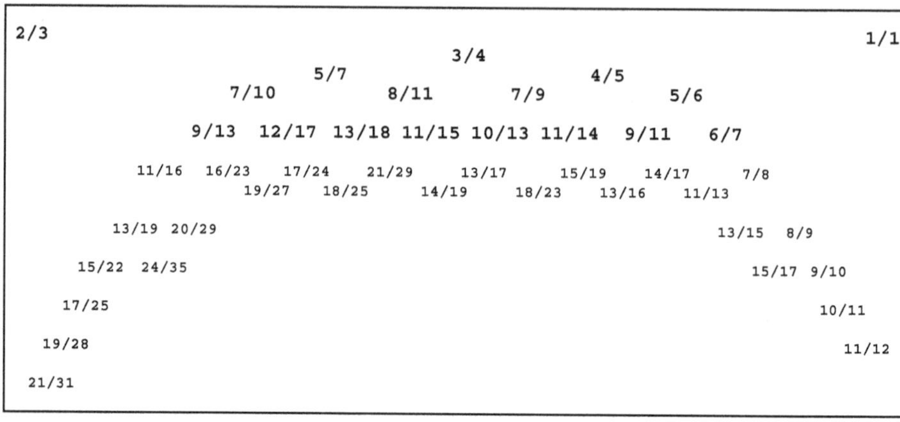

Figure 7.4: *Low-rank fractions in the Farey sequence between the Fibonacci fractions 1/1 and 2/3.*

5/8, and beyond towards τ, for higher Fibonacci fractions. It is shown in Figure 7.5 that all relevant points of intersection on the straight line from 1 to τ, that could refer to actual nuclides are generated by the Farey sequence that approaches 2/3. The value τ is approached at $Z = 102$.

All fractions beyond 3/4, at the centre of the Farey sequence, are further seen to define points of intersection between the curves of constant $A - 2Z$ and the straight line between coordinates of (14/19,0) and $(\tau), 102)$ at the intersection with the line $1 \rightarrow \tau$. The straight lines, so generated, are identical to the empirical limits of Figure 7.1 and automatically limit the maximum allowed atomic number of a stable element to 83. The limiting lines for nuclides of odd mass number lie inside the triangle of stability for even mass number [5], and may also be generated by the procedure described above. By definition, a proton:neutron ratio of unity is not defined for odd mass number. The maximum ratio is $Z/(A - Z) = 19/20$ that occurs for $A - 2Z = 1$ at $Z = 19$. The Farey sequence of fractions between 19/20 and 2/3 defines the limiting lines.

The range of all sequences determined by the procedure above, each with constant neutron excess $A - 2Z$ is shown as a function of atomic number Z vs $A - 2Z$, plotting as horizontal line segments in Figure 7.6. Each of the four sequences

$$(A - 2Z)(\text{mod}4) \equiv 0, 2, 1, 3 \tag{7.1}$$

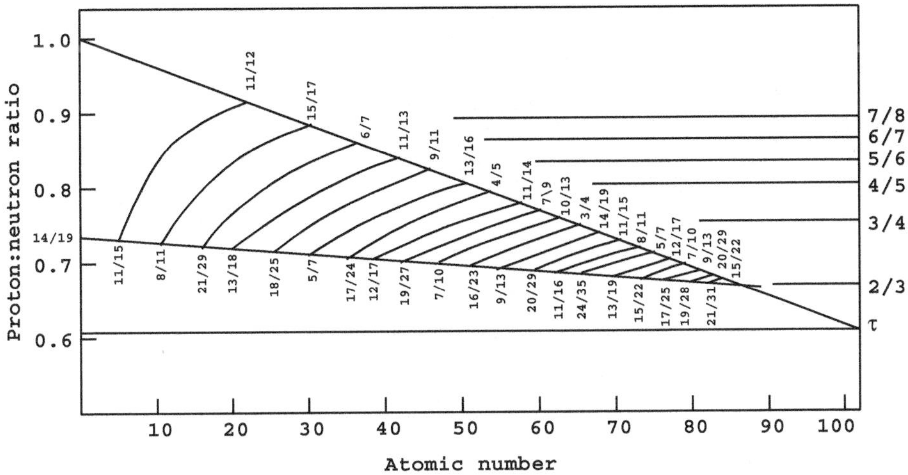

Figure 7.5: *Diagram to show points of intersection between curves of constant $A - 2Z$ and the straight lines that limit the occurrence of stable nuclides.*

consists of 11 segments. When collected into a single set the 264 stable nuclides occur on 44 segments of constant $A - 2Z$, ranging from 0 to 43, contained within the triangle of stability.

The modular form of the distribution (7.1) suggests that the 264 nuclides may be arranged into groups of 24, repeating as eleven periods, as conjectured before [5]. The origin of this periodicity remains to be established.

7.4.1 Nuclidic Periodicity

The regular recurrence of four isotopes of the same element generated by the four sequences, $A(\text{mod}4) \equiv 2, 3, 0, 1$ provides one possible basis for the proposed periodicity. Indeed, the first recurrence happens with the isotopes of Ti with neutron excess numbers of $A - 2Z = 2, 3, 4, 5$ – a uniform increase of four units over the neutron numbers of the starting set, and it coincides with the completion of the second set of 24 on the 24×11 matrix [5]. Two further exact coincidences occur with four isotopes of Sn ($A - 2Z = 14, 15, 16, 17$) and of Os ($A - 2Z = 34, 35, 36, 37$), closing periods 6 and 10, respectively.

These exact coincidences therefore happen when the period number, $P(\text{mod}4) \equiv 2$. Further recurrences are observed at

Mo(10,11,12,13)
Ru(10,11,12,13)

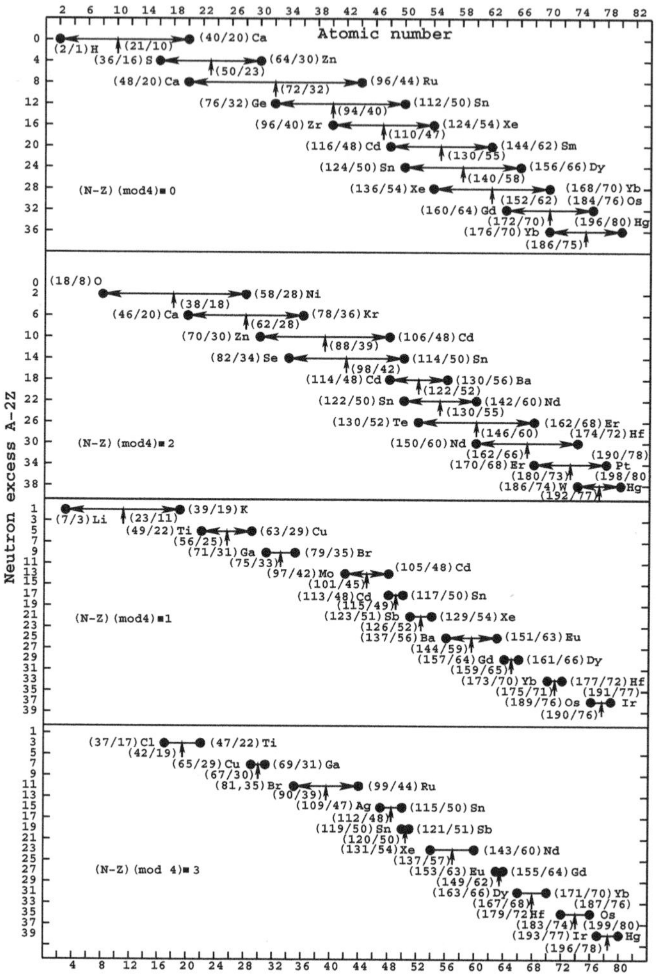

Figure 7.6: *Four plots of atomic number Z vs neutron excess $A-2Z$ for the nuclidic sequences $(A-2Z)(mod4) \equiv 0, 2, 1, 3$ respectively. Each plot consists of 11 segments centred at positions of mass and atomic number indicated by small arrows.*

Cd(14,15,16,17)
Ba(22,23,24,25)
Nd(22,23,24,25)
Gd(26,27,28,29)
Yb(30,31,32,33)
Hg(38,39,40,41)

The total of 11 observed recurrences does not necessarily define the 11 expected periods. The consecutive Mo and Ru recurrences, for instance, only show that none of the four sequences has been interrupted between these two elements by the generation of some radioactive nuclide. Furthermore, one or more of the isotopes defining a recurrence at the close of a period may be radioactive and hence does not feature in the table of stable isotopes. An example of such an incomplete coincidence is provided by the isotopes of Zn(6,7,8,-) that lacks the β-unstable nuclide ^{69}Zn to complete the expected quartet; its position being taken up by an isotope of ^{69}Ga. Extending this argument, the expected coincidence at the end of the fourth period may be viewed as having the nuclide ^{85}Rb replacing the missing ^{85}Kr isotope in the quartet Kr(10,11,12,-).

A plausible reconstruction, based on the preceding observations, that may reflect the 24-group periodicity of stable nuclides, is shown in Table 7.3. It serves to identify the linear segments that occur in each period. For comparison the nuclides at which each period has previously been found to terminate are underlined in Table 7.3. It is reassuring to find that a simple average over the mid-points of those segments grouped together as in Table 7.3, agrees well with the corresponding end member of the 11×24 matrix. This agreement, demonstrated in Table 7.4, makes good physical sense and follows the magic-number periodicity (Figure 6.18) fairly well.

7.5 Nuclear Stability

It is clear that nuclidic stability is some function of the proton:neutron ratio, $R = Z/(A - Z)$, and may be viewed as arising from an interaction that depends on the exchange of electrons between neutrons and protons. Only one stable nucleus, that of ^{3}He has $R > 1$. Nuclei of the $4n$ series with $Z < 20$ all have $R = 1$. Beyond $Z = 20$ all nuclei have $R < 1$ and this ratio decreases steadily with increasing Z. These observations imply that with increasing mass or atomic number a larger excess of neutrons over protons is needed to overcome coulombic repulsion and stabilize an atomic nucleus. The stability of the nucleus in a given sequence at constant $A - 2Z$, reaches an optimum at a certain point after which it steadily declines until an unstable nuclide is reached, at both ends of the line segment of constant neutron excess. At the lower end the unstable nucleus decays by β^- emission and at the higher end by β^+ and/or electron capture. This trend occurs in all four series $A(\text{mod}4) \equiv 0 \rightarrow 3$, but not necessarily in phase. If it is assumed that the maximum stability in each segment occurs near the mid-point, the average maximum for a set of four neighbouring segments in a common range of Z,

Table 7.3: *Eleven periods of nuclides.*

Period	End member	Recurrence	Enclosed segments
1	^{28}Si	Si(–,–,0,1)	0,1
2	^{48}Ti	Ti(2,3,4,5)	2,3,4,5
3	^{68}Zn	Zn(6,7,8,–)	6,7,8,9
4	^{85}Rb	Rb(–,11,–,–)	10,11
		Kr(10,11,12,–)	
		Ru(10,11,–,–)	
5	^{102}Ru	Ru(12,13,14)	12,13,14,15
		Sn(14,15,	
6	^{117}Sn	Sn(16,17,18,19)	16,17,18,19
7	^{132}Ba	Ba(20,–,22,23,24,25)	20,21,22,23
8	^{152}Sm	Sm(26,–,28,–)	24 → 29
		Gd(26,27,28,29)	
9	^{169}Tm	Yb(30,31,32,33)	
		Tm(–,31,–,–)	30,31,32,33
10	^{188}Os	Os(34,35,36,37)	34 → 39
		Hg(38,39,40,41)	
11	^{209}Bi	Bi(–,43)	

should then coincide with the common centre of gravity. Nuclidic periodicity as identified before may thus be related to a periodic fluctuation of nuclear stability as a function of A and/or Z. As shown schematically in Figure 7.7, in any period of 24, nuclear stability first decreases to a minimum, as each segment in turn terminates in a radioactive nuclide, after which the series continues with an increased neutron excess, building up to another maximum at the end of the period. Since the line segments at constant neutron excess in the odd series $A(\text{mod}4) \equiv 1, 3$, are in general shorter than those of the even series, the pattern of overlap gets out of step. Any number of segments may hence occur in different periods, but the average is four. A schematic representation of the proposed periodicity as a function of mass number is shown in Figure 7.8. In the middle of each period the stability falls below the limiting line beyond which β-type emission occurs.

Table 7.4: *Observed and calculated periodicities.*

		(N-Z)(mod4)					
Period	N-Z	0	1	2	3	Ave.	Obs
1	0	21/10				22/11	28/14
	1		23/11				
2	2			38/18			
	3				42/19	47/22	
	4	50/23					48/22
	5		56/25				
3	6			62/28			
	7				67/30	69/31	
	8	72/32					68/30
	9		75/33				
4	10			88/39			
	11				90/39	89/39	85/37
5	12	94/40					
	13		101/45			101/44	
	14			98/42			102/44
	15				112/48		
6	16	110/47					
	17		115/49			117/50	117/50
	18			122/52			
	19				120/50		
7	20	130/55					132/56
	21		125/52			131/55	
	22			122/52			
	23				137/57		
	24	140/58					
8	25		144/59				
	26			146/60			
	27				154/63	151/62	
	28	152/62					152/62
	29		159/65				
9	30			162/66			
	31				167/68	169/69	169/69
	32	172/70					
	33		173/70				
10	34			180/73			
	35				183/74		
	36	186/75				188/76	188/76
	37		190/76				
	38			192/77			
	39				196/78		
11	40	198/79					
	41		203/81				
	42			202/80		204/81	
	43				207/82		209/83
	44	208/82					

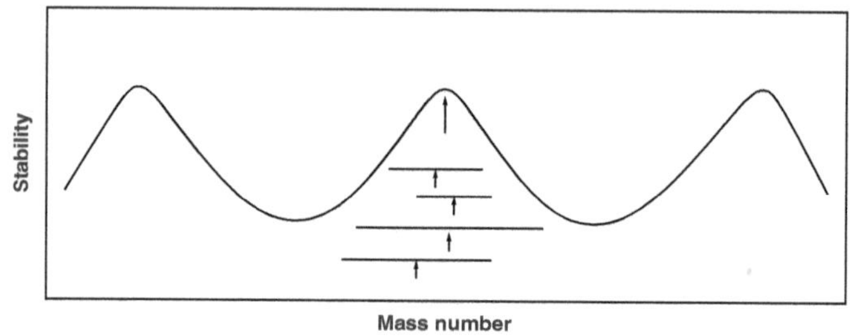

Figure 7.7: *Schematic diagram to envisage the periodic stability trend among stable nuclides and how it arises as an average over the mid points of nuclidic groups of constant neutron excess.*

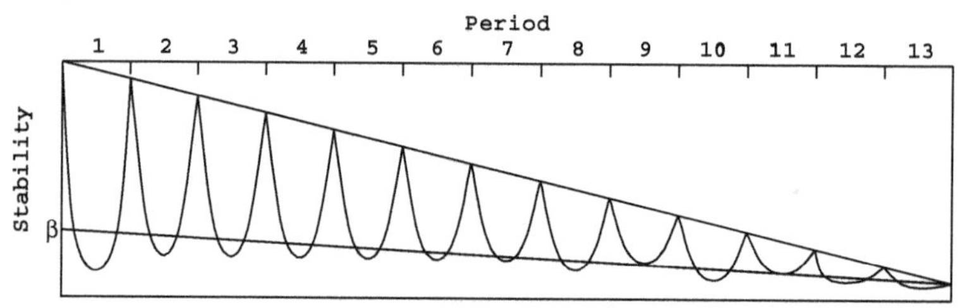

Figure 7.8: *Periodic trend proposed to derive from nuclidic stability. Whenever the periodic curve falls below the β line the corresponding nuclide will be unstable against $\beta^{\pm}$ decay.*

It now becomes clear why the recurrence of four consecutive isotopes tends to coincide with period termination. The latter occurs close to the point of maximum local nuclear stability for all four series, $A(\text{mod}4) \equiv 0 \rightarrow 3$, and hence maximum likelihood that each series should produce a stable isotope.

The question of periodic nuclear stability may be considered as solved in principle. It remains to work out the quantitative details that would predict, amongst other things, the abundances of stable nuclides and the distribution of α-emitters.

7.5.1 Nuclear Binding Energy

An immediate demonstration that the predictions of the previous section are essentially correct comes from an examination of experimentally measured

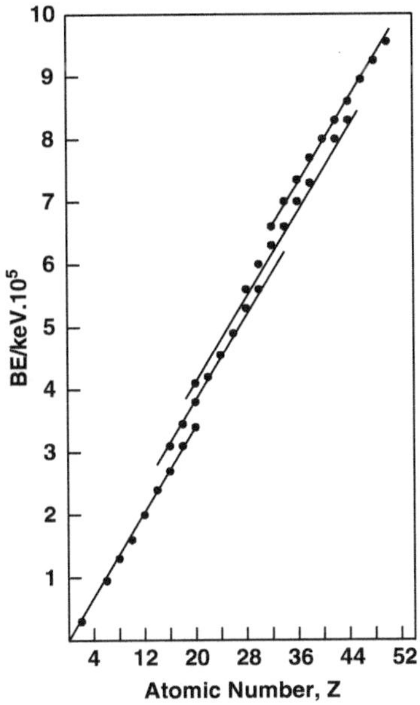

Figure 7.9: *Plot of experimental binding energies for nuclides of the same type as shown in Figure 7.3.*

nucleus binding energies (NBE). The measurement is invariably that of a mass defect.[2]

A plot of BE vs $(N-Z)/Z$ may, for instance, be scaled to appear virtually indistinguishable from a plot of $Z/(A-Z)$ vs Z. A plot of BE vs Z, for the same nuclides of Figure 7.3, is shown in Figure 7.9.

It may be scaled to near coincidence with the plot of A vs Z, shown in Figure 7.3. These relationships need considerably more study, but at first

[2]A possible source of confusion in the estimation of nucleus binding energies from measured mass excess (ME), is the arbitrary assignment of 12 a.m.u. to the mass of the isotope ^{12}C. The effect of this is that the ^{12}C nuclide seems to have a mass excess of zero by definition, although its binding energy must have some positive value. Zero binding energy must clearly be rather associated with ^{1}H, which on the ^{12}C scale has a ME of 7.29 MeV. The apparent minimum in a table of ME at about $Z = 50$, occurs for the same reason on the ^{12}C scale. To convert ME (on ^{12}C scale) to NBE (on ^{1}H scale) it is necessary to subtract the appropriate number (Z) of proton and (N) neutron masses,

$$NBE = ME - (Zm_p + Nm_n) \quad \text{where} \quad m_n = 8.07 \text{ MeV}$$

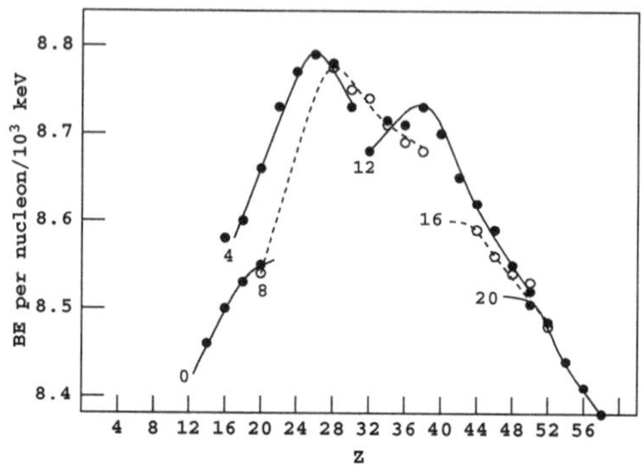

Figure 7.10: *Experimental binding energy per nucleon for the nuclides of Figure 7.9.*

glance they suggest that the qualitative stability measure of Figure 7.7 may be related to binding energy per nucleon. The variation of this quantity with atomic number is shown in Figure 7.10. There is no direct correspondence between the two graphs, but the central idea of maximum stability near the central nuclide in a series of constant neutron excess is clearly borne out.

7.5.2 β-Stability

The diagram shown in Figure 7.5 provides a summary of how naturally occurring nuclides may be identified by number theory and how their stability is controlled by the golden section. To find the theory that applies specifically to nuclides of mass number A and atomic number Z, it is necessary to identify the fractions along the vertical axis with $Z/(A-Z)$ and the numbers on the horizontal axis with $Z(A)$. Fractions that represent allowed nuclear compositions are generated by the Farey sequence in the interval $(0,1)$, as four sub-sets, at points $A(\text{mod}4) \equiv 0 \to 3$. Each sub-set further subdivides into segments of constant $A-2Z$. Curves connecting points of the even mass-number segments generate a straight line through a set of points on the curves, at coordinates corresponding to fractions generated by the Farey sequence, in an interval defined by successive non-zero Fibonacci fractions, e.g. $(1/1, 2/3)$. This straight line extends from the unit fraction at $Z=0$ to the irrational fraction $\tau = 1/\Phi$ at (atomic) number 102 and it limits the stability of nuclides against β^+ decay.

A sub-set of the fractions that define the limiting line, i.e. those with values less than 3/4, generates another straight line at second points of intersection with the curves, in the interval $(14/19, \tau)$. This second line limits the stability of nuclides against β-decay. The area between the limiting lines represents the triangle of nuclear stability against β-type decay. A smaller triangle of stability for the odd series is contained within the larger triangle for even mass-number. Its number-theoretic definition arises from the Farey sequence in the interval $(19/20, 2/3)$. The theory behind α-decay remains obscure.

Each of the modular sub-sets is represented in the triangle of stability by 11 segments of constant $A - 2Z$, adding up to 44 segments with $A - 2Z$ increasing from 0 to 43. Within each segment nuclear stability reaches a local maximum, half-way between β^- and β^+ unstable species, to define eleven periods of fluctuating nuclear stability, accommodating 24 nuclides in each period.

The observations of the previous paragraph are summarized in Figure 7.11 that shows a plot of all stable nuclides on axes $A - 2Z$ vs Z. The nuclides at which the 11 periods end are marked by squares and connected by approximate curves. The discontinuity in period 8 coincides with a large number of α-unstable nuclides in this region.

The profile that bounds the area of stability once more reveals a structure, based on a set of atomic numbers that features prominently in the

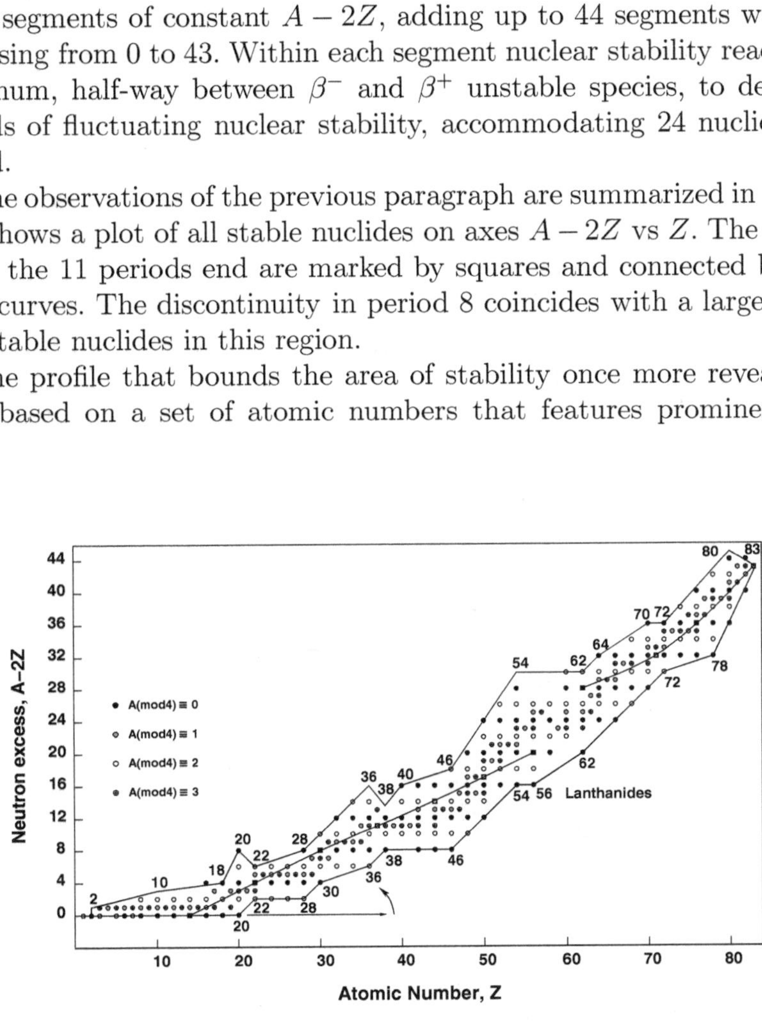

Figure 7.11: *Map of the stable nuclides in a field defined by the variation of neutron excess as a function of atomic number. The nuclides are grouped into four families defined by* $A(\mathrm{mod}4) = 0, 1, 2, 3$. *The eleven nuclidic periods repeat between the nuclides represented by squares.*

periodic classification of the elements, and first recognized [5] in a plot on axes $Z/(A-Z)$ vs Z. This structure establishes the important principle that the periodic law of the chemical elements is a subset of a more general law that governs the periodicity of stable nuclides.

Neutron excess is shown in Figure 7.12 as a function of mass number. The diagram confirms the discontinuity in the expected linear periodicity of the nuclear-stability function. As in Figure 7.11 the discontinuity coincides with the appearance of the lanthanide series. It sets in at the point where the sequence based on α-addition runs into the region of (β^-)-instability at ^{137}Ba and ^{139}La, shown in Table 7.1. Beyond the lanthanides the stability function becomes linear again. The stability profiles match the hem lines that define the assumed periodicity of 24 well.

Figures 7.11 and 7.12 demonstrate, once more, the important relationship that exists between the stable nuclides and the Farey sequence. To better appreciate this relationship any fraction m/n may be considered as a point (n, m) in the Cartesian plane. A fraction m/n is called *visible* iff the segment of the straight line connecting the origin with the point (n, m) contains no other grid points. Now imagine a ray along the positive x-axis, rotated counter-clockwise to mark all visible points, inside the first octant, whose x-coordinates does not exceed some number, N say. For example, if $N = 5$, the succession of visible fractions

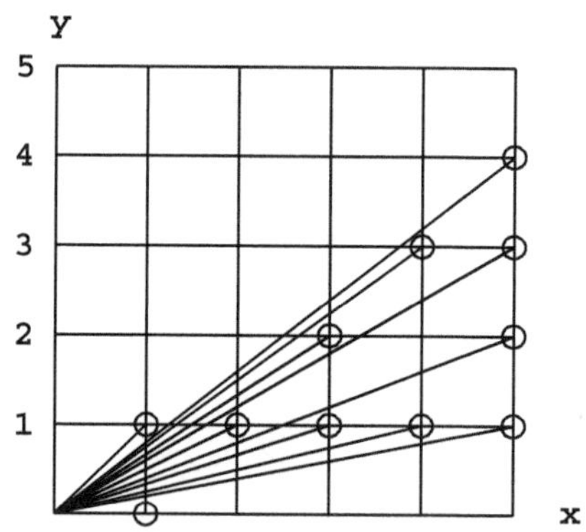

0/1, 1/5, 1/4, 1/3, 2/5, 1/2, 3/5, 2/3, 3/4, 4/5, 1/1, may be recognized as the Farey sequence, F_5. All other Farey series are obtained in the same manner. All the marked fractions (m/n) are proper $(m < n)$ and each fraction equals the slope of the line connecting the corresponding point with the origin. Therefore, they are traversed in ascending order. The rotating ray indicated in Figure 7.11 marks the visible points of the Farey sequence that maps the points of Figure 7.1. All other points are generated from the visible set by repetition at regular intervals, according to slope.

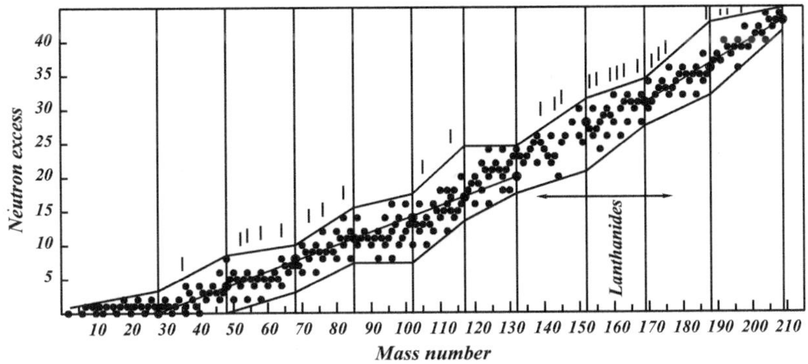

Figure 7.12: *Neutron excess of stable nuclides as a function of mass number. Short vertical markers indicate the position of some odd nuclei with high thermal-neutron cross sections.*

It is noted that the first indication of nuclidic periodicity [5] arose from analysis of the prime-number distribution, mapped onto the $6n \pm 1$ arms of a 24-stop number spiral with features reminiscent of extranuclear electronic configurations. Whereas this previous analysis relied primarily on an empirical match, no such assumption has been made in the number analysis. The two approaches lead to the same picture.

7.6 Golden Parabola

The central two-part curve of Figure 7.11 intersects each linear segment of constant neutron excess at, or close to the position of the nuclide of maximum binding energy on that segment. Each segment terminates at both ends in a rational fraction Z/N of the Farey sequence generated by Fibonacci numbers. The two resulting limiting curves both converge to τ. The function (2.8)

$$x^2 - x - n = 0$$

describes a parabola with largely the same property. It is a more general form of the expression, $x^2 - x - 1 = 0$ (2.10) that generates the golden mean.

The proposed function defines the parabola shown as an inset in Figure 7.13; it has a minimum $n = -\frac{1}{4}$ at $x = \frac{1}{2}$. A related parabola is obtained from the end members of the linear segments on plotting maximum and the negative of minimum atomic numbers for each section against neutron excess. The two parabolas coincide after scaling the theoretical x-coordinate by a factor $2\tau = 1.236$ and matching the neutron excess to n. The minimum point of the curve shifts to τ. The relationship between the two sets of axes is defined by $A - 2Z = 4n$. The resulting curve is described by the equation

$$x^2 - 2\tau x - 4\tau^2 n = 0$$

with solutions $x = \tau(1 \pm \sqrt{1 + 4n})$. The relationship between x and atomic number, after scaling is $Z = 10(2x - 1)$. These results establish the elusive general relationship that exists between neutron excess and atomic number. The scatter around the mean curve arises from two factors – the relatively lower stability of nuclides with odd mass number and the restriction of both Z and A to integer values. The nuclides enclosed within the two arms of the golden parabola are the same as those in the triangle of stability.

To obtain the golden parabola (Figure 7.13) that fits the stability bounds of nuclides it is necessary to scale the x-coordinate by a factor 2τ. Since the point $(x, n) = 0$ is not affected by the scaling, the resulting parabola is no longer centred at $x = \frac{1}{2}$, $(Z = 0)$, but at $x = \tau$, i.e. $Z = 10(1.236 - 1) = 2.36$.

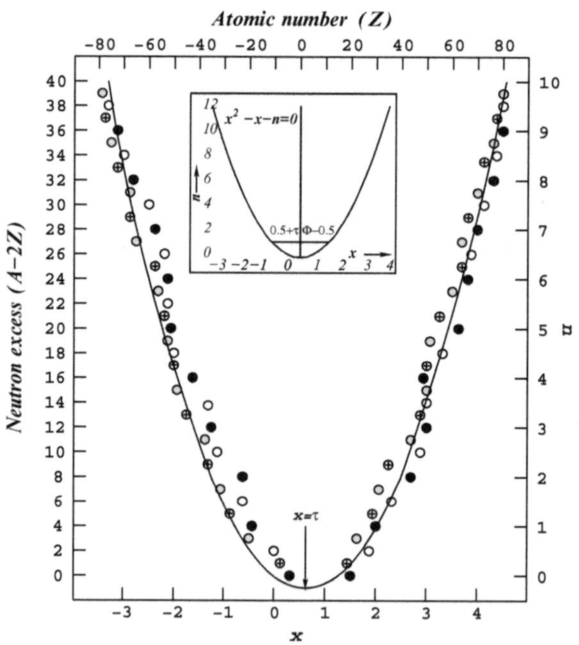

Figure 7.13: *Nuclides have a limited range of stability for each value of neutron excess, A-2Z. The maximum Z of these ranges plotted against A-2Z define the right arm of the parabola shown in the diagram. The left arm of the parabola is obtained by plotting minimum Z as negative integers. The curve is described by the equation* $x^2 - 2\tau x - 4\tau^2 n = 0$, *i.e.* $x = \tau(1 \pm \sqrt{1 + 4n})$.

This shift of the golden parabola has an effect on the way in which a periodic function can simulate the symmetrical periodic state of the elements. This state was found to occur at the ratio $Z/N \simeq 1.04$, and it follows the sine curve (6.2)

$$z = \sin 2\pi \left(\frac{i-3}{32} \right), \quad i = 0, 102$$

Appearance of the non-periodic part (Figure 6.11) at $Z < 3$, first noticed by Reynolds [35], is now seen to arise from recentring of the parabola from 0 to 2.36. The nearest integer at which the periodic section can start is at $Z = 3$.

Chapter 8

Properties of Atomic Matter

8.1 Periodicity

The conclusions of the previous two chapters suggest that, by virtue of their digital binary make-up, a valid physical model of neutral atoms, and of matter in general, follows directly from number theory. The number basis of nuclide periodicity is summarized in Figure 8.1. The point of convergence appears to be the same as in Figure 6.1, that shows the variation of proton:neutron ratio with atomic (or mass) number. Should the two points of convergence be identical it means that

$$\frac{N-Z}{Z}=\frac{Z}{N}$$

at that point. This equality rearranges to

$$Z^2+NZ-N^2=0$$

with solutions

$$Z=\frac{-N\pm\sqrt{N^2+4N^2}}{2}=N\tau$$

This result provides final proof that the limiting ratio Z/N for stable nuclides is, not approximately, but exactly equal to the golden ratio.

The numbers of importance in Figure 8.1 are A, Z, $N=A-Z$, $N_e=A-2Z$ and the irrational number $\tau=1/\phi$. The field of stability, when mapped as a function of A, converges to $N_e/Z=\tau$ at $A=262=100\phi^2$. The limiting lines intersect the vertical axis ($A=0$) at $\tau^2=0.382...$ and $\tau^2-0.5=0.5-\tau=-0.118...$ It is noted that the point $(A-2Z)/Z=\tau^2$ corresponds to $Z/N=Z/(A-Z)=1/(\tau^2+1)=(1+\phi)/(2+\phi)=0.723...$,

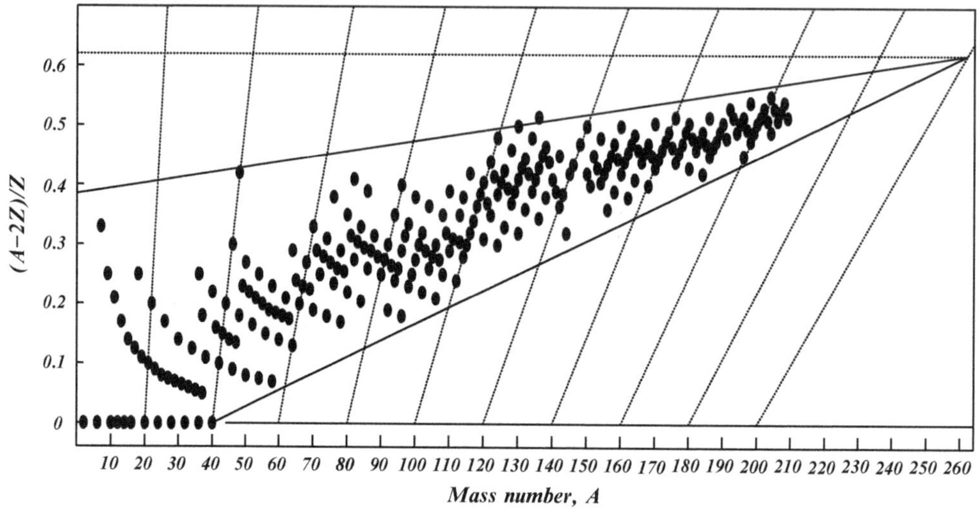

Figure 8.1: *Plot of the stable nuclides to show the ratio of neutron excess to atomic number as a function of mass number. The plot seems to converge to a limiting value of* $\tau = 0.618\ldots$ *at about* $A = 262$.

a number[1] that featured in many plots of Chapters 6 and 7. Isotopes of any single element occur as linear arrays between the points $A = 2Z$ and $A = Z\phi^2$ at $N_e/Z = 0$, and τ respectively, as marked for elements with $Z = 10n$. Assuming a point of convergence at $A = 262$ the number of stable elements is limited to 100, as first inferred from the number spiral,[2] Figure 1.2.

The relationship $A = Z\phi^2$ corresponds to $A - Z = Z(\phi^2 - 1) = Z\phi$, i.e. $Z/(A-Z) = \tau$ of Figures 6.1 and 6.4. Points $A = 2Z$ translate into $Z/N = 1$. The limiting point $A = 200$ accords with the inference that under cosmic conditions which favour a proton/neutron ratio of unity, the maximum number of stable nuclides is 100. Slopes of the limiting lines are fixed (Figure 7.5) by their intersection with the 44 N_e curves at fractional values of Z/N that follow the Farey sequences in the intervals (1,5/8) and (14/19,5/8).

The variation of neutron imbalance, N_x is shown as both N_e/Z and Z/N on the same diagram, Figure 8.2, as a function of A, for the isotopes of a

[1]This number is close to $\tan\left(\frac{\pi}{5}\right) = 0.7265$, at the trigonometric basis of the golden ratio.

[2]Symmetry arguments favour $Z = 102$ and maximum $A = 300$. The actual numbers of stable elements and nuclides are 81 and 264, respectively. The limiting lines in Figure 8.2 do not define the triangle of stability and have been inserted to emphasize the convergence.

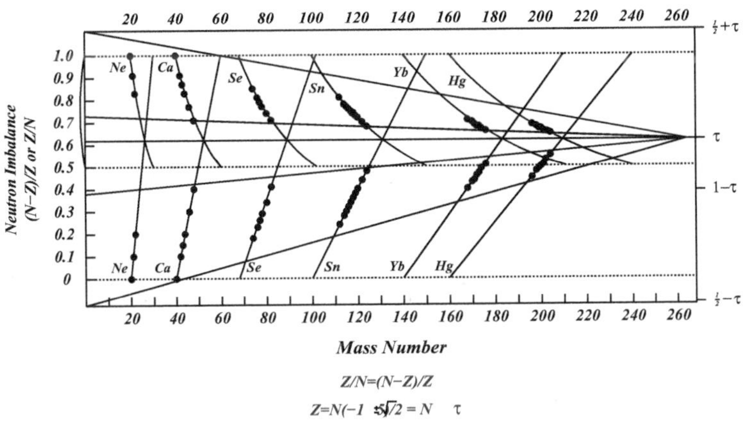

Figure 8.2: *Neutron imbalance as a function of mass number for the isotopes of selected elements.*

number of selected elements. As in Figure 8.1, the values of N_e/Z for the isotopes of a given element are mapped on straight lines between coordinates $N_x = 0, A = 2Z$ $(N = Z)$ and $N_x = 1, A = 3Z$ $(N = 2Z)$. For $(N-Z)/Z = \tau$, $A = N + Z = Z(2 + \tau) = Z\phi^2$. Similarly, the values of Z/N for a set of isotopes appear on smooth curves between coordinates $1, 2Z$ $(N = Z)$ and $\frac{1}{2}, 3Z$ $(N = 2Z)$. For $Z/N = \tau$, $A = Z(1 + 1/\tau) = Z\phi^2$. The intersection of straight lines and curves for the same element, always happens at exactly $N_x = \tau$, independent of assumption.

The limiting lines that define the triangle of stability, for maximum atomic number $Z = M$, are given by the following expressions:

$$\frac{A}{Z} = \frac{(5\tau - 3)A}{M} + (3 - \tau)$$
$$\frac{2A}{Z} = \frac{(9\tau - 5)A}{M} + (5 - 2\tau)$$

The correctness of these expressions is demonstrated by equating the two solutions for A/Z, i.e.

$$\frac{2(5\tau - 3)A}{M} + 2(3 - \tau) = \frac{(9\tau - 5)A}{M} + 5 - 2\tau$$

which simplifies to[3] $(A/M)(\tau - 1) = -1$ i.e. $A = M/\tau^2$. The most probable value of $M = 100$ finally confirms the validity of the assumption, originally based on the number spiral, that the number of stable elements never exceeds 100.

The empirically established periodicity of the stable nuclides and of the chemical elements has been demonstrated consistent with simple number theory and the golden mean. The inference that the properties of nuclear stability and solar abundance are functions of the same periodicity will be explored next.

8.2 Nuclear Stability

Nuclear stability is commonly defined in terms of either binding energy (BE) or binding energy per nucleon (BE/A). Periodic relationships that involve these quantities are not totally convincing, e.g. in Figure 7.10. To bring out the effect more clearly it may be necessary to introduce a binding energy term that relates to neutron excess. This is done by defining the contributions z and n of individual protons and neutrons to the total nucleus binding energy, such that

$$Zz + Nn = BE$$

and forming the difference

$$z - n = BE\left(\frac{N-Z}{NZ}\right) + \left[n\left(\frac{N}{Z}\right) - z\left(\frac{Z}{N}\right)\right]$$

[3]The limiting lines, (1) and (2), $y_{1,2} = (A - 2Z)/Z$ are given by

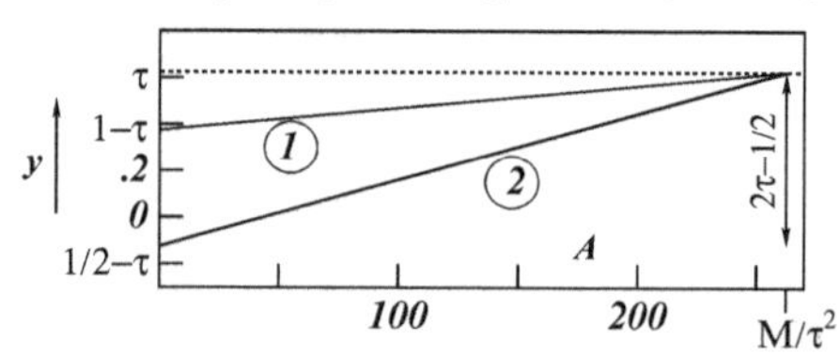

$$y_1 = \frac{(2\tau-1)\tau^2 A}{M} + 1 - \tau \qquad y_2 = \frac{(2\tau-\frac{1}{2})\tau^2 A}{M} + \frac{1}{2} - \tau$$

$$= \frac{(5\tau-3)A}{M} + 1 - \tau \qquad = \frac{(9\tau-5)A}{2M} + \frac{1}{2} - \tau$$

$$\frac{A}{Z} = \frac{(5\tau-3)A}{M} + 3 - \tau \qquad \frac{A}{Z} = \frac{(9\tau-5)A}{2M} + \frac{5}{2} - \tau$$

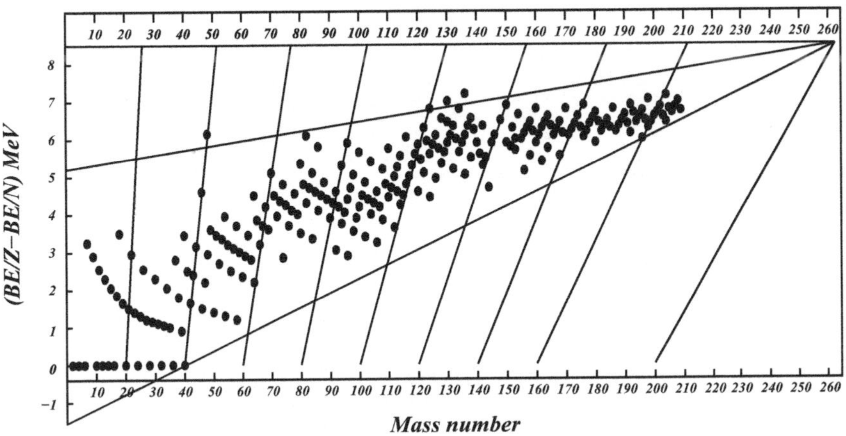

Figure 8.3: *Plot of* $BE(1/Z - 1/N)$ *vs* A.

The term in square brackets vanishes for $z/n = (Z/N)^2$. The ratio Z/N is known to vary between 1 and ϕ, which then implies either $z = n$ or $z = n\phi^2$, both of which are reasonable assumptions at the extreme ratios of the triangle of stability, as in Figure 6.1.

The quantity $BE(N-Z)/NZ$ is shown for all stable nuclides as a function of mass number in Figure 8.3. The remarkable resemblance that it shows with the plot of Figure 8.1 indicates that, since BE/N is not a constant, it must obey the periodic law of the nuclides. In fact, both BE/N and BE/Z are found to obey the periodic law, but with opposed trends which cancel in the combined plot of Figure 8.3. To emphasize the periodicity, plots of BE/N and BE/Z are shown in Figure 8.4 for the nuclides $A(\text{mod}4) \equiv 0$. Whereas the binding energy per proton varies periodically with mass number over a limited range, the binding per neutron decreases steadily, such that the difference converges to 8.5 MeV, the energy equivalent of τ. Several vital factors emerge from these plots:

1. Nuclear stability is a periodic function of mass number and neutron imbalance;

2. The returns on nuclide stabilization by an increased neutron excess diminish at high mass number and approach zero at about $A = 210$, the natural limit to nuclide stability.

3. If nucleogenesis happens in equilibrium processes, as implied by the assumed model of α-addition, nuclide abundance should obey the same periodic law as nuclear stability.

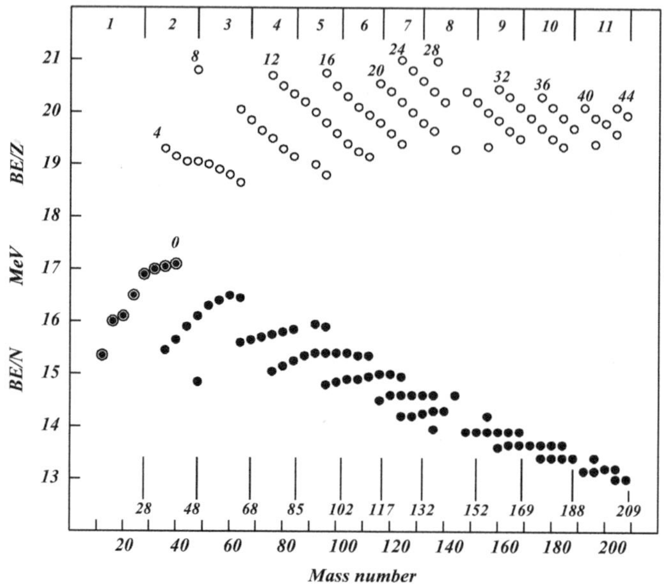

Figure 8.4: *Plots of binding energy per proton and binding energy per neutron vs mass number for nuclides with* $A = 4\,n$. *Each plot separates into 12 segments of constant neutron excess, as numbered. For* $N = Z$ *the plots coincide. Vertical hem line segments demarcate the 11 nuclidic periods.*

8.2.1 Cosmic Abundance

There is an important difference between the two rival theories of nucleogenesis which has not been investigated before. Only the equilibrium-synthesis model implies a direct relationship between nuclear stability and cosmic abundance. As pointed out before (Section 4.2.4, Nucleogenesis) there is nothing to suggest that the web of reactions assumed in non-equilibrium theories should produce nuclides in abundance related to their binding energies. Having established a periodic trend that dictates nucleus binding energies (Figure 8.4), nuclidic abundances can be tested against the same periodic law, to either confirm or rebut the equilibrium model.

Using Anders-Grevesse [61] abundance data, separate plots of solar abundance vs mass number A, for all sets of nuclides with common neutron excess, $N_e = N - Z = 0, 43$, are shown in Figure 8.5. For the sake of clarity the plots are grouped together according to $A(\text{mod}4) \equiv 0, 2, (1, 3)$. Despite agreed uncertainty around abundance data the plots for even mass number clearly show the expected periodicity inferred from Figure 8.4. Vertical hem lines

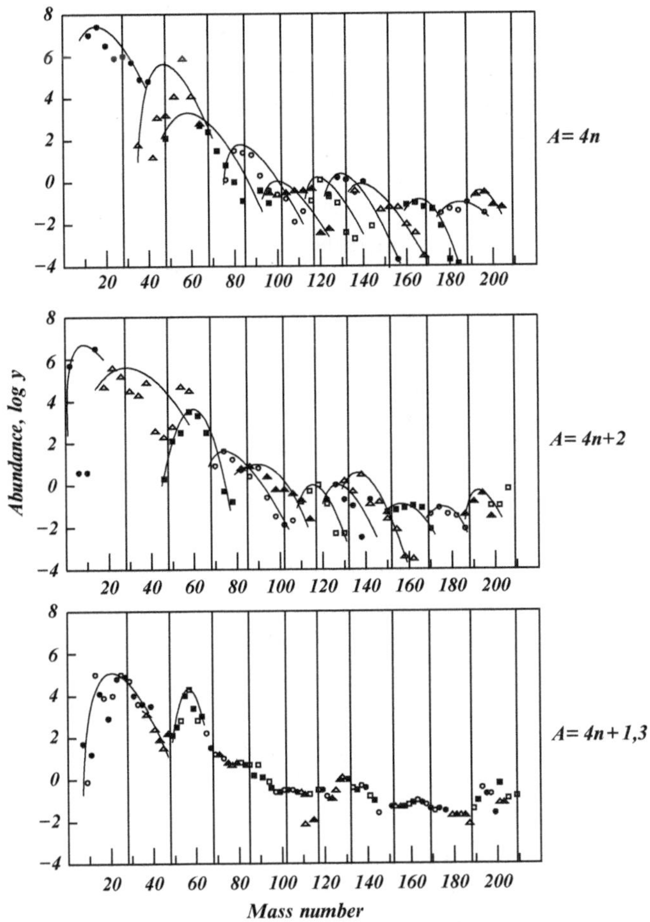

Figure 8.5: *Solar abundances [61] of stable nuclides plotted in subgroups of constant neutron excess and groups A(mod4).*

correspond to the periods of the 11 × 24 matrix defined before (6.2). The periodicity among odd mass number nuclides is not as striking, but equally convincing. It follows that the periodic law of Figure 7.7 applies to both nucleus-binding energies and nuclide abundance. The plots of Figure 8.5 identify several outliers, that may be re-examined.

It is instructive to note how the scatter plots of Figures 4.5 and 4.6 acquire new meaning on grouping the data points into 4 × 11 families of common neutron excess. The 11 families of each modular, $A(\mathrm{mod}4)$, group obey the 11×24 periodic law, first inferred from the number spiral, Figure 1.2. Empirical evidence, based on solar abundances therefore identifies neutron excess as fundamental to the periodic classification of the stable nuclides.

If nuclides derived from a variety of synthetic processes with different mechanisms, there cannot be a general rule that relates their abundance to binding energy. Current models of nucleogenesis therefore need serious revision. The bulk of stellar material must clearly originate from equilibrium processes that would yield products, after degradation, in quantities related to their relative stabilities.

8.3 Nuclear Structure

The demonstrated failure of the shell model with spin–orbit coupling to account for observed nuclear spins argues for an alternative definition of the shell structure of atomic nuclei. A vital clue on how to overcome the problem is provided by the observed periodicity of the nuclides. The mass numbers at which nuclidic periods close, show remarkable correspondence with the periodic table of the elements. The mass numbers 28, 48, 68, 85 and 102 run in parallel with the atomic numbers that close the electronic $3d(28)$, $5s(48)$, $4f(70)$, $6p(86)$ and $5f(102)$ levels. Although the neutron magic spectrum has been shown to arise in a non-central potential the total nucleon spectrum, on the other hand, is shown by its correspondence to the extranuclear electronic spectrum, to be of the central-field type. This observation suggests that protons and neutrons should be treated as equivalent within a common periodic law. In theory only charge and isospin differentiate between the two equivalent nucleon states. Although the observed decay of free neutrons by electron and neutrino emission may suggest that such leptons could contribute to nuclear spin, there is no interaction to stabilize such a distribution within the nucleus. Those electrons that may formally be considered to produce the charge balance in the nucleus, have no independent existence in that regime and disappear into the quark structure of the nucleus.

To describe the behaviour of the extranuclear electrons they are considered to exist in a central coulombic field which they cannot penetrate more deeply than to a well-defined limiting level that defines a ground-state energy for the most strongly bound electron. The energy spectrum of the nucleus is found to obey the same law as the extranuclear electrons, albeit at much higher energies. To rationalize this observation it is assumed that available energy levels for nucleons are defined by the same relationship that specifies extra-nuclear electronic energy levels, only with different boundary conditions. Nucleons are made up of quarks in *up* and *down* flavour and respective charges of $\frac{2}{3}$ and $-\frac{1}{3}$. Although quarks remain confined their polarization effects provide the strong interaction between nucleons, conveniently interpreted in terms of virtual pion exchange, e.g.

$$\left[\psi(uud) \equiv N^{+}\right] + \psi(\bar{u}d) \rightarrow \psi(ddu) \equiv N^{0}$$

which is formally equivalent to the Yukawa prescription

$$p + \left[\left(e^{-} + \bar{\nu}_{e}\right) \equiv \pi^{-}\right] \rightarrow n$$

The total cohesive energy must increase with mass number, but since quarks may be considered homogeneously distributed throughout the nucleus, the binding energy remains essentially constant at about 8.5 MeV per nucleon, the energy equivalent of τ. The packing of nucleons follows the same pattern as the stacking of extra-nuclear electrons in the central field of the nucleus. The tendency of atoms to have spherical symmetry is achieved by quenching of orbital angular momentum [11] and this leads to Hund's rule that links the number of unpaired electrons at any level to the orbital angular-momentum quantum number, l. The basic pattern of nuclear spin distribution is proposed to obey the same rule.

Nuclear interactions are on the edge of sub-quantum phenomena in the Bohm sense of Section 5.5. The neutrino, which essentially carries spin angular momentum only is probably no more than a vortex in the substratum, while electrons and quarks are the solitary waves at the source of electromagnetic charge and mass. It is futile to look for a quantum model of these entities; only their manifestations in familiar space behave quantum-mechanically.

Stability of the proton indicates interaction at the sub-coulomb level. The partons or quarks, postulated as constituents of the proton must therefore reside within the sub-quantum plenum. Their confinement is due to the coulombic field that only exists between points more than a Planck length apart. The quarks in a nucleon therefore stay together in order to avoid the Coulomb field.

8.3.1 Bound-state β^{-} Decay

Bound-state β^{-} decay is a weak decay mode in which the decay electron remains in a bound atomic state rather than being emitted into the continuum. This mode was observed for the first time [95] in the case of ^{163}Dy, which is stable as a neutral atom, but when fully ionized, it decays to ^{163}Ho with a half-life of 47 days. Another dramatic difference of this type, and one which could have serious implications for theories of nucleosynthesis and cosmic abundance, concerns the ^{187}Re – ^{187}Os pair [96].

Bound-state β^- decay of fully ionized ^{187}Re nuclei that circulate in a storage ring has been observed. The time-dependent growth of hydrogenlike ^{187}Os ions was measured and a half-life of 33 years for bare ^{187}Re could be determined, compared to 42 Gy for neutral ^{187}Re atoms. The decay electron remains in a bound state and the total decay energy is carried away by the simultaneously emitted antineutrino. The atomic charge state remains the same.

The driving force of the decay process is the enhanced attraction between the Os nucleus and the extranuclear electronic charge cloud, compared to Re. It is pointed out [96] that ^{187}Os^{76+} decays by capturing an electron from the continuum.

Bound-state decay accords with the quark model of the nucleus. The overall configuration of an atom depends on the total charge distribution. Imbalance due to ionization may hence be restored by electron flow from the core. Because of the large discrepancy in energy between core and extranuclear levels, equilibrium is strongly biased towards the core, making bound-state decay a rare event at extreme levels of ionization. However, like electron capture, it provides experimental evidence of continuity between the nucleus and extra-nuclear electrons, consistent with the observation that a common periodic law governs both regions.

8.3.2 Nuclear Spin

Sub-atomic particles with spin are confined to well-defined energy levels and sub-levels. For extranuclear electrons the resultant spin, called the multiplet, is non-zero for all partly-filled sub-levels. Normal multiplets arise from the spins of equivalent electrons when the sub-level is less than half full. Inverted multiplets arise when the sub-level is more than half full.

According to Hund's rule the multiplet of highest multiplicity for a given configuration is lowest in energy, and therefore favoured. In the case of atomic nuclei non-zero spin is observed mainly for nuclides with an odd number of nucleons. The effect is that for l-fold degenerate levels a total of only l non-zero multiplets are observed. In this case the spins of relevant nuclides are conveniently summarized by listing them in the same order as the allowed multiplets. In such a table the l successive multiplets per sub-level correspond to the observed spins of l successive odd nuclides, listed in order of increasing mass number. This is feasible since only for $A = 113$ are there more than one odd nucleus with the same mass number. An effort to understand and predict nuclear spins must involve the following steps:

1. Define the distribution that assigns nucleons, considered in order of increasing mass number to energy sub-levels defined by an appropriate periodic law.

2. Compare the implied multiplet structure with observed spins and establish the nuclear equivalent of Hund's rule.[4]

3. Consider the effect of secondary periodic laws, pertaining to individual types of sub-atomic particle, that may have an effect on the total spin carried by the nucleus.

The assignment of nuclear spins in terms of the 11×24 periodic law and central-field interaction is summarized in Figure 8.6 and Table 8.1. Since all, but a few, of the even mass number nuclei have zero spin, only the spins associated with odd-numbered nuclides are shown on the energy-level diagram as 2σ, where σ represents half-integer spins.

The energy-level assignment uses the one-electron spectroscopic notation and follows the structure of the observed periodic table of the elements as closely as possible, consistent with the 11×24 periodicity of the stable nuclides. Placement of the $6s$ level at the end of the third period could, for instance, ensure a perfect match with the table of elements (Figure 6.4), without materially affecting the argument. As in the Schrödinger problem, sub-level energies are assumed to increase in the order $s < p < d < f < g$ for any principal quantum number n. Multiplicity of the sub-levels is fixed by the degeneracy of $2l + 1$ for $l = 0 \rightarrow n - 1$. Five consecutive nuclei with an odd number of nucleons will, for instance, be involved with the stepwise occupation of the available energy levels at $l = 2$, with total spins of 1, 3, 5, 7, 9 respectively. Not more than five of these can be parallel and any spin observed in excess thereof must be of different origin. For convenience, the spin distribution of the consecutive odd nuclei is denoted by $nd(1 - 5)$ and treated as separate sites, or blocks in a set of five.

Individual nuclear spins may be assigned in terms of Hund's rule that specifies maximum parallel spins in degenerate sub-levels. By this procedure one third of the spins are predicted correctly. It is noted that in a significant

[4]This procedure differs from the spin assignment according to the shell model, which assumes that the total angular momentum ($j = l + s$) is carried by the single nucleon at the highest energy level in the nucleus [97].

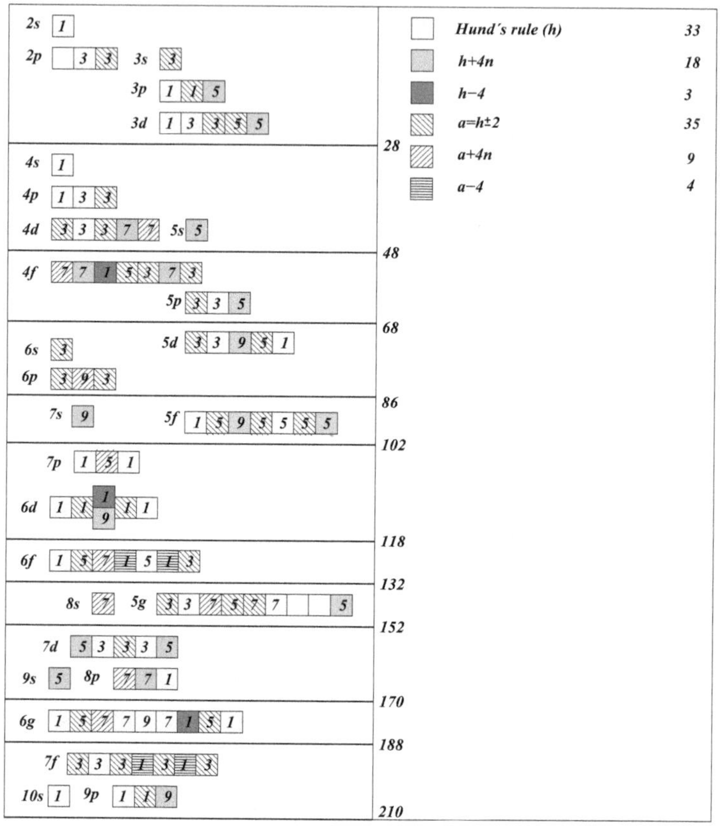

Figure 8.6: *Reconstruction of the 102 odd-nucleon nuclear spin values* $(2 \times \sigma)$ *in terms of a modified Hund's rule.*

number of cases Hund's rule underestimates the observed spin by $4n$ units, i.e. $\sigma = h + 4n$, $n = 1, 2$; e.g. 5/2 or 9/2 for predicted 1/2, etc. It will be argued that these additional spins arise in the orbital angular momentum of distorted non-spherical nuclei. In an even larger number of cases observed spins differ from Hund's rule values by two units, $h = 4n \pm 2$, $n = 0, 1, 2$. Overestimation in this case is assumed due to extra pairing of nucleon spins. Spin assignment according to these two schemes is defined in Table 8.2. It is of interest to note how the boldly outlined central part of the alternative Table 8.2(b) that describes spin pairing beyond the Hund prediction, fits into Table 8.2(a). All nuclear spins are assigned correctly on the basis of these generalizations although it is not possible to predict when each of the rules should apply.

Table 8.1: *Assignment of nuclear spins in terms of Hund's rules and nuclear distortion.*

	A	Z	N	2σ	
2s	3	2	1	1	
	7	3	4	3	
2p	9	4	5	3	
3s	11	5	6	3	
	13	6	7	1	
3p	15	7	8	1	
	17	8	9	5	*
	19	9	10	1	
	21	10	11	3	
3d	23	11	12	3	
	25	12	13	5	
	27	13	14	5	* e
4s	29	14	15	1	
	31	15	16	1	
4p	33	16	17	3	
	35	17	18	3	
	37	17	20	3	
	39	19	20	3	
4d	41	19	22	3	
	43	20	23	7	*
	45	21	24	7	*
5s	47	22	25	5	* e
	49	22	27	7	*
	51	23	28	7	*
	53	24	29	1	
4f	55	25	30	5	
	57	26	31	3	
	59	27	32	7	*
	61	28	33	3	
	63	29	34	3	
5p	65	29	36	3	
	67	30	37	5	* e
	69	31	38	3	
	71	31	40	3	

	A	Z	N	2σ	
5d	73	32	41	9	*
	75	33	42	5	
	77	34	43	1	
6s	79	35	44	3	
	81	35	46	3	
6p	83	36	47	9	*
	85	37	48	3	e
7s	87	38	49	9	*
	89	39	50	1	
	91	40	51	5	
	93	41	52	9	*
5f	95	42	53	5	
	97	42	55	5	
	99	44	55	5	
	101	44	57	5	* e
	103	45	58	1	
7p	105	46	59	5	*
	107	47	60	1	
	109	47	62	1	
	111	48	63	1	
6d	113	48	65	1	
	113	49	64	9	*
	115	50	65	1	
	117	50	67	1	e
	119	50	69	1	
	121	51	70	5	
	123	51	72	7	*
6f	125	52	73	1	
	127	53	74	5	
	129	54	75	1	
	131	54	77	3	
8s	133	55	78	7	* e
	135	56	79	3	
	137	56	81	3	

	A	Z	N	2σ	
	139	57	82	7	*
5g	141	59	82	5	
	143	60	83	7	
	145	60	85	7	
	151	63	88	5	* e
	153	63	90	5	*
	155	64	91	3	
7d	157	64	93	3	
	159	65	94	3	
	161	66	95	5	*
9s	163	66	97	5	*
	165	67	98	7	*
8p	167	68	99	7	*
	169	69	100	1	e
	171	70	101	1	
	173	70	103	5	
	175	71	104	7	*
	177	72	105	7	
6g	179	72	107	9	
	181	73	108	7	
	183	74	109	1	
	185	75	110	5	
	187	76	111	1	e
	189	76	113	3	
	191	77	114	3	
	193	77	116	3	
7f	195	78	117	1	
	197	79	118	3	
	199	80	119	1	
	201	80	121	3	
10s	203	81	122	1	
	205	81	124	1	
9p	207	82	125	1	
	209	83	126	9	* e

As a first step towards identification of the alternatives to Hund's rule it is noted that all observed spins ($1/2 \rightarrow 9/2$) can be stated in the form of the modular equivalence

$$2\sigma \equiv (h + 4n)(\mathrm{mod}10)$$

where h is the number of parallel spins predicted by Hund's rule. A simple pattern that summarizes alternatives to Hund's rule in terms of n is shown in Table 8.2(c). This table transforms (a) into (b) by modular addition.

Predicting the Spin

The prediction of individual spins requires an understanding of the mechanism whereby spins are generated by the nuclear equivalent of Hund's rule symmetry, and interact with spins contributed by asymmetry effects. There appears to be periodic factors that relate to both atomic and neutron number, superimposed on the grand periodicity that relates to mass number. One interpretation is that the overall periodicity depends on the packing of nucleons whereas additional factors arise from nuclear charge distributions in the

Table 8.2: *(a) Hund's rule spin (b) Alternative spin (c) Excess spin as multiples of n, in modular notation.*

a

s					1				
p				1	3	1			
d			1	3	5	3	1		
f		1	3	5	7	5	3	1	
g	1	3	5	7	9	7	5	3	1

b

s					3				
p				3	1	3			
d			3	1	3	1	3		
f		3	5	3	5	3	5	3	
g	3	5	3	5	7	5	3	5	3

c

				3				
			3	2	3			
		3	2	2	2	3		
	3	3	2	2	2	3	3	
3	3	2	2	2	2	2	3	3

nucleus. These are not smoothly varying functions of mass number and may have an effect on nuclear spin.

Neutron excess changes in steps of unity within each of the four modular series $A(\text{mod } 4)$ of nuclides. Breaks occur where the build-up of a series by α-particle addition reaches a β-unstable nuclide, e.g.

$$ {}^{A}_{Z}X - \beta^{+} \rightarrow {}_{Z}{}^{A}_{-1}\,Y $$

Interplay of the two factors are demonstrated in Figure 8.7. Arrows indicate breaks in the regular sequence that also indicate switching between an even and odd number of positive charges on the nucleus, i.e. between odd neutron or proton number. In the case of the $4f$ sub-level, the $4f(k)$ multiplets separate into two groups. For odd k, the nuclei are from the $A = 4n+1$ series, have an even number of charges and a-type spins. For $5f(k)$ the opposite is true. Even-k nuclei are now from the $4n - 1$ series, with even charges and a-type spin. In both cases the remaining nuclei behave differently, but not uniformly so. Evidently the spin type depends on the interplay of at least three factors: the mass-number series $(4n \pm 1)$, the even–odd charge distribution and placement within the sub-level, k. Each of these indices can be represented as either plus or minus one and the product of the three interpreted as indicating either h or a spin type. Whereas mass and charge indices are fixed, definition of the sub-level index is less obvious. An important clue is provided by the observation (Figure 8.8) that a sub-level index which oscillates between positive and negative values down the row of ns sub-levels, predicts

	A=4n+1							A=4n+1					
	9				3	2p(3)		113	o	6d(3)	1		
	13		3p(1)	1				113	e	6d(3)	9		
	17		3p(3)	5			→	117	e	6d(5)			
	21		3d(2)	3			→	121	o			5	6f(2)
	25				5	3d(4)	→	125	e			1	6f(4)
	29		4s	1				129				1	6f(6)
	33		4p(2)	3			→	133	o			7	8s
→	37	o			3	4d(1)	→	137	e	5g(2)	3		
	41				3	4d(3)	→	141	o			5	5g(4)
	45				7	4d(5)	→	145	e	5g(6)	7		
→	49	e			7	4f(1)		149					
	53				1	4f(3)	→	153	o	7d(1)	5		
	57				3	4f(5)	→	157	e			3	7d(3)
	61				5	4f(7)		161		7d(5)	5		
→	65	o	5p(2)	3			→	165	o			7	8p(1)
	69				3	5d(1)		169		8p(3)	1		
→	73	e	5d(3)	9			→	173	e			5	6g(2)
	77		5d(5)	1				177		6g(4)	7		
→	81	o			3	6p(1)	→	181	o	6g(6)	7		
	85				5	6p(3)		185				5	6g(8)
	89		5f(1)	1			→	189	e			3	7f(1)
	93		5f(3)	9			→	193	o			3	7f(3)
→	97	e			5	5f(5)		197				3	7f(5)
	101		5f(7)	5			→	201	e	7f(7)	3		
	105				5	7p(2)	→	205	o	9p(1)	1		
→	109	o	6d(1)	1				209		9p(3)	9		

	A=4n−1							A=4n−1					
	3		2s	1				107		7p(3)	1		
→	7	o	2p(2)	3			→	111	e			1	6d(2)
	11				3	3s		115				1	6d(4)
	15				1	3p(2)	⇉	119	e	6f(1)	1		
	19		3d(1)	1			→	123	o			7	6f(3)
	23				3	3d(3)		127		6f(5)	5		
	27		3d(5)	5			→	131	e			3	6f(7)
	31		4p(1)	1				135				3	5g(1)
	35				3	4p(3)	→	139	o			7	5g(3)
	39		4d(2)	3			→	143	e			7	5g(5)
→	43	e	4d(4)	7				147					
	47		5s	5			→	151	o	5g(9)	5		
→	51	o	4f(2)	7			→	155	e	7d(2)	3		
	55				5	4f(4)	→	159	o	7d(4)	3		
	59		4f(6)	7			→	163	e	9s	5		
	63				3	5p(1)		167		8p(2)	7		
→	67	e	5p(3)	5				171		6g(1)	1		
→	71	o	5d(2)	3			→	175	o			7	6g(3)
	75				3	5d(4)	→	179	e	6g(5)	9		
	79				3	6s		183		6g(7)	1		
→	83	e			9	6p(2)		187		6g(9)	1		
	87		7s	9			→	191	o	7f(2)	3		
	91				5	5f(2)	→	195	e			1	7f(4)
	95				5	5f(4)		199				1	7f(6)
	99				5	5f(6)	→	203	o	10s	1		
→	103	o	7p(1)	1			→	207	e			1	9p(2)

Figure 8.7: *Nuclides of the* $4n \pm 1$ *series assigned to the energy levels of Figure 8.6 and showing radioactive breaks.*

the correct spin in all cases, with two provisos: The $1s$ entry does not describe a composite nuclide and is included only for completeness. Secondly, ^{3}He(2s) is the only nuclide with an excess of charged nucleons and the charge index is therefore inverted from even to odd. Starting from each s and moving to the right, the index oscillates through consecutive sites on contiguous sub-levels. At the fifth level however, the sequence along sub-levels with common n, and hence the index oscillation, is interrupted by the insertion of sub-levels with higher n. The $5d$ sub-level is anomalous because of the superposition of another periodic factor as discussed below. It does not affect the other levels as $5f$ falls into step with $6p$. The sequence $7s \rightarrow 7p \rightarrow 6d \rightarrow 6f \rightarrow 8s \rightarrow 5g$ follows regular oscillation. $7d$ picks up from either $7p$ of $8s$. After level 8, the np sub-level no longer follows directly on ns, but rather on $(n+1)s$, e.g. $9s \rightarrow 8p \rightarrow 7f$.

As shown in Figure 8.8, the three-index products as defined above, predict the correct spin type (h or a) in 83 out of 102 cases.

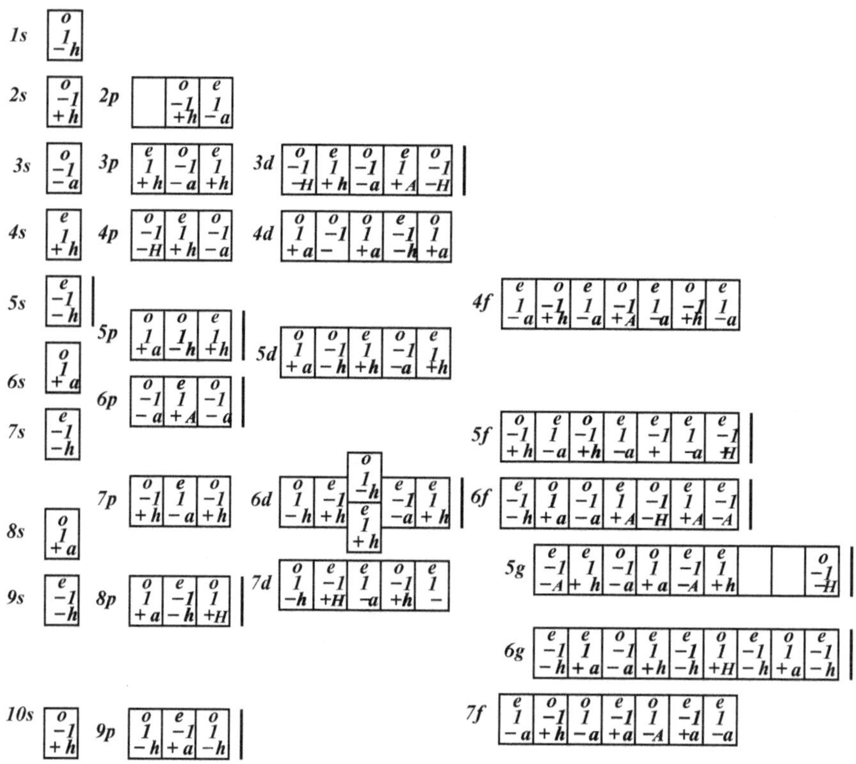

Figure 8.8: *Predicted spin arising from the interplay between three main indices and two periodic functions. The charge index is shown as either e(ven) or o(dd), the mass number index as* ± 1, *and the multiplet index by* $+$ *or* $-$.

A regular pattern which is related to the discrepancies suggests that additional factors are at work in fixing these nuclear spins. The first, most obvious, is the observation (Figure 8.6) that all, but two of the periodic groups end at a nuclide of h-type, irrespective of the three-index product. Spin types that have been adjusted in terms of this ad hoc rule are identified by H labels in Figure 8.8. The two spins, at $6p(3)$ and $6f(7)$, not adjusted, are subject to the next, stronger rule.

Most of the remaining discrepancies are observed to bear a close relationship with the relative proton–neutron distribution predicted by Figures 6.8 and 6.9. The relevant atomic numbers 4, 10, 16, 24, 32, 38, 48, 54, 56, 64 and 78 correlate well with observed discrepancies at $A = 7(2p^3)$, $19(3d^1)$, $31(4p^1)$, $53(4f^4)$, $73(5d^3)$, $85(6p^3)$, $113(6d^3)$, $125 - 131(6f^{4-7})$, $135(5g^1)$, $155(7d^2)$, $195(7f^{4-6})$. In most cases an inversion of predicted spin type occurs for these nuclides. At $4f(3)$, $6d(3)$, $6f(4,6)$, $7f(4,6)$ and $6g(7)$ unit spins, rather than the predicted five occur, without affecting spin type. It will be argued that spherical symmetry for nuclei that would normally distort, is restored by the closure of the charge-energy shell. At $5d(3)$ and $6d(3)$ the effect is an $(h+8)$-type spin at the site of maximum multiplicity, compared to a-type for all other d, f and g sub-levels. This case is closely analogous to the appearance of dynamic Jahn-Teller distortion in molecular systems.

The strong preference for spins of type $a = h - 2$ at sites of maximum multiplicity in d, f and g levels overrides the three-index rule where the latter predicts h-type spins at such sites on $4f$, $5g$ and $6f$. At $6f$ the discrepancy coincides with the periodic effect causing all spins from $6f(4)$ to $6f(7)$ to be inverted. At $6p(3)$ an a-type spin at the end of period 4 is induced. Since the closure of both charge and mass periods coincides at this point, the combined effect carries through to the previous nuclide, $6p(2)$. Related to this feedback and to the inverted sequence at $6f(4-7)$, the spin at $3d(4)$ is inverted by a related knock-on effect that stops with an a-type spin on the central site of the sub-set. The spin type of only three nuclides, those with configuration $4d(2)$, $5f(5)$ and $7d(5)$ remain undetermined.

The final unresolved issue is identification of the nuclei with an excess of $4n$ on the spins predicted by the scheme of Table 8.2. As shown in Figure 8.6, many of these occur near the end of the mass-number periods, suggesting that the excess may coincide with an arrangement which is sensitive to distortions of the Jahn-Teller [98] type. This set includes nuclides with $A = 17$, 27, 45, 47, 67, 87, 101, 133, 151, 161, 163, 209 and excludes those nuclides at the end of sub-levels, also affected by the completion of the charge-energy levels, i.e. $2p^3$, $4p^3$ and $6f^7$. The opposing charge–nucleon effect is best illustrated at $4p^3$ that closes between the charge level at $A = 33$ and the neutron level at $A = 37$. The relatively high thermal-neutron cross section of ^{35}Cl at 44 barn will be shown to relate to this coincidence. According to the seminal table of Figure 6.6 three periods end at a nuclide of odd mass number, i.e. 85, 117 and 169. This oddity has been glossed over before for the sake of uniformity, but it now appears to be important. It is precisely at these positions that the $h + 4n$ rule breaks down and applies to an immediately preceding nucleus instead. The $6f$ and $6g$ discrepancies are discussed below.

The distortion of atomic nuclei is analogous to the Jahn-Teller distortion of inorganic coordination complexes where a ligand field is of higher symmetry than the electronic charge density on a central metal ion. A typical example is the Cu(II) ion with d^9 configuration in an octahedral ligand field. To compensate for the symmetry mismatch, the coordination geometry is distorted in a mode that enhances electronic interaction with the central ion. In the case of nuclei the arrangement of nucleons in general and of neutrons and protons individually, is dictated by different periodic laws. A completely filled nucleon energy level represents a special high-symmetry arrangement. However, should this arrangement coincide with an incomplete neutron or proton level, the high-symmetry is lowered spontaneously, exactly as in the case of molecular Jahn-Teller distortion. The resulting distortion of the nucleus is responsible for the release of orbital angular momentum that would remain quenched in the undistorted high-symmetry configuration.

The reconstruction of nucleon interaction within each mass-number period, in terms of total nucleons and effective pions (neutron and proton periods), is shown in Table 8.3. The completion of neutron and/or proton levels immediately after mass levels, causing Jahn-Teller distortion, is clearly evident. Completion of proton and neutron levels at $A = 33(16)$ and $37(20)$ respectively, cause distortions in opposite sense at $A = 35(4p^3)$, that carries no extra spin. The sub-levels $6f^7$, $5d^5$ and $7p^3$ are not affected by partially completed neutron or proton levels.

At period 4 that ends at $A = 85$, the immediately preceding odd nucleus, as well as the next, have the predicted spin + 8. Near the end of period 6 ($A = 117$) extra spins appear at $A = 113$, at the end of another magic sub-level. The closure of period 9 ($A = 169$) coincides with a magic sub-level 2B(6) that runs from $A = 159 - 169$ and the four nuclides at $A = 161 - 167$ all have spins increased by four. Another straight run of five augmented spins occurs along magic period $2A(4)$ ($A = 37 - 47$) at the end of period 2. Of the remaining four augmented spins, three appear at the end of magic periods $1B(4)(A = 59)$, $1C(2)(A = 73)$ and $3A(2)\,)(A = 139)$. The increased spin at $A = 93$ can only be due to its position, near the beginning of the magic sub-level $1C = (91 - 100)$.

Most of the regularities outlined above apply more faithfully to periods 1–6 than to higher periods. The reason for this trend becomes obvious from

Figures 7.11 and 7.12. The regular trend, represented on those diagrams by a mean stability curve over the early periods, breaks down abruptly at the end of period 7 with the appearance of the lanthanides. It re-appears over the last two periods, but with different slope. Uncertainty over $6g(5)$, $6g(9)$ and $7f(7)$ may therefore just as well be treated as special cases since the sample is too small to establish an alternative pattern.

Summary

Nuclear spin has resisted prediction for a long time because a large number of observed values exceed expectation in terms of the odd number of either protons or neutrons responsible for the spin. The most successful prediction to date has been based on a scheme of spin–orbit coupling, formulated without the support of a reasonable physical model. The model has been reworked here with the benefit of the larger number of measured spins now available, and careful assignment (Table 6.3) of both proton and neutron levels to ensure the best fit. The total of 62 out of 102 spins correctly predicted, may be considered as optimal because of the almost random distribution of the remaining discrepancies. The most negative aspect of the scheme is the fact that an adequate explanation of the assumed strong spin–orbit coupling has never been provided.

The spin–orbit scheme performs better than standard Russel-Saunders coupling because it provides a mechanism to rationalize the observation of spin, almost uniformly higher than expected. Splitting of the sub-levels defined by the quantum number l into $2l$ values of $l - \frac{1}{2}$ and $2l + 2$ values of $l + \frac{1}{2}$, allows more flexibility for matching the empirically observed magic numbers. The formulation of Table 6.9 should therefore be seen as matching magic numbers rather than the Goeppert-Mayer scheme.

Tight packing of nucleons should promote quenching of orbital angular momentum to ensure maximum approach to a spherically symmetrical nucleus and affect the occupation of degenerate energy levels exactly as in the case of extranuclear electrons. The expected result is multiplet structure according to Hund's rule [11]. Being more massive than electrons, nucleons may however, be expected to behave differently in situations of high spin, because of the higher kinetic energy required for the outermost nucleon to orbit the tightly packed kernel. The effect would be exactly the opposite of spin–orbit coupling. The first and end members of a degenerate sub-level are expected to follow Hund's rule better than those at the centre where increased spin pairing must be expected. The observations embodied in Table 8.7 confirm this conjecture.

The analysis of nuclear spin is seen to depend exclusively on the behaviour of odd-numbered nucleons in the outermost nuclear shell, floating on a tightly packed spherically symmetrical kernel. Whatever the nature of the strong interaction, the spherical core may be approximated by a central field for the mobile outer layer of nucleons. The energy spectrum should therefore be closely analogous to that of the extranuclear electronic charge cloud. This argument justifies the energy-level sequence of Figure 8.6, defined to match the periodicity of the nuclides as specified in Figure 6.6. It is reassuring to find an almost 2:1 preference for Hund's rule spins at the beginning and end of degenerate levels and a similar preference for reduced spin at the central sites.

Deviations from the qualitative picture are plentiful, but rational. Apart from the grand periodicity of nucleons, protons and nucleons follow different, but related periodic laws which interact with the former and the modular composition of each nuclide. Systematic consideration of all these factors results in a unique definition of virtually all nuclear spins. Of overriding importance is the recognition that the observed excess over Hund's rule spins in about one quarter of all cases may be rationalized by spontaneous distortion of atomic nuclei.

8.3.3 Packing of Nucleons

Because of the interplay between periodicities nuclear structure must be such as to satisfy at least two different periodic functions at the same time. Nucleon periodicity is defined by Table 6.6 and Figure 8.6, neutron periodicity by Table 6.5 and proton periodicity in Table 6.4. The schematic diagram in Table 8.3 shows a simple model solution to the overall problem. Solid vertical lines represent shells of closely packed nucleons, held together by pions,

Table 8.3: *Numerical details of periodicities in atomic nuclei.*

Period	1	2	3	4	5	6	7	8	9	10	11
Total (A)	28	48	68	84	100	116	132	152	168	188	208
Nucleons	14 14	10 10	10 10	8 8	8 8	8 8	8 8	10 10	8 8	10 10	10 10
Neutrons	14	12	12	10	8	10	10	14	10	12	14
Total (A−Z)	14	26	38	48	56	66	76	90	100	112	126
N−period	16	28	38	50	56	64	76	90	100	112	126
Z	14	22	30	36	44	50	56	62	68	76	82
+−period	16	24	32	38	48	54	58	64	70	78	
N_x	0	4	8	12	12	16	20	28	32	36	44

shown as stipled lines. The neutron (π^-) excess increases with an average of four units per period as the ratio $Z/(A-Z)$ gradually decreases from 1; converging to τ. The neutron and proton levels, although obeying entirely different periodic functions, stay largely in step with nucleon periodicity. The mismatch is responsible for deviations of the observed nuclear spins from the predicted Hund's-rule values.

Comment

The model outlined here has reached the same level of development as the periodic table of the elements in the hands of chemists around 1920. At the time, it provided the main impetus for a formulation of quantum theory that provided a theoretical basis for the observed periodicity. An important advantage on that occasion was that the required eigenvalues for electronic energy and angular momentum were known from the periodic law, and the potential field from the nuclear model of the atom. In the present instance an energy spectrum is provided by Table 6.5 and the potential field relates to the annular charge distribution. An important parameter that specifies the potential field is the neutron excess, that varies in step with the proton/neutron ratio, from an initial value of 1 and coverging to τ. A simple one-partricle model locates an effective unit of negative charge within a spherical annulus and a total positive charge of N/Z on the walls. The calculation is left as an exercise to the reader.

Geometrical Model of the Nucleus

The quark model of atomic nuclei does not provide an inequivocal account of internucleon interaction. Accepting that the main interaction is of the dispersion type, it is most unlikely that a static arrangement of nucleons would occur, but a probability nucleon density that, on the average, resembles crystalline close packing is not excluded. Close-packing models of atomic nuclei, the most famous of which is that of Pauling [72], have been proposed and from the observed distribution of nuclear spins, such proposals are not totally unreasonable.

The appearance of $4n$ excess spin has been correlated convincingly with the mismatch between nucleon periodicity and magicity. At the same time the appearance of these excess spins occurs at surprisingly regular intervals of mass number. Most of them repeat at intervals defined by multiples of 8, i.e. $8n$, $n = 1, 9$, along two related number sequences that remain out of phase across the entire range, by 10 mass numbers. A scale model is shown in Table 8.4. In five special cases, repetition after four mass numbers is observed.

Table 8.4: *Periodic recurrence of excess spin along two mass-number series that stay out of phase by 10 units.*

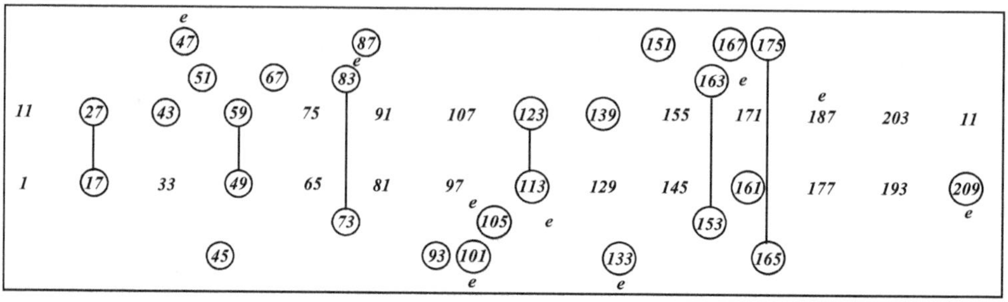

The observed regularity appears related to the packing of nucleons. The phase difference is generated at the outset and then it persists. A two-dimensional model of parallel layers that exhibit such a property is shown in Figure 1.7. A three-dimensional analogue of such a model arrangement is required to explain the observed trend. It bears no relationship to the close-packed model of the nucleus proposed by Pauling [72].

According to the two-dimensional analogy the packing of nucleons occurs in two parallel layers that spiral together from the centre outwards. To remain in phase it is necessary that nucleons should occupy progressively more space as their distance from the origin increases. The literal meaning of this requirement is that nucleons at the centre are compressed, relative to those on the outside. This compression may be compared with the size effect of Rydberg atoms that expand in rarefied environments. Compression in the nucleus is due to the strong interaction.

The diagram in Figure 1.7 is based on Fibonacci phyllotaxis, distorted to emphasize the occurrence of parallel layers. In the botanical analogue (e.g. in a sunflower head) the arrangement assures closest packing of self-similar florets that grow in size, with retention of shape, from the centre, outwards. If the packing of protons follows the same pattern in three dimensions they may be viewed as spreading along three-dimensional Fibonacci spirals (compare Figure 2.14), interspersed by layers of neutrons that spiral in the opposite sense. This phyllotaxis predicts the ratio of protons to neutrons to be the ratio of successive Fibonacci numbers, which is known to converge to the golden ratio at high mass number.

The proposed packing is related to the mysterious role of the golden ratio in all periodic relationships in and between atoms. It explains why neutron imbalances should converge to τ and why heavy nuclei become unstable, even before they reach the limiting size. The strong grip on the outer layer of nucleons, inflated in size, is no longer adequate to prevent fragments in the form of α-particles to escape from the nucleus.

8.3.4 Nuclear Size and Shape

One measure of nuclear size and shape is the thermal-neutron cross section of atomic nuclei. Up to mass number 20 these are all in the millibarn range, (1 barn $= 10^{-24}\,\mathrm{cm}^{-2}$). For mass numbers $20 < A < 139$ the cross sections are typically less than 10, with only about 15% exceeding this limit. In the range $139 < A < 176$, corresponding to the lanthanides, 52% exceed 10 b and 20% exceed 100 b. In the range $176 < A < 209$, 20% exceed 10 b and 7% exceed 100 b. The nuclides of highest cross section are $^{113}\mathrm{Cd}(2 \times 10^4\mathrm{b})$ and $^{164}\mathrm{Dy}(2 \times 10^3\mathrm{b})$. The distribution for odd-mass nuclei is highlighted in Figure 7.12 that shows the clustering of high cross sections in the region identified before in terms of anomalous stability properties. A large number of α-emitters occur in this region. Around $A = 190$ high cross section again correlates with α-instability. Since the likelihood of nuclear distortion must increase with size, it may be inferred that those nuclei proposed to acquire additional spin by non-spherical distortion, should have large neutron cross sections. This correlation is explored in Table 8.5. With few exceptions, the nuclei predicted to have high spin due to distortion, are those with high neutron cross sections. Those odd-mass nuclides with normal spin and high cross section are interpreted to have large polarizable but symmetrical nuclei, not subject to distortion by other periodic factors. Two even-mass nuclides are included in Table 8.5 to emphasize the observation that the two groups

Table 8.5: *Relatively high thermal-neutron cross sections of odd-mass nuclei with high nuclear spins.*

High cross section	*High spin*	*High cross section*	*High spin*	*High cross section*	*High spin*	*High cross section*	*High spin*	*High cross section*	*High spin*
^{17}O (.5mb)	5*	^{59}Co 21	7*	^{103}Rh 11	1	^{145}Nd 47	7	^{171}Yb 53	1
^{27}Al (.23)	5*	^{65}Cu 66	3	^{105}Pd 22	9*	^{151}Eu (10^3)	5*	^{173}Yb 16	5
^{35}Cl 44	3	^{67}Zn 7	5*	^{113}Cd,In $2x10^4$	9*	^{153}Eu 300	5*	^{175}Lu 16	7*
^{43}Ca 6	7*	^{71}Ga 5	3	^{123}Sb (.02)	7*	^{155}Gd 61	3	^{179}Hf (50)	9*
^{45}Sc 10	7*	^{73}Ge 15	9*	^{127}I 6	5	^{159}Tb 23	3	^{187}Os 200	1
^{47}Ti (1.6)	5*	^{77}Se 42	1	^{129}Xe 22	1	^{161}Dy 600	5*	^{191}Ir 660	3
^{49}Ti (2)	7*	^{83}Kr 183	9*	^{131}Xe 90		^{163}Dy 120	5*	^{193}Ir 40	3
^{50}Cr 21		^{87}Sr 16	9*	^{133}Cs (3)	7*	^{164}Dy $2x10^3$		^{197}Au 99	3
^{51}V 5	7*	^{93}Nb (1)	9*	^{137}Ba 5	3	^{165}Ho (58)	7*	^{201}Hg <60	1
^{53}Cr 18	1	^{95}Mo 13	5	^{139}La 9	7*	^{167}Er 700	7*	^{203}Tl 11	1
^{55}Mn 13	5	^{101}Ru 5	5*	^{143}Nd 330	7	^{169}Tm 8	1	^{209}Bi 11	9*

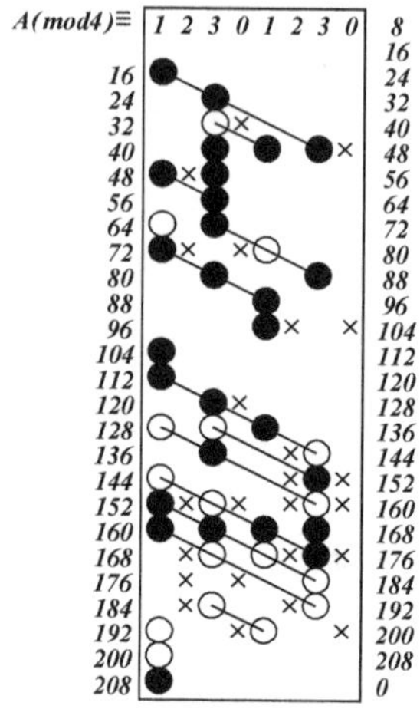

Figure 8.9: *Diagram to illustrate the periodic relationship between thermal-neutron cross section, nuclear size, polarizability and high-spin nuclides.*

of high-spin nuclides, $A = 43 - 51$ and $A = 161 - 167$ occur in regions of uniformly large polarizable nuclei. To emphasize the common periodicity of nuclear size and polarization related spin, all affected nuclei are shown in Figure 8.9, that collates the nuclides into modular groups of A(mod4).

Black circles represent polarizable odd-mass nuclei of high spin. Open circles represent large odd-mass nuclei not subject to distortion. Crosses identify even-mass nuclei of high cross section. Because of symmetry these nuclei cannot distort to a more stable state and several nuclides in this mass region are α-unstable. The surmized packing regularity of layers that are out of phase by ten units, is highlighted by thin lines connecting odd-mass nuclei. To a large extent even-mass nuclei follow the same trend. It has been pointed out before [97] that collective properties of nuclei may arise from collective motion, in which many nucleons contribute cooperatively to nuclear properties. Both vibrational and rotational collective motions may, for instance, be recognized in the nuclear magnetic moment they generate by the circulation of the charge carried by protons. Because of these effects many nuclei are permanently deformed. It has been a problem to understand how shell-model orbits, calculated from a spherical potential, result in a non-spherical nucleus. Figure 8.11 suggests a possible solution.[5] Nuclei enclosed

[5]Although an accurate description of the effect, it is not an explanation.

in the circles, centred between major magic numbers, are those most severely deformed and also those that occur far from filled neutron and proton shells.

The observed location of deformed nuclei corresponds well with the observed trends in neutron scattering cross sections, shown in Figure 7.11 and correlates with the lanthanide break in the mean stability curve of Figure 7.11. The entire argument for assignment of excess over Hund's rule spins and the analysis of nuclear packing and shape are obviously based on the same idea of cooperative interactions and nuclear distortion.

Neutron Scattering Length

Another well-documented property that relates to the size of nuclides is neutron scattering length. It provides a rough measure of the maximum distance at which a neutron which has been scattered off a nucleus still interacts with the nuclear field.

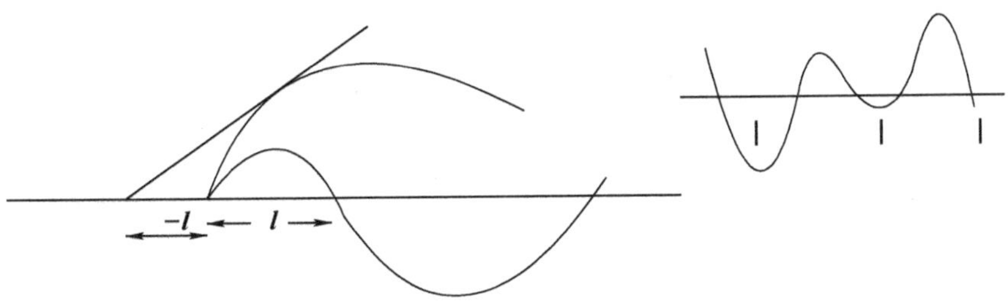

As shown in the sketch, a negative scattering length, derived by back extrapolation of a divergent scattering track, may be interpreted to indicate an impervious closed shell of nucleons. The strongest attraction on an external neutron is expected to be exerted in the vicinity of closed shells. A number of high and negative pairs of scattering length, such as ^{36}S-^{48}Ti; ^{45}Sc-^{53}Cr; ^{58}Fe-^{62}Ni; ^{81}Br-^{83}Kr; ^{111}Cd-^{113}Cd; ^{164}Dy-^{168}Yb; and ^{174}Yb-^{186}W, may be associated in this sense with the closed nucleon shells at $A = 48$, 48, 68, 85, 117, 168 and 188 respectively. Closer scrutiny of all available data suggests that in a family of nuclides with common neutron excess, scattering length varies periodically with mass number, to reach minima close to or at the completion of packing-energy shells as sketched on the right. Plots for nuclides of mass number $A = 4n + 2$ follow the proposed pattern particularly well, as shown in Figure 8.10. The plot of odd-mass nuclei has no breaks because there is little overlap between curves for different neutron excess.

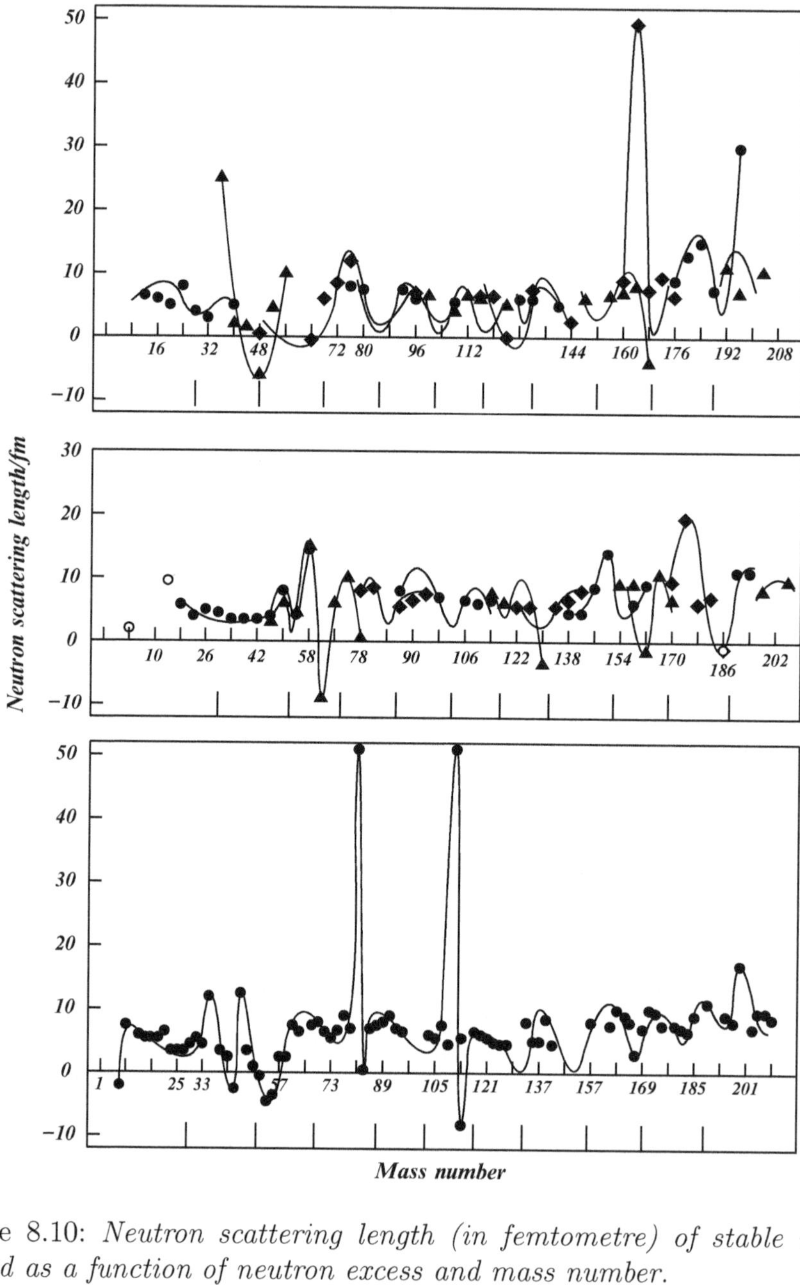

Figure 8.10: *Neutron scattering length (in femtometre) of stable nuclides, plotted as a function of neutron excess and mass number.*

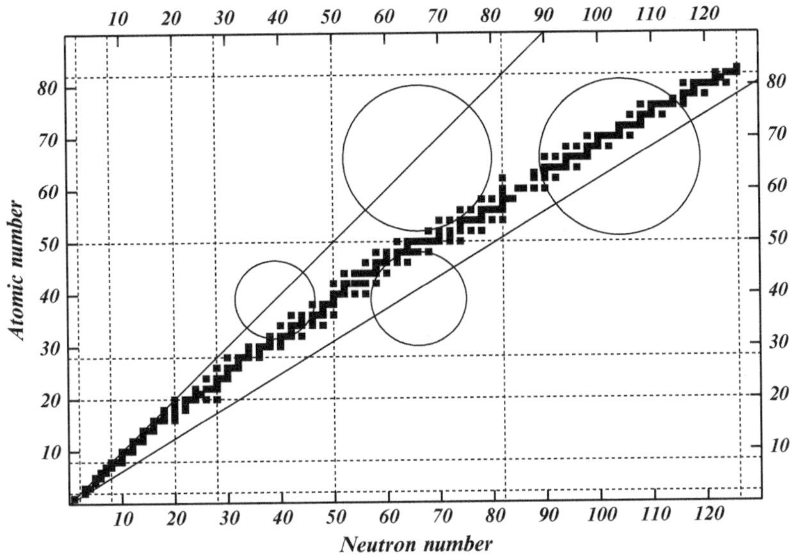

Figure 8.11: *Distribution of stable nuclides relative to major magic numbers, after [97]. Those nuclei far removed from closed shells by magic number are subject most to nuclear distortion The two straight lines represents Z/N ratios of unity and τ, respectively.*

8.3.5 Parity

The success in accounting for the observed ground-state assignment of nuclear parity has been claimed [97] as a great triumph for the shell model. Indeed, up to about $Z = 50$ the assignment of $\pi = (-1)^l$ appears almost flawless, but at higher mass numbers it goes seriously wrong. Like spin, parity is considered a property of the entire nucleus. The parity operation causes an inversion of all coordinates through the origin. Since observable properties depend quantum-mechanically on ψ^2 it follows that, if $V(\boldsymbol{r}) = V(-\boldsymbol{r})$, then $\psi^2(\boldsymbol{r}) = \psi^2(-\boldsymbol{r})$, and $\psi(-\boldsymbol{r}) = \pm\psi(\boldsymbol{r})$. The case $\psi(\boldsymbol{r}) = +\psi(\boldsymbol{r})$ is known as positive or even parity, while the case $\psi(-\boldsymbol{r}) = -\psi(\boldsymbol{r})$ is negative or odd parity. In three dimensions, the parity operation applied to spherical harmonics, $Y(l, m_l)$ gives a phase $(-1)^l$:

$$Y(l, m_l) = (-1)^l Y(l, m_l)$$

Central potentials, which depend only on the magnitude of $\boldsymbol{r}$, are thus invariant with respect to parity, and their wave functions have definite parity, odd for l odd and even for l even.

The wave function for a system of many particles is formed from the product of the wave functions of individual particles. In practice such a procedure is not possible and there is no theoretical relationship between spin and parity. Scrutiny of experimental assignments of nuclear parities (Table 8.6) suggests the existence of a periodic relationship, much simpler than the spin distribution function. Although parity assignments for odd-neutron and odd-proton nuclei have many aspects in common for low mass number, this trend breaks down for N or $Z > 32$. Unlike spin, the variation of nuclear parity can therefore not be formulated as a function of A, but rather as two separate functions on N and Z, respectively. The best candidate function is defined in table 6.5, mindful of possible deviations in the proton case.

On a first trial the energy spectrum of Table 6.5 predicts nuclear parity in complete correspondence with experimental values for odd-neutron nuclides. In order to formulate parity as $\pi = (-1)^l$, where l is some quantum number related to the energy levels of Table 6.5, the arrangement shown in Table 8.7 predicts all parity inversions correctly with l defined as the sum $l = k + n$. n may be considered as a principal quantum number and l as an auxiliary number that defines energy sub-levels and their degeneracies, shown in parentheses. Using the same quantum numbers, the parities of all odd-proton nuclides are readily interpreted in terms of the same

Table 8.6: *Experimental nuclear parity of odd nuclei and its variation with magic number.*

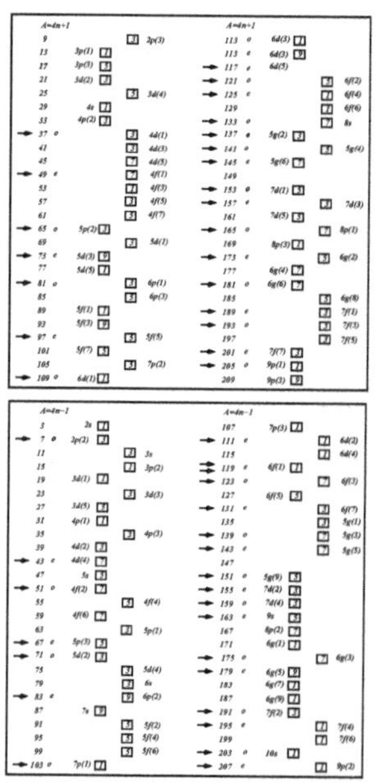

Table 8.7: *Assignment of nuclear parity in terms of the energy levels defined in Table 6.5.*

k \ n	1	2	1	1	3	2	2
1	+ 1A(2)	− 2A(2)		+ 1C(2)	+ 3A(2)		− 2C(2)
2	− 1A(4)	+ 2A(4)	− 1B(4)		− 3A(4)	+ 2B(4)	
3	+ 1A(8)	− 2A(8)		+ 1C(6)	+ 3A(8)		− 2C(6)
4			− 1B(6)				+ 2B(6)
5			+ 1B(10)	+ 1C(8)		− 2B(10)	− 2C(8)
7				+ 1C(12)			

$\pi = (-1)^l, \quad l = k + n$

	Neutron levels	m		Proton levels	m
+	1A(2)	2	+	1A(2)	2
−	1A(4), 2A(2)	8 (6)	−	1A(4), 2A(2)	8 (6)
+	1A(8), 2A(4)	20 (16)	+	1A(8), 2A(4)	20 (16)
−	2A(8), 1B(4)	32 (28)	−	2A(8), 1B(4)	32 (28)
−	1B(6)	38	−	2C(8)	40
+	1C(2), 1B(10)	50 (40)	+	1C(2) − 1B(6)	48 (42)
+	1C(6), 1C(8)	64 (56)	+	1C(6),1C(8)	64 (56)
+	1C(12),2B(4),3A(2)	82 (76,80)	+	3A(2), 2B(4) − 2C(2) + 1B(10)	82 (66,70,72)
−	2C(2),2C(6),3B(4)	94 (84,90)	−	4A(2)	
+	2B(6) − 4A(2), 2B(10)	112 (100,102)			
−	3B(6), 2C(8)	126 (118)			

scheme. The only discrepancy with observed parity occurs at $2B(4)^1$, i.e. ^{165}Ho, which has odd parity. A corresponding discrepancy at $2B(6)^3$ occurs in the neutron assignment for ^{163}Dy, with odd parity. A second discrepancy in the neutron assignment occurs at $2B(10)^5$, i.e. ^{179}Hf, with odd parity. What these nuclides have in common are high spin, long coherent scattering lengths, and high neutron cross section, with excitation, shown in parentheses in Table 8.5. There is sufficient evidence to conclude that the three exceptions occur because of real physical effects, and not because of wrong assignment.

An important feature of Table 8.6 is the way it ties together purely theoretical and purely experimental observations on a one-to-one basis. The neutron and proton shell structures that occur in the table may be accepted as complete and final. It has hence now been established beyond reasonable doubt that the energy spectra of neutrons and protons in the nucleus are not identical, but complementary.

It has already been shown in Table 6.6 how the mass numbers at the closing of the 11 nucleon shells are approximated by nuclear compositions containing neutron and proton numbers that represent completion of these respective shells. As an extension of this observation it can be readily demonstrated that several atomic numbers which represent the completion of proton levels, also describe stable nuclides with a magic number of neutrons. As a matter of interest, the resulting nuclides are ^{4}He, ^{12}C, ^{16}O, ^{32}S, ^{48}Ca, ^{60}Ni, ^{70}Ge, ^{72}Ge, ^{88}Sr, ^{90}Zr, ^{92}Mo, [^{104}Cd], ^{112}Cd, ^{132}Ba, ^{154}Gd, ^{162}Dy, ^{168}Er, ^{172}Yb, ^{188}Os, ^{208}Pb. Reference to Tables 8.1 and 8.5 shows that in a remarkable number of cases, odd nuclei, contiguous to these closed-shell nuclides have spins in excess of Hund's rule values, and/or exceptionally high neutron cross sections, characteristic of deformed nuclei. The remainder of the nuclides with excess spin can, in fact be shown to include a magic number of either protons or neutrons, or lies one mass unit away from an even mass nuclide with this property.

Spin assignment by the spin–orbit shell model makes no distinction between normal and high spins and predicts the latter rather poorly. The regular appearance of high spins for off closed-shell nuclides therefore has no special meaning. In the scheme presented here, this feature identifies a real physical basis of the assignment.

8.3.6 α-Instability

Radioactivity in the form of β-type decay or electron capture is a well-defined periodic property of atomic nuclei. Any nuclide with a proton/neutron ratio that falls outside the triangle of stability decays by β^- emission, if on the low ratio side, and by positron emission or electron capture on the high-ratio side. According to their values of neutron excess, those naturally occurring β-emitters, not generated by cosmic-ray activity (4.1) do not belong with any of the four modular series summarized in Table 7.1 and hence cannot be products of the postulated equilibrium process of Sections 6.1 and 7.2. These nuclides must be of more recent origin in stellar synthesis as proposed by Hoyle [67].

A number of nuclides, predicted stable on the proposed criterion, decay by α-emission. At first glance the distribution of these nuclides appears to be almost random. Viewed against the distribution of distorted nuclei and large cross sections, a more regular pattern emerges. An obvious common

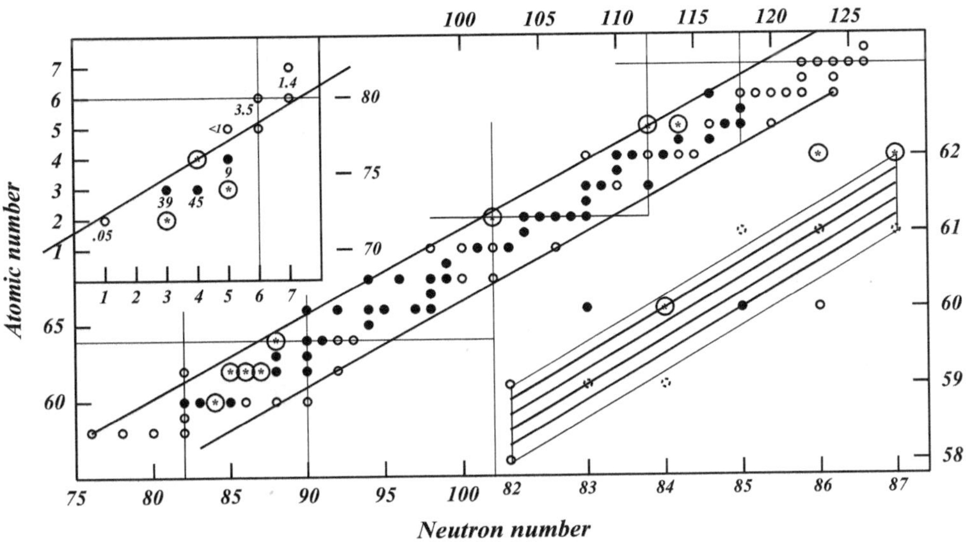

Figure 8.12: *Diagram to show the coordinates of α-unstable nuclides, marked by asterisks enclosed in larger circles, with respect to filled proton and neutron energy shells.*

property of the α-emitters is their proximity to a straight line of slope 3/5 that passes through the coordinates of several stable nuclides plotted on a Z/N diagram, as shown in Figure 8.12. All nuclides shown in the plot, lie on a family of straight lines with the same slope, only two of which have been drawn in. The affected nuclides are on the line closest to the stability limit. Black circles on the diagram identify nuclides of exceptionally high neutron scattering cross section and expected to be distorted, following Figure 8.11.

An additional common feature is the proximity of both proton and neutron energy-level lines. It means that these nuclides have configutations close to completed energy levels for both protons and neutrons. It has been argued that symmetrical conformations close to filled levels are likely to distort, in order to gain stability. However, when both proton and neutron levels are involved, distorting effects are likely to work in opposite directions. It may, in fact, prevent the asymmetric distortion that could have enhanced the packing energy. Lacking such stabilization, these nuclides may disintegrate radioactively.

The same reasoning explains, for the first time, why ^{8}Be does not exist. The corner inset on Figure 8.12 shows the environment of this nuclide on a

$Z - N$ plot, relative to the proton and neutron energy levels. Some other (synthetic) α-emitters are also shown. Compared to the heavy α-emitters, the nuclides shown here are much lighter and the stated neutron scattering cross sections (mbar) may appear negligible in comparison. On a local scale however, the stable nuclei amongst the α-emitters have relatively large cross sections. The situation is fully analogous to that in the lanthanide region where the radioactivity is also due to an inability of nuclei to lower their energy by distortion.

Chapter 9

The Grand Pattern

9.1 The Golden Ratio

This investigation has touched on many aspects, starting from the supposition that reality is a reflection of the number system. Towards the end of the quest this is the only conjecture to survive. No other preconceived idea could be upheld. Not a single result was correctly anticipated, but always comprehensible in a new guise. In retrospect, the central conclusion that supports the entire edifice is the magic of the golden ratio. As soon as its importance appears to be fully understood, it offers a new surprise at a deeper level. The use of Fibonacci numbers to order stable nuclides is the most recent surprise. Although it has been demonstrated that the triangle of stability is defined by a sequence of Fibonacci fractions, the deeper significance was not fully appreciated. It can now be shown that any consecutive triplet of Fibonacci numbers from the sequence

$$1, 1, 2, 3, 5, 8, 13, 21, \text{ etc.}$$

gives a valid classification of the nucleonic composition of all stable nuclides.

The entire analysis, up to this point has been conducted in terms of the first triplet that maps the nucleon numbers $\{Z, N, A\} \mapsto \{1, 1, 2\}$. In practice this means that by adding the Fibonacci numbers, in sequence, to any set $\{Z, N, A\}$ that represents a stable nuclide, a feasible composition of the next possible stable nuclide is generated. In order to characterize the α-emitters it was necessary to use the mapping $\{Z, N, A\} \mapsto \{3, 5, 8\}$. In an almost trivial exercise it is easily demonstrated that the mapping $\{Z, N, A\} \mapsto \{2, 3, 5\}$ provides another valid description. The difference between the various mappings lies in the slope of the Z/N relationship, that changes from $1/1$, $2/3$, $3/5$, etc., in a sequence that approaches the golden ratio. The closer it gets to the golden ratio, the more rigorous is the classification of nuclides. Examples of modular classifications in terms of the triplets $\{3, 5, 8\}$ and $\{5, 8, 13\}$ are shown in Figure 9.1.

Along with increased rigour, the visual impact of the higher classifications deteriorates very rapidly. Compared to the modular classification $A(\text{mod } 4)$ shown in Figures 6.2, 7.1 and 7.3, the modular families $A(\text{mod } 8)$ and $A(\text{mod } 13)$ appear on straight lines that are spaced much more closely and virtually unrecognizable to the unaided eye. The situation is exactly equivalent to the recognition of spirals in Figures 2.13 and 2.14. The more fundamental single

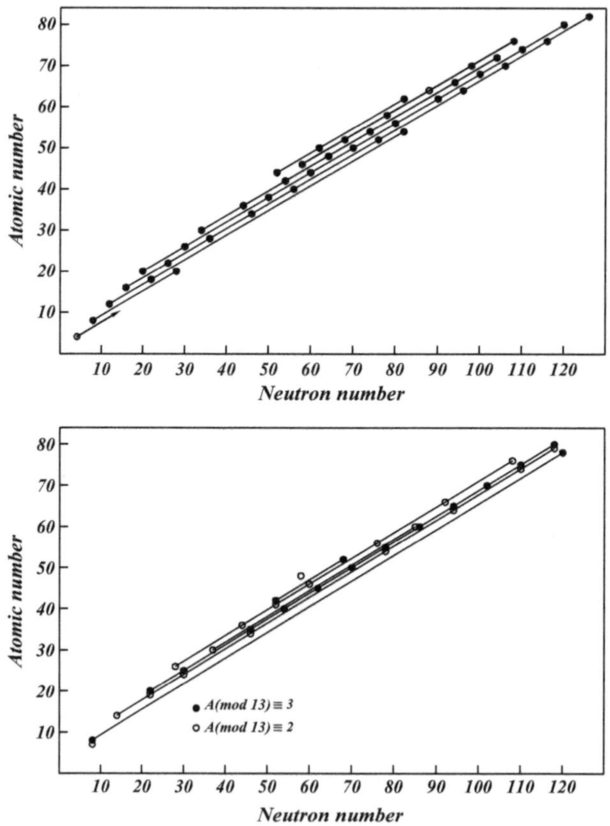

Figure 9.1: *The top diagram shows a plot of all stable nuclides with* $A(\textit{mod } 8) \equiv 0$, *along straight lines of slope 3/5. In the bottom part the nuclides with* $A(\textit{mod } 13) \equiv 2, 3$ *are shown in the same style. Interspersed, but not shown, among the nuclides shown on the diagrams, are all the other modular groups of each decomposition.*

spiral of Figure 2.13, although it navigates through the same points as those of Figure 2.14, is impossible to recognize without guidance.

The important conclusion is that the distribution of protons and neutrons in atomic nuclei, although superficially governed by sets of Fibonacci numbers, is in fact fixed by the golden ratio. The implications of this observation are far reaching.

9.2 Nuclear Structure

Of primary importance are the consequences for the geometrical distribution of nuclear matter. What is directly implied is that, although the actual proton:neutron ratio in a given nucleus may be far from τ, the major interaction arises in extended regions with the exact τ distribution. As in Figure 8.12 all modular groups can be interpreted to occur on lines with slope τ and non-zero intercepts. One interpretation is that the proton excess is concentrated in a cluster, leaving the correct ratio that maximizes proton–neutron interaction elsewhere. The same reasoning was used before to argue for a three-dimensional spiral distribution of nucleons.

The crude picture that may be drawn from the previous arguments is of a proton-enriched cluster, associated with protons and neutrons in golden ratio, spiralling out in opposite directions to constitute a three-dimensional analogue of botanical phyllotaxis. Strong interaction is mediated by the exchange of virtual pions that also serve to neutralize coulombic repulsion. In a formal sense each neutron corresponds to a proton, which is associated with negative charge and compensating angular momentum, which on disintegration occurs as neutrinos. There is no suggestion that leptons or pions have independent existence in an atomic nucleus. In the same way that sodium metal loses its well-known chemical properties when it enters into chemical combination with chlorine to form salt, the leptons when swallowed into the nucleus disappears into newly created entities, called neutrons. Even individual nucleons lose their identity when they combine to establish an atomic nucleus. However, in the same way that ionic positions in a sodium chloride crystal provide a useful working model for simulating the physical properties of the crystal, a formal geometrical model has heuristic value.

In this sense, the statement that protons and neutrons spiral in opposite directions does not imply the existence of separate spirals. There is only one grand spiral of the type shown in Figure 2.13. Apparent secondary spirals are like the perceived spiral arms, $F_5 + F_6$, in Figure 2.14. The total construct is therefore more like the space-filling chiral helicoid shown in Figure 5.8. The helicoidal structure determines the chirality of matter, as opposed to anti-matter, its enantiomeric form. Nuclear parity may be related to the dominant macroscopic Fibonacci spirals associated with protons and neutrons respectively. This interpretation is in line with the ease whereby nuclei may switch parity at the completion of nuclear energy levels.[1]

9.3 The Five Domains

The essence of the golden ratio is hidden somewhere in fivefold symmetry, and manifests itself in the details of atomic structures, in a way that can only be guessed at. It is no accident that prominent periodic relationships are encountered at Z/N ratios of $\simeq$ 1.0, 0.6 and 0.2, at intervals of 0.4 apart. In the extended closed interval [1.0,−1.0], these limiting values define five equal domains, but experience shows (Figure 6.12 and Eq. 6.2) that actual atomic periodicities arise in the closed interval [1.02,−1.02], shown in Figure 9.2. The symmetrical two-dimensional reconstruction contains all the

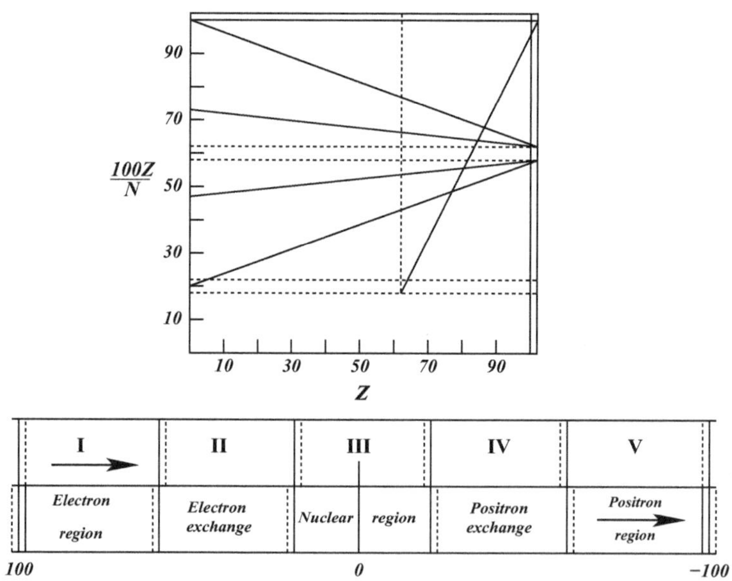

Figure 9.2: *Elements of symmetry that relate to the golden ratio and dictate the arrangement of sub-atomic matter.*

[1]The spirals of Fibonacci phyllotaxis always involve two adjacent Fibonacci numbers, F_n and F_{n+1}. For n even, the dominant spiral has opposite parity from $n+1$. For n odd, the situation reverses. The switch happens at the completion of nuclear energy levels, where parity is reversed, until the next level is reached.

information about observed periodic relationships and is obtained by placing the quantity $100Z/N$ on the same scale as atomic number and reflecting the familiar triangle of stability in the line $Z/N = \tau$. The significance of $Z/N = 0.2 \pm .018$ immediately becomes apparent. It represents the limit between nuclear and extranuclear regions. The region $0.2 \rightarrow 0.582$ may be interpreted as both positron distribution in anti-atoms, or as the region of electron capture by the nucleus. The τ-equivalent in atomic number is 62. A vertical line through this point intersects the triangle of stability at the onset of the lanthanide region of nuclear distortion, where the average nuclear-stability trend breaks down (compare Figures 7.11 and 9.1) and cuts through the cluster of α-emitters. Remarkably, the line between (102,102) and (62,18) coincides with the limiting hem line of nucleon period 11, Figure 6.7. It separates stable nuclides from the trans-bismuth α-emitters.

The emphasis on Z, rather than N or A, happened for historical reasons. There is little doubt that equivalent analyses of stable *nelements* (elements defined in terms of neutron number) and *melements* (defined in terms of mass number) would produce the same type of golden symmetry. As an example, the nelement equivalent of Figure 8.2 is shown in Figure 9.3. It demonstrates

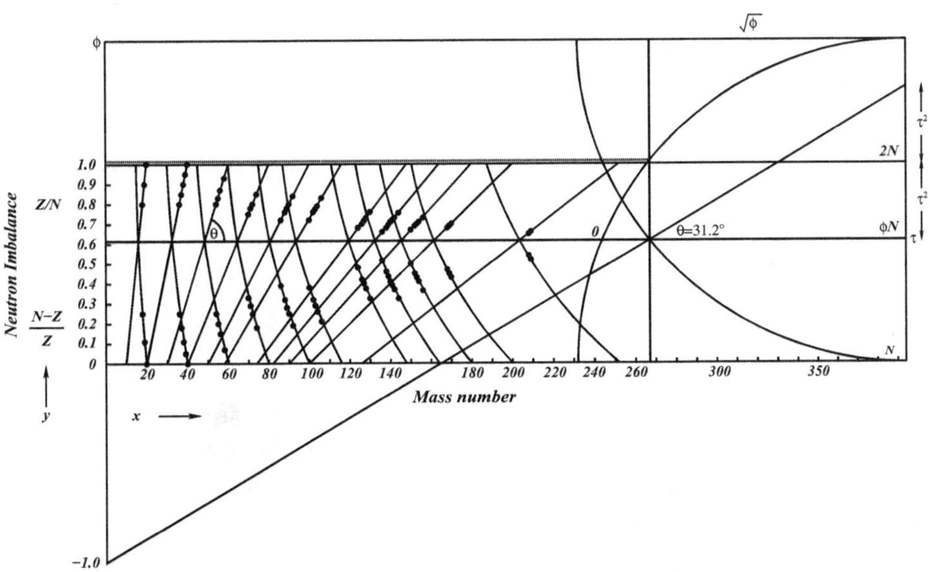

Figure 9.3: *Nuclides of selected nelements with* $N = 10, 20, 30, 40, 50, 58, 74, 82, 90, 100$ *and 126, plotted as functions of mass number and neutron imbalance, both as* Z/N *and* $(N-Z)/Z$*. Many important aspects of nuclide symmetry and periodicity, shown on the diagram, are explained in the text.*

the complementary nature of elements and nelements. Where the isotopes of the elements plot as straight lines, the isotones of the nelements occur along smooth curves, and vice versa. The linear relationship between isotones in the Z/N region defines straight lines that meet at neutron imbalance coordinate, $y = -1.0$. They intersect the lines $y = 1$ and $y = \tau$ at mass number coordinates, x of $2N$ and ϕN, respectively. The slopes of these lines are hence given by

$$\tan\theta = \frac{(1-\tau)s}{(2-\phi)N} = \frac{s}{N}$$

From Figure 9.2 the value of the scale factor is assumed $s = 100$ and this factor has been used in preparation of the diagram to ensure the same measure being used on both axes. The predicted angles correspond to their measured values, e.g.

$$\theta(40) = \tan^{-1}\left(\frac{100}{40}\right) = 68^\circ$$

With the axes properly scaled it was further noticed that the isotones in the $(N-Z)/Z$ region actually lie on circles, centred at points that approach $(394, \phi)$ as $N \to 165$. This observation has the important implication that the limiting atomic number for natural nuclides, i.e. the value to which the triangle of stability converges, is indeed 102 and not 100. This conclusion follows from the observation that the circle of radius ϕ, centred at $(394, \phi)$, intersects the line $Z/N = \tau$ at the mass number coordinate 267, and noting that $102\phi = 165$ and $102\phi^2 = 267$, the limiting values for N and A respectively. It needs to be re-iterated that the convergence to τ and identification of the point of convergence, have both now been established on the basis of internal evidence, without assumption. A second circle, centred at $(394, 0)$ and radius ϕ intersects the limiting line $x = 267$ at coordinates (267, 1.02). The isotone circles are centred at points along this second circle, starting from the point marked 0 and approaching the limit defined before. This result is equally important and confirms the structure assumed in Figure 9.2. The periodicity of the elements has already been worked out on this basis. The straight line (shown extended to $y = -1$) and circle, through the hypothetical limiting nuclide, $N = 165$ are shown, intersecting at $(267, \tau)$, with slope $\theta = \tan^{-1}\left(\frac{100}{165}\right) = 31.2^\circ$.

The periodicities of nelements and melements must clearly follow the same pattern, i.e. on the basis of 165 and 267 members, respectively. Details of the nucleon periodicity by mass number are summarized for the stable nuclides in Table 8.6 and the preliminary periodic table for neutrons and protons in the nucleus is in Table 8.7.

9.3.1 A Golden Diagram

Despite the superficial resemblance between isotope and isotone plots, there are significant differences, and both of them bristle with golden symmetry. Re-examination of Figure 8.2 shows that the isotopes of each element also occur on circular segments in the Z/N region. In fact, the generating circles are readily demonstrated to be centred at regular points along another circular segment, as shown in Figure 9.4. Isotope curves for more elements have been added to the modified diagram by noting that each such curve intersects the lines at imbalance coordinates, $y = 1.0, \tau$ and 0.5, at $x = 2Z, \phi^2 Z$ and 3Z respectively. As shown on the inset, the central points define a smooth function on Z. The curves for the limiting cases, $Z = 102$ and 0 are also shown. These circles are centred at coordinates of (330[7-6τ], 2.38[3-τ]) and (165[102ϕ], 0.75) respectively. The curve for $Z = 102$ intersects, as expected at $A = 204$ with $y = 1.0$ and at $A = 267$ with $y = \tau$. It is tangent to $y = 0.5$ and is the limiting curve beyond which no stable nuclides can occur. The generating circle that defines the centring of the isotope curves, is itself centred at coordinates $(447, -0.38[\tau - 1])$. The value of this x coordinate derives

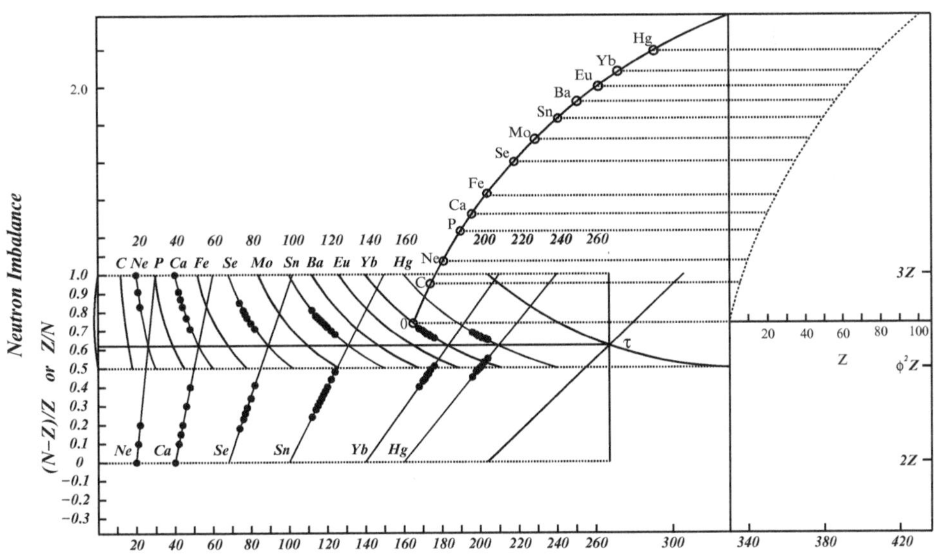

Figure 9.4: *Figure 8.2, modified to show the construction of circular segments that connect the isotopes of a given element in a plot of Z/N vs A.*

from the identity $2(2-\tau)\phi = 4.47$. On the mass number scale the large rectangle in the diagram therefore measures $(3-\tau)+(1-\tau) = 276 \times 447$, which is a golden rectangle.

The lines that connect isotopes in the lower half of the imbalance plot have slopes of $100/Z$ and converge at $(x, y) = (0, -2)$. The Sn line, for instance, is inclined at $\theta = \tan^{-1} 2 = 63.4°$. The $Z = 102$ limiting line is inclined at $\theta = 44.4°$.

Isotopes of given Z and isotones with $N = Z$ map on to lines with the same slope, but displaced by $\Delta x = Z$. The respective x-coordinates at $y = 1.0, \tau$ and 0 are marked on the right in Figures 9.3 and 9.4. These two diagrams illustrate that periodic classifications in terms of Z and N are equally fundamental. The reason why only the former has gained relative prominence is because of its direct link to extranuclear electron distributions that control chemical properties. The fact is that a systematic search for observable periodic trends amongst nelements or melements has never been considered. One topical issue that could benefit by redefinition as a periodic function of mass number, rather than atomic number, is superconductivity, to be addressed, as a practical application, in the penultimate chapter of this monograph.

9.4 Matter Transformation

The five domains of Figure 9.2 correlate with five possible states of atomic aggregation, or perhaps with three states of matter and three states of antimatter, with one region in common as shown in Figure 9.5. This interpretation is consistent with previous conclusions and even more compelling when the regions are mapped on the Möbius strip of Figure 6.11. This operation links regions I and V through an involution and produces an additional five antimatter regions. There are in fact two Möbius twists in this construction which is closed in the direction of both atomic number $(0 \mapsto 102)$ and proton:neutron ratio $(1 \mapsto -1)$. This orthogonality implies periodicity as a func-

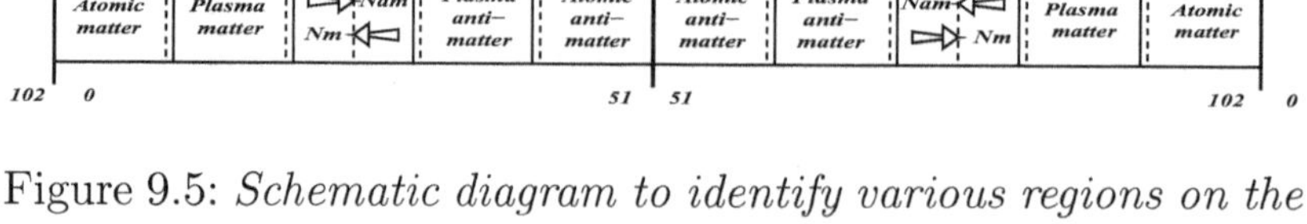

Figure 9.5: *Schematic diagram to identify various regions on the double cover of the Möbius band, obtained by identifying* $Z = (0, 102)$ *and* $Z = (51, 51)$. *Nm = nuclear matter.*

tion of number, as well as the continuity of the different forms of matter, and can only be understood on a closed Möbius surface, which is a projective plane [11].

To map the surface of a Möbius band on a single two-dimensional euclidean strip, the double cover is opened at a single point only and split along the interface between individual elements of the double cover. The double cover, opened up, and after unwinding, is shown in Figure 9.5. It shows the surface, stretched between identical points with a common coordinate of unity. The left- and right-hand halves of the diagram lie on opposite sides of the interface in the intact Möbius band. Discontinuity between matter and antimatter states seems to occur at the two points labled 0. These positions are on opposite sides of the interface. The regions of nuclear matter may hence be joined by cutting through the interface and continuing on the reverse side as indicated by the arrows. The same applies to regions of nuclear antimatter. The combined operation that amounts to the interpenetration of two intact surfaces is impossible in three-dimensional space, but feasible in four. This effect is illustrated by the impossible construction of either a Klein bottle or a Roman surface as models of the projective plane in three dimensions.

The interpretation of Figure 6.12 in terms of Figure 9.5 is straight forward. Recognizing the region between ± 0.22 as characteristic of nuclear matter explains why the intersection of periodic hem lines at $Z/N = 0.18$ and 0.22 yields details of nucleon distribution and why the spectrum at $Z/N = 0$ is a purely neutron spectrum. The properties of atomic matter are encoded in the field demarcated by $Z/N = 0.62$ and -0.62. Two periodic tables are generated by intersecting hem lines at these coordinates, being the observed and purely theoretical tables, respectively. The inverted counterparts of these tables are generated near $Z/N = 1$, where nuclear synthesis by an equilibrium process has been assumed to occur. The matching thermodynamic conditions are that of high temperature and pressure. Atomic nuclear levels are revealed at $Z/N = 0$, which on the cosmic scale are encountered in neutron stars. By analogy with theories on the formation of neutron stars [52] it may be argued that conditions of high pressure and low temperature prevail in this region. The regions marked, plasma, are intermediate between atomic and nuclear regions. This is also the region of lepton exchange between the nucleus and the extranuclear charge cloud. In one direction it embodies the process of electron capture and in the opposite direction, bound-state β^--decay [Section 8.3.1]. On the cosmic scale it requires low pressure and high temperature.

9.4.1 The Cosmic Phase Diagram

The conjectural conclusions reached here point at a new interpretation of Figure 9.2 as a cosmic phase diagram. It represents a torus which is closed with a twist, therefore a projective plane embedded in four dimensions. By the theory of periodic functions (Section 2.7.4) it may hence be inferred that the distribution of matter in the cosmos is self-similar with the periodicity of atomic matter. In principle, the diagram therefore maps the topology of space-time and regulates transformation between different states of matter that occur with periodic regularity.

Convergence to the golden ratio probably relates to the geometry of the cosmos, as a balance between general curvature and the overall size of the universe, as measured by Hubble's constant. The observation [2] that spiral galaxies display the structural golden ratio in the same way as biological growths, implies aspects of self-similarity and evidence of the general curvature of space.

Although cosmic periodicity is not evident in astronomical observation, there is evidence out there of neutron stars, plasma activity, extreme conditions in black holes and quasars, empty space and moderate regions, to fit the phase diagram. The apparent lack of observable periodicity between these isolated objects is not too surprising. Self-similarity only becomes apparent with appropriate scaling. As an example, the symmetry between electrons in an atom, planets in a solar system and stars in a spiral galaxy, is easily overlooked because of differences in scale.

At the same time there are some observations that sit uneasily with standard cosmography, but find ready explanation in terms of the ideas developed here. It was suggested before [5] that inherent red shifts associated with quasar light [99] find a simple explanation in terms of variable electronic configurations of atoms in strong gravitational fields. A related phenomenon that raises serious controversy is the observation of anomalous Fraunhofer lines in quasar light [100]. According to some theorists this observation implies that the fine-structure constant varies with time. However, an alternative explanation is now also available if the electronic configuration of atoms, and hence their electronic spectra, should be a function of their thermodynamic environment. The observed deviations from expected absorption frequencies therefore do not relate to time, but to differences in the local curvature of space–time.

The discovery of isotope periodicity [5], that later revealed the hidden symmetry of periodic classification [101], was made possible only by exploring the relationship between atomic structure and the prime-number cross [102]. The observed isotope periodicity has been shown to be, not only consistent with the known electronic configuration of atoms and the energy spectra of nuclei, but also a reliable predictor of unexpected relationships between atoms of matter and anti-matter. The most significant aspect of this discovery is that elaborate details of atomic and nuclear structure, strictly results of quantum theory before, have been derived here without the use of higher mathematics. The conclusion that an intimate link exists between natural numbers and the physical world is almost inescapable.

Chapter 10

The Golden Excess

10.1 Introduction

It has been demonstrated that nuclear properties such as spin, parity and stability are functions of more than atomic number, characterizing these as properties of nuclides rather than elements. Other properties, usually thought of as functions of atomic number only, may well be of the same type. One such property is superconductivity, first observed as a property of certain metals at low temperature. Periodic variation of superconductivity with atomic number has been analyzed by many authors [103, 105, 106]. A number of trends appear to be well established. Pure elemental superconductivity is restricted to metals, but many semi-metals and non-metals become superconducting under special conditions such as cooling at high pressure or after structural modification. Whereas the critical temperature for non-metals increases with applied pressure, the inverse is often observed for superconducting metals. Superconductivity cannot be induced in ferromagnetic or anti-ferromagnetic metals.

Most of the superconducting metals with critical temperatures $T_c > 1$K occur as either single isotopes (Al, V, Nb, Tc, La and Ta), or twin isotopes, (Ga, In, Re and Tl). All of these are odd mass-number isotopes. More of the elements that have been claimed to be superconductors at a lower level are also of this type: P, As, Y, Cs, Lu; Br, Sb, Ir. In other cases isotope enrichment towards lower atomic mass results in increased T_c. This latter observation features prominently in the BCS theory of superconductivity [107] that invokes the coupling of electrons to lattice phonons. Statistical analysis [108] of the correlation between normal-state properties and superconductivity however, casts serious doubt on the validity of the conventional BCS theory. From another point of view these isotope effects may be interpreted to define metallic superconductivity as a property of nuclides rather than of elements. To better understand the phenomenon it needs to be characterized in terms of nuclidic rather than elemental periodicity.

10.2 Nuclide Periodicity

The most stable arrangement of atomic nuclei is predicted to occur as the ratio of protons to neutrons approaches the golden ratio, $Z/N \rightarrow \tau = 0.6180\ldots$. The lightest nuclei have $Z/N = 1$ and this ratio decreases steadily with increasing Z, approaching τ. However, not a single nucleus ever achieves this ratio, which is projected to occur at $Z = 102$, since all nuclides with $Z > 83$ are unstable, albeit for different reasons. Still, the nucleon distribution $Z/N = \tau$ plays a prominent role in the stabilization of all nuclei by condensation of the proton excess into a positive surface layer. The proton excess has a unique value for each nuclide, i.e.

$$x = Z - \tau N$$

A plot of x/Z vs A, the mass number, is shown in Figure 10.1 and reveals the periodicity of nuclides as a function of A. The boundary profile that defines the region of stability is readily drawn to divide the 264 stable nuclides into 11 groups of 24. The nuclear properties mentioned above echo this same periodicity. When Figure 10.1 is redrawn in Figure 10.2 as a function of atomic number rather than mass number, the periodic profile is preserved and the 11 hemlines slant away from the vertical to intersect the Z-axis at

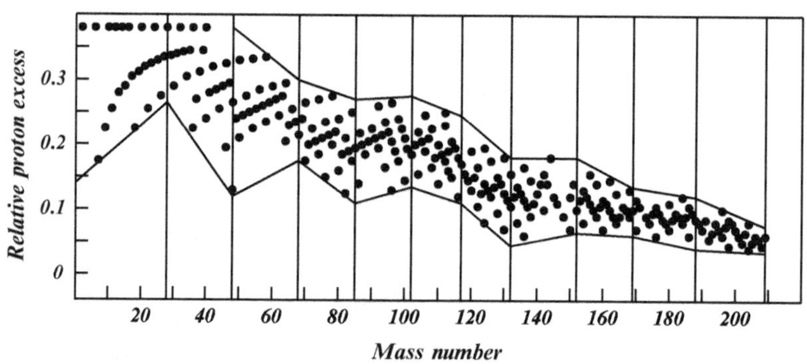

Figure 10.1: *Mass-number periodicity of relative proton excess based on golden ratio packing of nucleons.*

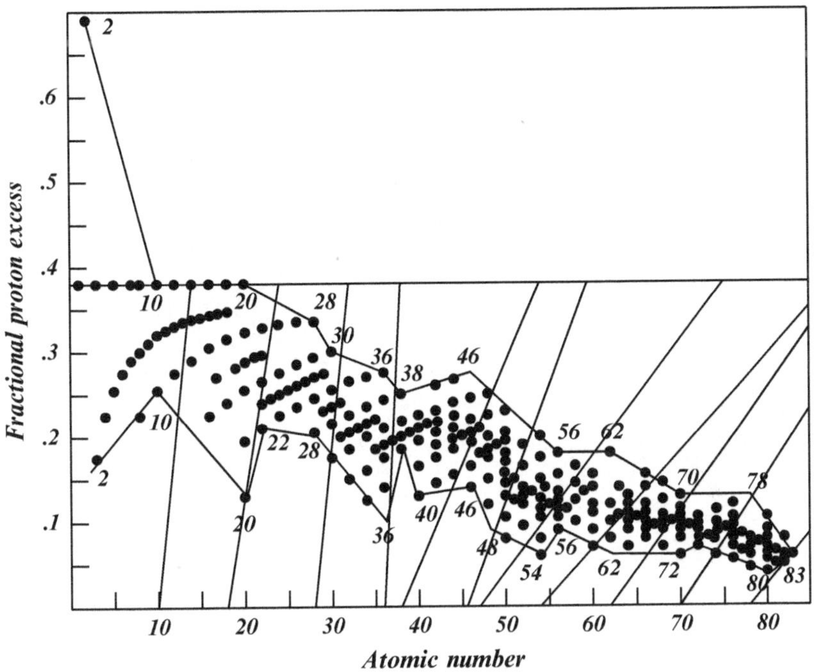

Figure 10.2: *Atomic-number periodicity derived from golden-ratio packing of nucleons.*

points that define the periodic closure of eleven prominent electronic energy shells, $Z = 10, 18, 28, 36, 38, 46, 48, 54, 62, 70, 78$ (compare Figure 6.4).

The periodic table of the elements is thereby defined as a subset of nuclidic periodicity. Characterization of superconductivity in terms of the more fundamental classification and the implied structure of atomic nuclei will be examined here.

10.2.1 Superconducting Nuclides

In the first instance an empirical match between the superconducting properties of the elements and the periodic table of the nuclides is observed on plotting x directly against A to show the actual number of excess protons for each nuclide. This number never exceeds 13, reaches a maximum at $A = 106$ ($Z = 48$) and approaches zero at $A = 267$ ($Z = 102$). Note that $267 = 102/\tau^2$. The nuclides at which each of the eleven periods comes to a close are circled in Figure 10.3. It is observed as an empirical rule of thumb that the maximum proton excess that allows superconductivity in

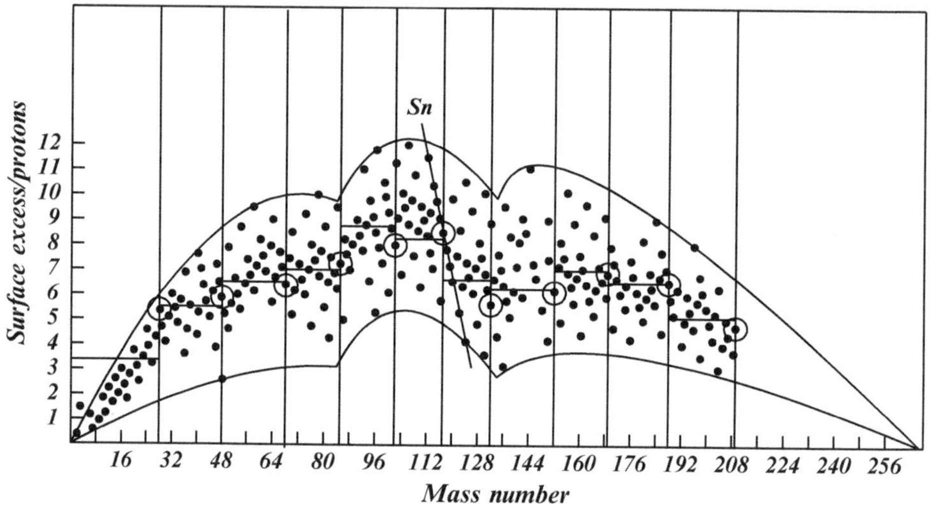

Figure 10.3: *Diagram for discussion of the correlation between elemental superconductivity and nuclear proton excess.*

each period lies close to that of the circled nuclei. Following this trend an empirically defined upper x limit for each period, shown as horizontal lines, can be chosen to differentiate between potential superconductors and non-superconductors. In a few cases, nuclei such as ^{27}Al, ^{118}Sn and ^{133}Cs that occur close to periodic hem lines, could better be considered part of the contiguous group. A summary of all nuclides, predicted to be potential superconductors, is shown in Table 1.

Entries for non-metals are shaded. None of these turn superconducting on cooling, except under high pressure, as indicated. Some nuclides in this group which are present in low abundance have not yet been demonstrated to be superconductors. Some others (^{9}Be, ^{53}Cr, and ^{108}Pd) require some structural modification or high pressure (^{7}Li, ^{142}Ce and ^{196}Pt) to become superconducting. Magnetic nuclides such as ^{64}Ni and the lanthanides are excluded on other grounds. Iron transforms into a non-magnetic phase under pressure. The radioactive isotope ^{87}Rb with 28% abundance and proton excess of 6.0, in period 5, suggests that the Rb metal could be a superconductor under appropriate conditions. The only serious exception is ^{174}Yb.

Table 10.1: *Elemental superconductivity tabulated as a periodic function of mass number. Radioactive isotopes are marked by asterisks.*

Period	Nuclide	Excess	Abundance	Tc/K p/GPa
1 (x<3.4)	6Li3	1.14	8	20 at 50
	7Li3	0.52	92	
	9Be4	0.9	100	10 (mod)
	10B5	1.9	20	11 at 25
	11B5	1.28	80	
	16O8	3.04	>99	.5 at 100
2 (x<5.5)	27Al13	4.32	100	1.20
	28Si14	5.32	92	7 at 12
	29Si14	4.70	5	
	31P15	5.09	100	5.8 at 17
	34S16	4.85	4	17 at 160
	41K19	5.36	7	
	44Ca20	5.12	2	
	46Ca20	3.88	<1	
	48Ca20	2.64	<1	
	48Ti22	5.87	74	0.4
	49Ti22	5.26	5	
	50Ti22	4.64	5	
3 (x<6.3)	51V23	5.64	100	5.4
	53Cr24	6.02	10	3 (mod)
	54Cr24	5.4	2	
	58Fe26	6.16	<1	1 at 10
	64Ni28	5.68	1	$3d^8$
	68Zn30	6.45	19	0.85
	70Zn30	5.19	<1	
4 (x<7.4)	71Ga31	6.2	40	1.1
	73Ge32	6.59	8	5.4 at 4.5
	74Ge32	5.95	36	
	75As33	6.96	100	0.5 at 20
	78Se34	6.73	24	6.8 at 13
	80Se34	5.47	50	
	82Se34	4.25	9	
	81Br35	6.48	49	90GPa

Period	Nuclide	Excess	Abundance	Tc/K p/GPa
5 (x<8.8)	87Rb37	6.00	28	
	86Sr38	8.25	10	4 at 5
	87Sr38	7.64	7	
	88Sr38	6.99	83	
	89Y39	8.0	100	2.3 at 11
	91Zr40	8.4	11	0.6
	92Zr40	7.76	17	
	94Zr40	6.52	17	
	96Zr40	5.28	3	
	93Nb41	8.77	100	9.25
	96Mo42	8.53	17	0.92
	97Mo42	7.90	10	
	98Mo42	7.27	24	
	100Mo42	6.05	10	
	99Tc*43	8.28		7.8
	101Ru44	8.67	17	0.49
	102Ru44	8.05	32	
	104Ru44	6.82	19	
6 (x<8.3)	108Pd46	7.54	26	3.3 (mod)
	110Pd46	6.35	11	
	113Cd48	7.68	12	0.52
	114Cd48	7.10	29	
	116Cd48	5.86	7	
	115In*49	8.08	96	3.41
	118Sn50	7.85	24	3.72
7 (x<7.1)	120Sn50	6.6	33	3.72
	122Sn50	5.35	5	
	124Sn50	4.1	6	
	123Sb51	6.38	43	3.5 at 8.5
	126Te52	6.14	19	5 at 3.8
	128Te52	4.89	32	
	130Te52	3.64	34	
	127I53	7.1	100	HP
	133Cs55	6.66	100	2 at 12
8 (x<6.2)	137Ba56	5.77	11	2 at 5.5
	138Ba56	5.15	72	
	139La57	6.16	100	5

Period	Nuclide	Excess	Abundance	Tc/K p/GPa
	142Ce58	5.92	11	2 at 10
	148Nd60	5.46	6	$4f^4$
	150Nd60	4.2	6	
9 (x<7.0)	154Sm62	4.96	23	$4f^5$
	158Gd64	5.70	25	$4f^7$
	160Gd64	4.48	22	
	163Dy66	5.87	25	$4f^{10}$
	164Dy66	5.21	28	
10 (x<6.6)	173Yb70	6.16	16	
	174Yb70	5.53	32	
	176Yb70	4.27	13	
	175Lu71	6.53	97	0.1
	178Hf72	6.26	27	
	179Hf72	5.69	14	0.35
	180Hf72	5.04	35	
	181Ta73	6.06	100	4.47
	183W74	6.44	14	0.02
	184W74	5.77	31	
	186W74	4.59	28	
	187Re*75	5.56	62	1.7
	188Os76	6.54	13	0.66
11 (x<5.1)	190Os76	5.32	26	0.66
	192Os76	4.1	41	
	193Ir77	5.08	63	0.11
	196Pt78	4.84	25	<1 (mod)
	198Pt78	3.57	7	
	201Hg80	4.96	13	4.2
	202Hg80	4.36	30	
	204Hg80	3.12	7	
	203 Tl81	5.35	30	2.38
	205 Tl81	4.13	70	
	206Pb82	5.08	24	7.2
	207Pb82	4.51	22	
	208Pb82	3.85	52	
	209Bi83	4.90	100	8 at 9

Non-metals, known to superconduct, are included in Table 10.1, which strictly refers only to metals, as the effect of pressure has not been considered.[1] The only class of non-metal where superconductivity has never been demonstrated are the noble gases, which are excluded, albeit for reasons

[1]Pressures in excess of 100 GPa are often required to produce superconducting states. The effect of such pressures may not be limited to structural transformation and remains largely conjectural.

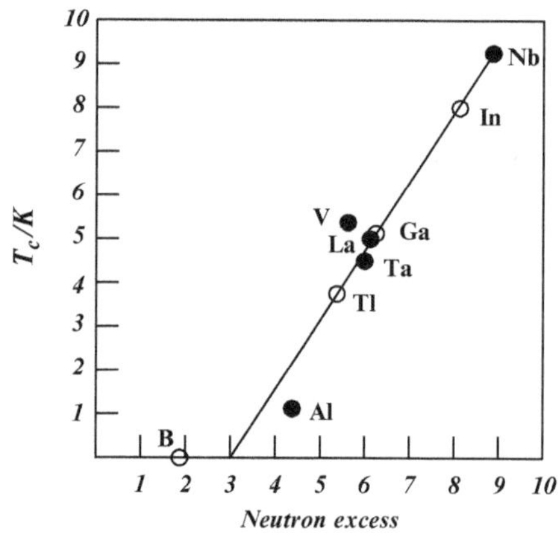

Figure 10.4: *Linear plot of T_c as a function of proton excess for single isotope group 3 metals.*

unknown.[2] The most successful non-metals that turn superconducting under pressure are sulphur and phosphorus. Based on proton excess superconductivity could very likely be induced in solid halogens. Weak effects at very high pressure have been observed only for Br and I.

The chemically alike single-isotope metals Al, V, Nb, La and Ta exhibit an almost linear relationship between nuclear proton excess and T_c, shown in Figure 10.4. The isotopes of lowest allowed mass number of the other non-magnetic elements (B, Ga, In, Tl) of the same elemental periodic group are shown as open circles, assumed to obey the same linear relationship. The corresponding values of T_c at normal pressure are predicted as 0, 5.2, 8 and 3.8 K respectively. It may be inferred that the preparation of metals in single-isotope form has the potential of substantially increasing critical temperatures, especially when conditions such as modified crystal structure that render the lower mass-number isotopes superconducting, can be identified.

10.2.2 Periodic Effects

The onset of superconductivitty is clearly conditioned by some periodic relationship that becomes evident in Figure 10.3. Previous efforts to find such a periodic relationship have failed, as these were invariably based on atomic

[2] Xe has been transformed into a metallic state without turning superconducting under high pressure.

number periodicity. The immediate conclusion is that superconductivity is not a chemical, but a nuclear phenomenon, directly linked to the proton excess.

Closer scrutiny of Figure 10.3 shows that the limiting value of x for each period lies close to half-way between the points of intersection of the closing vertical hemline, at the end of the period, with the two limiting curves. Given the approximate nature of the limiting curves the match is remarkably close. The situation is reminiscent of the break in chemical properties for a series of elements, with the same electronic valence shell, that occurs at half occupation of the sub-shell. In this instance the effect arises from Hund's rule that stipulates parallel spins for the first $2l + 1$ electrons in a sub-shell with angular momentum quantum number l.

We postulate a related rule for the layer of excess protons. For those nuclei with all spins parallel in the excess layer, alignment of these spins on neighbouring nuclei in the bulk material, must generate a powerful magnetic effect. Such alignment will be inhibited as soon as spin pairing in the excess layer becomes necessary.

Any nuclide with a number of excess protons, less than half the capacity for its periodic group, is a potential superconductor. The critical temperature, which is a measure of this tendency, increases with increasing x, until spin pairing commences. For a given element the number of neutrons in the nuclei of its isotopes decreases with decreasing mass number, while the proton excess increases. Isotope enrichment towards lower mass number is therefore predicted to raise the critical temperature – once more up to the isotope mass number where spin pairing starts.

The nuclear magnetic field which is caused by the alignment of excess spins is directed against external, and atomic, magnetic fields. Applied magnetic fields, below some critical value, will be expelled from superconductors. Supercritical fields destroy the superconducting state. Critical fields are predicted to be proportional to the proton excess of a superconducting nuclide.

These conjectures will next be compared with the phenomenon of superconductivity as observed experimentally and with known nuclear spins.

10.3 Superfluidity

A phenomenological analysis of the related phenomena of superfluidity and superconductivity, due to Bohm and Hiley [109], is based on the observation that at low temperature many substances enter a state in which currents (of either atoms or electrons) flow indefinitely without viscosity or resistance. This state is stable only up to a certain temperature at which the property disappears. To explain superflow in ^{4}He, made up of bosons, all particles are described by a single condensate wave function

$$\Psi_0 \propto \exp\left[i\boldsymbol{K}\cdot\sum_j \boldsymbol{r}_j\right]$$

and assumed to move at the same velocity $\boldsymbol{v} = \boldsymbol{K}/m$, $\boldsymbol{K}$ in units of $\hbar$ and $\boldsymbol{r}_j$ represents the coordinate of the jth particle. To maintain stable superflow no particle may go at a slower velocity, $\boldsymbol{v}' = (\boldsymbol{K} - \boldsymbol{q}_0)/m$, by giving up energy. This means that external perturbation cannot change the momentum of a single particle in the system from $\boldsymbol{K}$ to $\boldsymbol{K} - \boldsymbol{q}_0$ without affecting the other particles in such a way as to bring about a net increase in energy of the whole system.

Without this condition it should be possible for the momentum of the ith particle to increase, giving rise to the excited wave function

$$\begin{aligned}\Psi_i &= \exp[-i\boldsymbol{q}_0\cdot\boldsymbol{r}_i]\Psi_0\\ &= \exp\left[i\boldsymbol{K}\cdot\sum_{j\neq i}\boldsymbol{r}_j\right]\exp\left[i(\boldsymbol{K}-\boldsymbol{q}_0)\boldsymbol{r}_i\right]\end{aligned}$$

Although this is a possible solution of Schrödinger's equation, for bosons, the wave function has to be symmetrical and the lowest state of excitation corresponding to a loss of momentum $\boldsymbol{q}_0$, will be

$$\Psi = \sum_j \exp\left[-i\boldsymbol{q}_0\cdot\boldsymbol{r}_j\right]\Psi_0 \tag{10.1}$$

This wave function is defined in the configuration space of all particles,

$$i\hbar\frac{\partial\Psi}{\partial t} = \left[-\frac{\hbar^2}{2m}\sum_j\nabla_j^2 + V\right]\Psi$$

Writing $\Psi = R\exp(iS/\hbar)$ and defining $P = R^2 = \Psi^*\Psi$, this gives

$$\frac{\partial S}{\partial t} + \sum_j\frac{(\nabla_j S)^2}{2m} + V + V_q = 0$$

where

$$V_q = -\frac{\hbar^2}{2m}\frac{\left(\sum\nabla_j^2\right)R}{R}$$

The quantum potential V_q contains R in both denominator and numerator, so that it does not necessarily fall off with distance. This means that each particle can depend on distant features of the environment, and also, that all particles remain coupled at long distances. The behaviour of each particle may depend non-locally on the configuration of all others. The effect of the wave function on the particles is determined by the phase S and the quantum potential, both of which depend only on the form of the wave function and not on its amplitude. In this sense the wave function is regarded as containing active information.

A small perturbing potential which is a symmetric function of all particles changes the wave function (10.1) and the probability density only in the neighbourhood of the perturbation. Since the perturbed wave function vanishes at infinity, all particles will eventually return to their initial velocity. It follows that the liquid as a whole flows around the obstacle and reconstitutes its flow in the original direction. The quantum potential of the wave function (10.1) provides the active information that keeps the particles moving together in spite of perturbations that would otherwise scatter them.

Superconductivity is described in exactly the same terms, except that the charge carriers are electron pairs which act like bosons. Small obstacles in their path may be overcome by going around them and to reform without scattering, in the same way as happens for ^{4}He. Under special conditions the guidance principle and the quantum potential have the ability to organize the activity of an entire set of particles in a way that depends on a pool of information common to the whole set. What remains is to recognize the conditions under which the non-local cooperation of charge carriers happens.

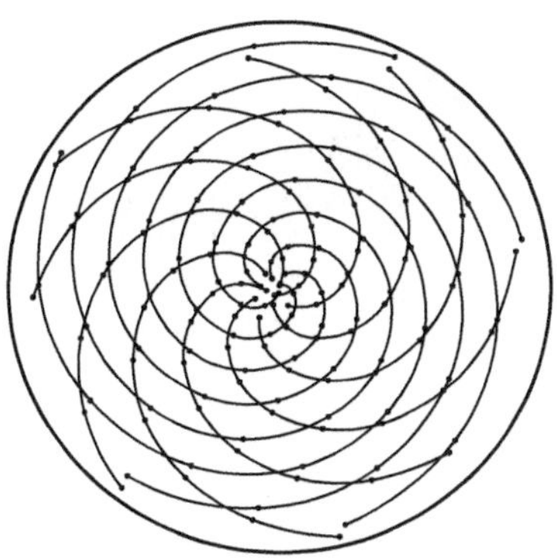

Figure 10.5: *Simple model of an atomic nucleus based on Fibonacci phyllotaxis.*

10.4 Structure of the Nucleus

The mechanism whereby the stacking of nucleons stabilizes an atomic nucleus is demonstrated by considering a hypothetical nucleus with $Z/N = 1$, built up according to the principle of Fibonacci phyllotaxis in three dimensions. Suppose that protons are spaced along five clockwise spiralling arms and neutrons on eight anticlockwise spiralling arms. Unlike botanical phyllotaxis the two sets of spirals are not coplanar and spread three-dimensionally in opposite sense to fill a spherical volume, while accommodating the same number of nucleons. When the supply of neutrons ($N = Z$) runs out only $5Z/8$ protons have been placed, such that $Z(1-5/8)$ are left over to be spread around the surface, as shown schematically in Figure 10.5. For $N \neq Z$ the excess $x = Z - 5N/8$, which is shown for all stable nuclides in Figure 10.6 as a function of mass number, A. All but one (^{3}He) of the nuclides occur in this plot on a 44×44 grid. The grid lines of positive slope correspond to the 44 possible values of neutron excess $(N - Z)$, perpendicular to 44 values of $2Z - N$. The same mapping of the nuclides are obtained by plotting Z vs $3Z - 2N$ in directions respectively defined by the vector sum and difference of $N - Z$ and $2Z - N$.

It is noted that the three equivalent mappings of the nuclides represent the surface excess in terms of three different Fibonacci ratios: $Z-\frac{1}{2}N$, $Z-\frac{2}{3}N$, and $Z - \frac{5}{8}N$, as functions of $N - Z$, Z and A respectively. The vector sums of $(2Z - N) + (3Z - 2N)$ and of the orthogonal pair $(N - Z) + Z$ define the axes that map the surface excess $Z - \frac{3}{5}N$ as a function of N. This mapping is shown as a separate plot in Figure 10.7.

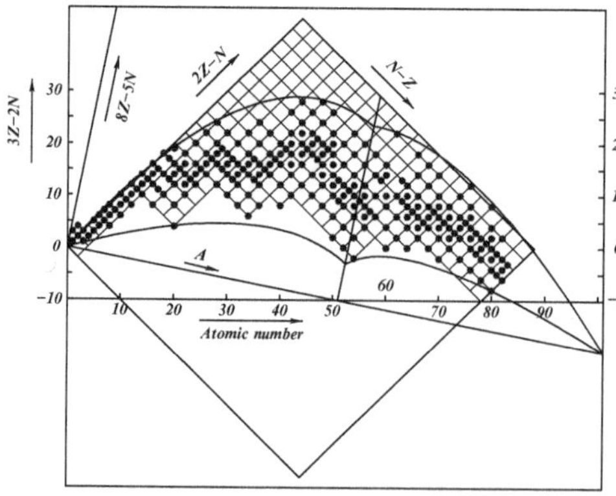

Figure 10.6: *Proton surface excess for 2/3 three-dimensional Fibonacci phyllotaxis of nucleons.*

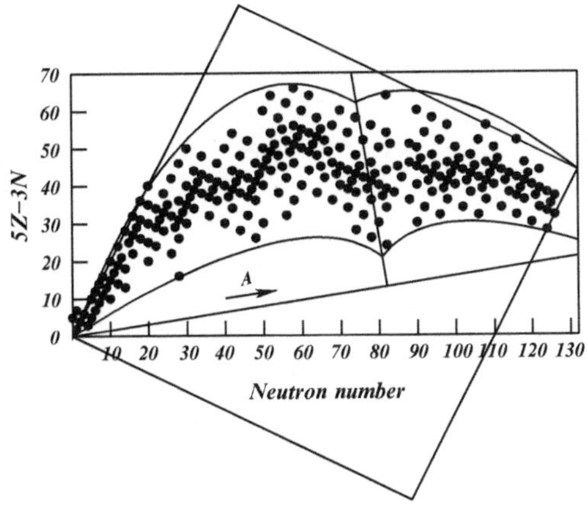

Figure 10.7: *Surface excess as a (3/5) function of neutron number.*

The relationship between the various maps follows directly from the Fibonacci series and the fractions defined by successive terms of the series:

$$\begin{array}{ccccccccccccccc} 0 & & 1 & & 1 & & 2 & & 3 & & 5 & & 8 & & \cdots \\ & \frac{0}{1} & & \frac{1}{1} & & \frac{1}{2} & & \frac{2}{3} & & \frac{3}{5} & & \frac{5}{8} & & \cdots & \tau \end{array}$$

i.e.

$$\begin{aligned} Z - \frac{1}{2}N &\leftrightarrow N - Z && (i) \\ Z - \frac{2}{3}N &\leftrightarrow Z && (ii) \\ Z - \frac{3}{5}N &\leftrightarrow N && (i) + (ii) \\ Z - \frac{5}{8}N &\leftrightarrow A = Z + N \end{aligned}$$

Figure 10.3 defines the map $Z - \tau N$ at convergence,[3] to be called the *golden excess*.

The proton/neutron ratio adjusts from 5/8 to τ by a modest increase of the spacing between protons along the spirals and concentration of the excess within the positive shell on the nuclear surface.

The profile shown to enclose the map of Figures 10.6 and 10.7 has a distinct break along the perpendicular bisector of $A_{max} = 267$ that also bisects

[3]According to the Fibonacci expansion $Z - \tau N$ is orthogonal to $A + \tau N = A(1 + \tau^2)$; hence $x \perp A$.

$N_{max} = 165$ and $Z_{max} = 102$ at this point. $A_{max} = N_{max} + Z_{max} = N_{max}/\tau = Z_{max}/\tau^2$; the same break as observed before with respect to numerous other nuclear properties. A less prominent break near $Z_{max}/3$, not emphasized before, is seen in Figure 10.3 and reflected in the superconductivity of the nuclides.

10.5 Superconductivity

It has been known for almost 100 years that the conduction properties of certain metals change abruptly at temperatures near absolute zero when their resistivities become vanishingly small. It was found by Kamerlingh Onnes (1911) that a current of several hundred amps, started in a lead ring under these conditions persisted indefinitely. This remarkable property is known as *superconductivity*. Such currents are generally started by magnetic induction, but above a certain critical magnetic field strength the superconductivity disappears. The conductor is placed in a supercritical magnetic field, which is then reduced gradually. Electric currents are induced when the field becomes subcritical and these currents which persist for extended periods may be detected through their external magnetic effects.

The magnetic properties of superconductors are as dramatic as their electrical properties. When a specimen is placed in a magnetic field and then cooled through the critical temperature, the magnetic flux is expelled from the specimen, then said to exhibit perfect diamagnetism. The magnetic field need not be of external origin and may be the result of electric current flow in the conductor. This, so-called Meissner effect provides a useful diagnostic for superconduction, often demonstrated by the levitation of a small magnet over a superconducting system.

The onset of superconductivity is associated with a second-order phase transition, characterized by a discontinuity in heat capacity without latent heat.

10.5.1 The Phase Transition

The sharp transition in many (type 1) superconductors has been traditionally interpreted to mean that electrons behaved coherently and interacted non-locally [109]. At the transition point, a fraction of the conduction electrons, condensed into a phase with reduced free energy, is supposed to give rise to superconductivity. At $T = 0\,\mathrm{K}$ all conduction electrons are assumed to be in the superfluid phase. As the temperature is raised, the number of electrons in the condensed state decreases until at $T = T_c$ all electrons are normal.

The existence of an energy gap in this two-fluid model has been verified experimentally. The absorption of electromagnetic radiation was found to increase abruptly when the frequency was increased to a critical value that matches the energy gap. For aluminium the critical frequency falls in the microwave region.

In the BCS theory the condensed phase is ascribed to the formation of electron pairs by the exchange of virtual lattice phonons with wave vector $\boldsymbol{k}$, between electrons of opposite spin and equal momenta. The total momentum of any electron pair is

$$(\hbar\boldsymbol{k} + m\boldsymbol{v}) + (-\hbar\boldsymbol{k} + m\boldsymbol{v}) = 2m\boldsymbol{v}$$

As in the two-fluid model, more electrons condense into this phase as the temperature is lowered.

To rationalize the creation of the energy gap it is necessary to assume that the normal density of states as a function of energy is modified when the metal becomes superconducting [110]. The occupancy of states with energy between $E_F + \Delta$ and $E_F - \Delta$, where E_F is the Fermi energy of the normal metal, is now forbidden. To keep the total number of states below The Fermi level the same, a sharp increase in density of states near the energy gap is assumed, as shown in Figure 10.8. At $T = 0\,\mathrm{K}$ all electrons are coupled as pairs and occupy the states with energy $E < E_F - \Delta$. As the temperature is raised, some pairs decouple into normal electrons by excitation to states with energy $E > E_F + \Delta$.

In a normal conductor the collisions of electrons with phonons and lattice irregularities result in a transfer of kinetic energy to the lattice. This process is the origin of resistance and joule heating. In a superconductor collisions still occur between pairs and the lattice. However, if one of the electrons of a pair is to suffer a loss in kinetic energy, decoupling of the pair occurs with an

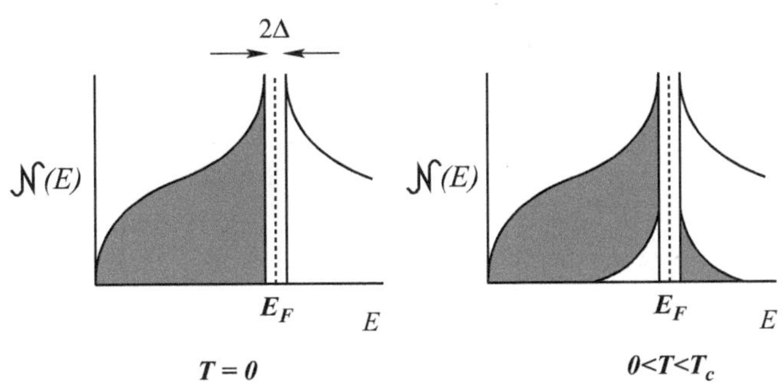

Figure 10.8: *Density of states in a superconductor according to BCS theory.*

increase of 2Δ in potential energy. If the loss in kinetic energy by scattering is less than 2Δ the electron pair will not decouple and the common velocity of the pair will remain the same.

Phase transitions are caused by spontaneous breakdown of symmetry when a system rearranges from a situation of high symmetry to a degenerate situation of lower symmetry. In the case of superconductivity, symmetry breaking of a gauge field in the form of a macroscopic wave function defines a conserved current that predicts the Meissner effect as well as zero electric field and resistivity [73]. Symmetry breaking, although consistent with BCS theory, does not directly imply phonon interaction and the physical rearrangement that triggers the transition remains unspecified.

Since the BCS model does not apply to all forms of experimentally known superconductivity and correlates poorly with related normal-state properties, it is not unreasonable to look for another type of lattice involvement in the process, different from the creation of Cooper pairs by electron–phonon coupling. One alternative is a rearrangement that involves the lattice at T_c and leads to the disappearance of the periodicity which is responsible for the forbidden bands that cause the scatter of charge carriers. The existence of a nuclear surface excess of protons, which has been shown empirically to correlate well with superconduction, offers such an alternative.

Nearly-free electron models assume periodic potentials to arise from the promotion of valence electrons into a conduction band, leaving behind a periodic array of positive ion cores. In fact, there is no energy gap or conduction band without a periodic potential. This paradox may be eliminated by ascribing the periodicity to positive charges at the nuclear surface.

The mathematical formalism of band structure in solids is unaffected by the alternative assumption, but the physical picture changes dramatically. The temperature dependence of conduction properties can now be related directly to modification of the atomic lattice and cooperative effects involving the positive surface layers. At the characteristic temperature an applied (sub-critical) magnetic field causes alignment of excess proton spins. The resulting magnetic field eventually expels the applied flux and the electric lattice potential is smoothed out to a constant level,[4] thereby eliminating the periodicity responsible for the scatter of charge carriers. The electronic quantum potential, no longer inhited by scattering centres, now extends non-locally through the entire crystal and the electron gas becomes superfluid. There is no need of a special effect to mediate the formation of boson pairs.

[4]Equivalent to the original ideal Drude model of metallic conduction.

10.5.2 The Critical Temperature

Transition to a superconducting phase happens when the protonic surface layers on atomic nuclei line up to form structured layers of lower symmetry. This transition is temperature driven and depends on two factors: the magnitude and spin of the surface charge and the separation between atomic nuclei. The first factor is a function of both proton and neutron count in the nucleus and the second depends on crystal structure.

Surface alignment would clearly be counteracted by a mixture of isotopes in a specimen. Each isotope of characteristic surface excess requires a different internuclear spacing for optimal alignment of spins. Hence the presence of even small amounts of different isotopes could seriously affect the critical temperature of a pure metal. An important factor that increases T_c of pure metals therefore is the production of isotopically pure materials.

For many nuclides (single isotopes) with a favourably large surface excess the natural internuclear spacing may be inappropriate to allow the superconduction transition. Various modes of structure modification could have an important effect in such cases.

10.5.3 Crystal Chemistry

There are three distinct types of compound that turn superconducting on cooling of the bulk material – elements (especially metals), binary compounds (alloys) and a variety of complex structures. The proposed alignment of positively charged surface-excess layers can only be a common factor, responsible for the onset of superconductivity, if linear chains of closely spaced active nuclei occur in all of these different types of compound. To explore this possibility the different types of superconductor are examined in turn.

Superconducting Elements

The most prominent superconducting metals are listed in order of decreasing T_c in Table 10.2, not including those with $T_c < 1$ K in the bulk form. Linear chains of the type shown in Figure 10.9 are readily identified for the structure types A1, A2 and A3, along the [110], [111] and [100] directions respectively. In the rhombohedral structure of Hg the channel that contains the linear chain is distorted, such that $\alpha = 70°$ rather than 90°. At $T < 13°$C Sn has the diamond structure. Although not perfectly linear, continuous chains of the same type identified for the other metals, run through the lattice. The tetragonal unit cell of In (c/a = 1.08) is pseudo-cubic close packed, A1. Atomic chains in Ga are non-linear, with alternating interatomic separations,

Table 10.2: *List of metals that turn superconducting on cooling below T_c. The structure types A1, A2 and A3 are commonly known as cubic close packed (ccp), body-centred cubic (bcc) and hexagonal close packed (hcp) respectively. Parameter d denotes the distance of closest approach in the chain.*

Metal	T_c/K	Excess	Structure type	Space Group	d/Å
Nb	9.25	8.77	A2	Im3m	2.86
Tc	7.80	8.28	A3	$P6_3/mmc$	2.70
Pb	7.20	5.08	A1	Fm3m	3.50
La	6.00	6.16	A1	$P6_3/mmc$	3.74
V	5.40	5.64	A2	Im3m	3.62
Ta	4.47	6.06	A2	Im3m	2.86
Hg	4.15	4.96	A10	$R\bar{3}m$	3.00
Sn	3.72	7.85	A4	Fd3m	2.80
In	3.41	8.08	A6	I4/mmm	3.24
Tl	2.38	5.35	A3	$P6_3/mmc$	3.42
Re	1.70	5.56	A3	$P6_3/mmc$	2.74
Al	1.18	4.32	A1	Fm3m	2.89
Ga	1.08	6.20	A11	Cmca	2.42

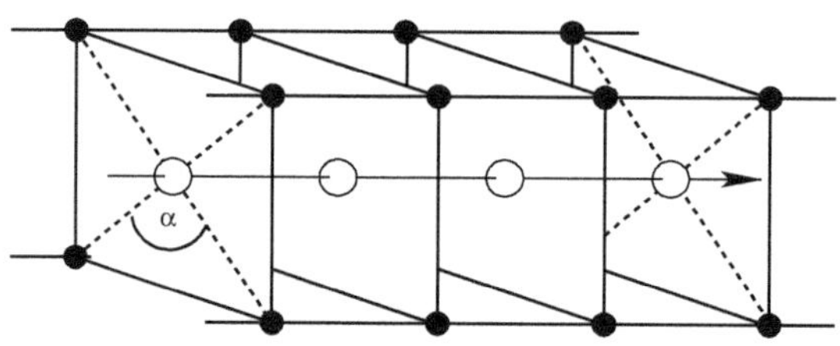

Figure 10.9: *Linear chain of atoms that occurs in most metal structures.*

more characteristic of atomic pairs, typical of non-metals. In amorphous film form the critical temperature has been raised as high as 8.6 K [106]. The effect of this quenching procedure on crystal structure is not known.

In metals of cubic symmetry there is no unique direction for the flow of persistent superconduction currents, unless these are initiated by a magnetic field. Non-metals become superconducting only at pressures which are sufficiently high to cause metallization. Hydrogen with its single proton can never go superconducting.

Superconducting Alloys

An amazing number of superconducting alloys and binary compounds of composition AB, AB_2, AB_3 and of more complicated or non-stoichiometric formulation are known. One of the components is invariably a potential elemental superconductor, with Ag_2F as the only possible exception. A subgroup of AB compounds with close-packed cubic (NaCl – Fm$\bar{3}$m) or hexagonal ($P6_3mc$ or P$\bar{6}$m2) structures is of special relevance for this discussion.

The compounds of type MX, listed in Table 10.3, are interstitial solid solutions of C and N in the metals. All of the metals are known superconductors and must be assumed responsible for the observed superconductivity of the compounds. In the cubic structures the metal atoms appear as linear chains along [110] directions with M–M distances somewhat longer than in the structures of the elementary metals. The enhanced critical temperatures, compared to the free metals, must therefore be due to another structure-related factor. All of these metals have the bcc structure, which is not close-packed. In this structure type only 68.1% of the available space is occupied by touching spheres, compared to 74.1% in ccp structures. Tight confinement of the conducting chain therefore appears to be as important as interatomic spacing in the superconducting state.

This confinement principle is of decisive importance in the case of superconducting intercalation compounds of C_{60} fullerene. Indications [111] that superconductivity (T_c = 0.55 K) occurs in single-walled carbon nanotubes, define C as a potential superconducting element. Fullerene itself is not

Table 10.3: *Superconducting binary compound AB with B1 structure.*

		MC		MN	
M	T_c/K	T_c/K		T_c/K	
d/Å	d(M–M)	d(M–M)	d(M–X)	d(M–M)	d(M–X)
Mo	0.92	14.3		13	
	2.72	3.02	2.14	(hex)	
Nb	9.25	11.5		16	
	2.86	3.16	2.23	3.11	2.20
Ta	4.47	10.3		13	
	2.86	3.15	2.23	3.06	2.17
W	0.02	10			
	2.74				
Zr	0.6			10.7	
	3.20			3.24	2.29

superconducting, but becomes a superconductor on intercalation by the alkali atoms K, Rb and Cs. Critical temperatures in these compounds have been shown [112] to scale monotonically with the M_3C_{60} lattice constants upon alkali–metal substitution and as a function of pressure, providing conclusive proof that the superconductivity relates to conduction along chains of fullerene units, rather than alkali atoms. Further confirmation comes from the crystal structure of K_3C_{60} [114]. The closest K–K distance at 6.17$\mathring{A}$ is much larger than in metallic potassium. The most important feature of the intercalation structure, compared to pristine C_{60} is the way in which interstitial K atoms lock the fullerene units into a more ordered arrangement. The rotational disorder of C_{60} is reduced to only two possible orientations, sufficiently alike to constitute a superconducting string. As the average size of interstitial alkali atoms increases with progressive substitution of K by Rb and/or Cs atoms, the state of order and the critical temperature increase accordingly. As a result T_c varies inversely with lattice constant, which explains the negative effect of pressure and the failure of Na_3C_{60} to show superconductivity.

The most striking group of binary superconductors is the AB_3 family of compounds with structure type *A*15, space group Pm$\bar{3}$n. Details of the structure type are shown in Figure 10.10. Each A atom is coordinated by 12 atoms of the B type and each B atom is tetrahedrally surrounded by four atoms of type A. In all cases, shown in Table 10.4, the B atoms are known

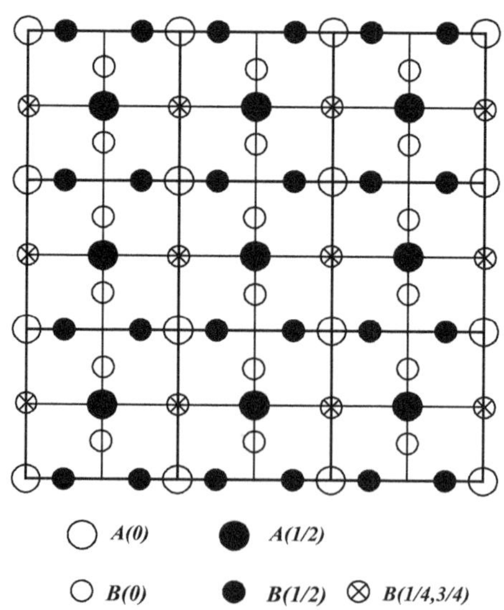

Figure 10.10: *The A15 structure type of* AB_3 *compounds.*

Table 10.4: *Superconduction critical temperatures (K) and d(B-B)/Å of AB_3 compounds of structure type A15.*

M ↓ m →	MTi_3	MV_3	MCr_3	MMn_3	MZr_3	MNb_3	MMo_3	MTa_3	MW_3
Al		11.8				18.0 2.59	0.58 2.48		
Si		17.1 2.36		12.5 2.86		19.0 2.58	1.4 2.45		
Ga		16.8 2.41				14.5 2.58	0.78 2.47		
Ge		11.0 2.39				23.2 2.58	1.43 2.47	8.0	
As		0.20 2.38							
Tc							15		
Ru			3.3 2.34				10.6		
Rh		0.38 2.39	0.3 2.39			2.64 2.56		10	
In		13.9				9.2 2.65			
Sn	5.8	3.8 2.49			0.92 2.82	18.05 2.65		8.35 2.64	
Sb	5.8 2.61	0.8 2.47						0.72 2.63	
Re							15.0		
Os			4.68 2.34			1.05 2.57	12.7 2.48		
Ir	5.4 2.51		0.45 2.35			1.9 2.57	9.6 2.48		
Pt	0.58 2.52	3.0 2.41				10.9 2.58	2.49	0.4	
Au		0.74 2.44			0.92 2.74	11.5 2.60			
Pb					0.76			17	
Bi						4.5 2.66			
O									3.35
Self	0.4 2.92	5.4 2.62	mod 2.50	-	0.6 3.20	9.25 2.86	0.92 2.72	4.47 2.86	0.02 2.74

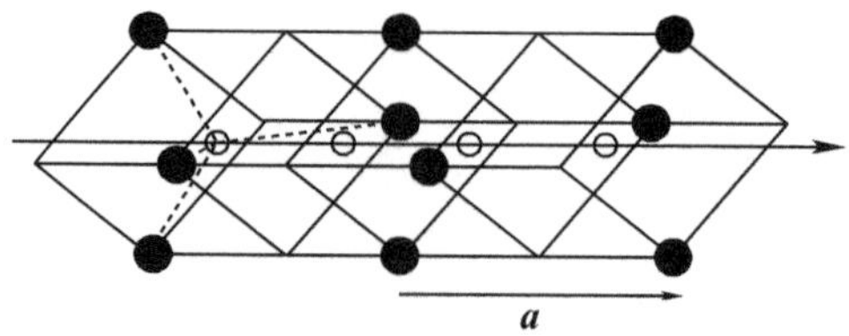

Figure 10.11: *Perspective drawing of the superconducting chain in A15 structures. An AB_4 tetrahedral unit is outlined.*

superconductors. The projective drawing of Figure 10.11 shows the linear array of B-type atoms at an interatomic separation of $d(\text{m–m}) = a/2$, within a channel of A-type atoms. By comparison, $d(\text{m–M}) = (\sqrt{5}/4)a$ and $d(\text{M–M}) = (\sqrt{3}/2)a)$, are both larger than $d(\text{m–m})$. The m-m chains of superconducting atoms must clearly be identified as the active element in the structure. Critical temperatures and m–m distances for this family of compounds are shown in Table 10.4. Interatomic distances in the active chains are uniformly shorter, compared to the parent metals. The atoms that constitute the supporting channel have an obvious effect on T_c. In the series of Nb compounds a substantial increase in T_c is observed for its compounds with Al, Si, Ga, Ge and Sn – all of these are superconductors with relatively high proton excess. In, Pt, and Au have little or no effect while Rh, Os, Ir and Bi have a negative influence. Apart from In, these elements have no, or only weak, superconducting properties.

The pattern for V is essentially the same. In this case In has a positive effect, Sn and Pt are neutral, with As, Rh, Sb and Au all negative.

The high performance of several other compounds from Table 10.4 as superconductors is noted, but the samples are too small to draw general conclusions. The relevant compounds are: $OsCr_3$, $SiMn_3$, (Tc,Ru,Os,Ir)Mo_3, (Sn,Sb,Ir)Ti_3, and (Ge,Rh,Sn,Pb)Ta_3.

Some other binary compounds that deserve special mention, because of their relatively high critical temperatures, with T_c in brackets, include: RuW (7.5), MoRu (10), $RhZr_2$ (11), CMo_2 (12.2), LaC_3 (11.0), $MoTc_3$ (15.8), Y_2C_3 (11.5) and MgB_2 (39.0). In all of these cases, and for the many non-stoichiometric compounds, superconductivity is readily understood on the basis of the principles already outlined, except for the last compound on the list.

Boron is not a superconductor at normal pressure. This is readily understood from its structure, made up of close-packed icosahedral units, shown in Figure 10.12. Under pressure, links between the units may form, and atomic chains that support superconductivity so established. The hexagonal

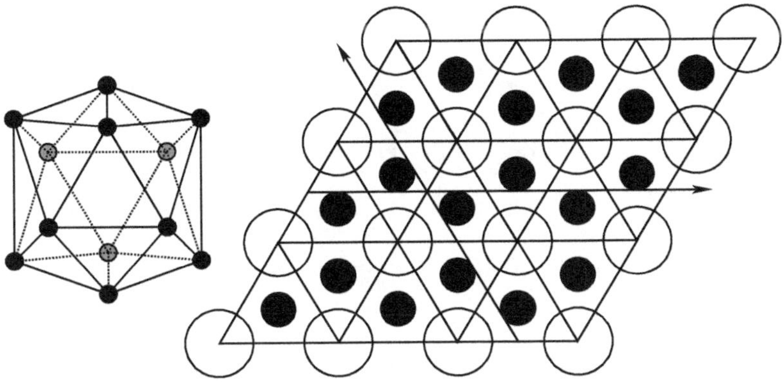

Figure 10.12: *Left: Perspective drawing of the icosahedral cluster of boron atoms in the metal. Right: The hexagonal structure of MgB_2. Close-packed layers of Mg atoms (open circles) are overlaid by atomic networks consisting of zig-zag chains of close-packed B atoms running in the directions indicated by arrows.*

structure of MgB_2 [115] in (001) projection is shown in Figure 10.12. The distance of 1.78 Å between boron atoms in magnesium diboride is the same as in the metal. The zig-zag chains, in either direction, span all of the boron atoms in a layer. The superconductivity may therefore be interpreted as a function of the entire lattice of boron atoms in the crystal. This remarkable compound [116], which is intermediate between traditional and high-temperature superconductors, provides indisputable evidence that all superconductors function by the same mechanism and that a single theory should account for their properties.

The compound $Ba(Bi_{0.2}Pb_{0.8})O_3$ with $T_c = 13.2\,K$ [117] is another intermediate, in a structural sense, between low and high-T_c superconductors. This compound has a slightly distorted perovskite ($CaTiO_3$) structure type ($E2_1$, $Pm\bar{3}m$) and exhibits the same linear atomic (Ba) chains that characterize all high-T_c superconductors. The perovskite structure is shown in Figure 10.13.

High-T_c Superconductors

The discovery of superconductivity ($T_c = 36$ K) in the La-Ba-Cu-O system [118] inaugurated the search for high-temperature superconducting ceramics, which led to the identification [119] of the distorted perovskite $YBa_2Cu_3O_7$ with $T_c = 93$ K. The idealized structure [120] of this compound, which is

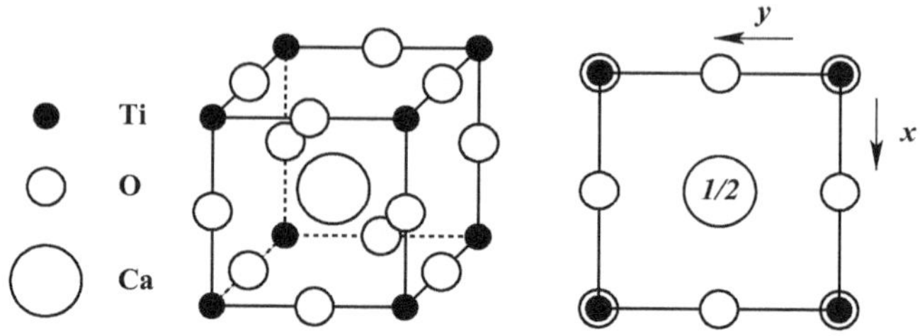

Figure 10.13: *The perovskite structure.*

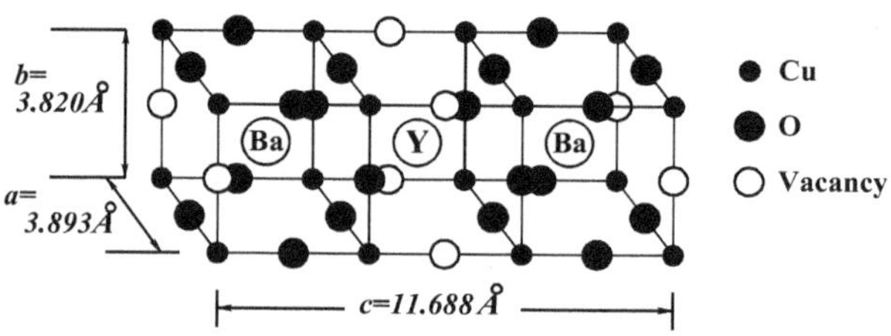

Figure 10.14: *Idealized defect perovskite structure of 1-2-3 superconductors.*

characteristic of a whole family of (1-2-3) $M(Ba,Sr)_2Cu_3O_{9-x}$ superconductors, is shown in Figure 10.14.

This layered structure was soon [122] likened to the chain of Nb atoms in $GeNb_3$, the traditional superconductor with maximum T_c, at the time. Noting that the Y chain could be replaced by magnetic lanthanide atoms without affecting the activity of 1-2-3, the active element was identified as the coordinated Cu polyhedra. This theme has been taken up by all subsequent authors, so that the chains of Ba and/or Sr atoms were never implicated as the superconducting entities. Quite remarkably, not a single high-T_c ceramic has been synthesized without Ba and/or Sr at the alternate body centres of the perovskite units, as in Figure 10.14. The interatomic distances in the resulting chain along [010] are, on average, equal to 3.82 Å. This separation is considerably shorter than $d(\text{Ba–Ba}) = 4.34$ Å and $d(\text{Sr–Sr}) = 4.30$ Å in the parent metals. Given the proton excess of 5.15 for ^{138}Ba and 6.99 for ^{88}Sr, enhanced T_c for both metals is predicted immediately.

The other group of superconducting copper oxide ceramics, $La_{2-x}M_xCuO_4$ with M = Ce, Sr, Ba, and structures, formally related to that of K_2NiF_4 [123], can also be considered as derivatives of a distorted perovskite structure. The structure of an authentic superconducting single crystal in this class has never been determined. All indications are that only a small fraction of each sample has bulk superconductivity and that the crystals show a range of T_c values [124]. The La and (Sr, Ba) atoms have been determined to be distributed at random over the equivalent sites of the same position, although the possibility of short range ordering could not be ruled out [125]. The distance within a chain of atoms, independent of space group, converges to about 3.76 Å, at 10 K, equivalent to d(La–La) and considerably less than d(M–M) for Ba and Sr in the free metals. Evidence of phase ordering near 10 K [127] strongly supports the possibility of segregated chains, like those in 1-2-3, to be responsible for the observed superconductivity.

Organic Superconductors

The most common component of organic superconductors is *bis*-ethylenedithia-tetrathiafulvalene, codenamed ET [128]:

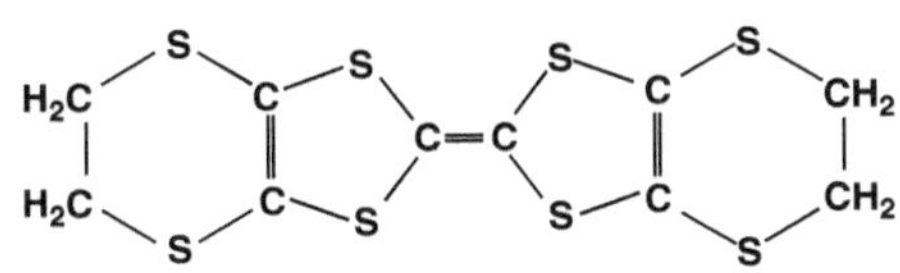

whose complex $(ET)_2Cu(NCS)_2$ has $T_c = 11.4$ K. The common structural feature in a variety of related superconductors, often with Se replacing S, is the appearance of segregated stacks of planar molecules, such that multiple linear strings of S or Se atoms occur in the stacking direction, at separations close to van der Waals contact distances. The interpretation is obvious.

10.5.4 An Alternative Mechanism

On the basis of an exhaustive analysis of the correlations between normal-state properties and the superconductivity of materials Hirsch [108] concluded that

> Hall coefficient is by far the normal-state property most closely associated with the existence or nonexistence of superconductivity among all properties considered. The quantities most closely associated with BCS theory, electronic specific heat, Debye temperature, and electrical conductivity, rank consistently low in their degree of association with superconductivity according to these measures.
>
> *Within the conventional theory of superconductivity there is no correlation between existence of superconductivity in a material and the sign of its Hall coefficient.*

> The conventional theory of superconductivity has been firmly in place for several years, and had been thought to describe superconductivity in all materials. In recent years, new families of materials have been discovered that do not seem to fit the conventional framework, and scientists have been working in developing new theoretical frameworks to describe the new materials. However, the applicability of the conventional theory to the "conventional" materials has not been called into question. Perhaps this is the time to do so.

In view of these strong indications that the conventional theory of superconductivity is in trouble, the alternative mechanism that developed from the notion of a golden excess of protons on atomic nuclei, will be outlined and assessed against the Hirsch criteria.

The Assumptions

1. Atomic nuclei are constructed on the basis of three-dimensional Fibonacci phyllotaxis. In a nucleus, consisting of Z protons and N neutrons, nuclear growth terminates when all neutrons have been consumed, leaving a golden excess of $x = Z - \tau N$ protons as a surface layer.

2. The golden excess is a periodic function of A, Z and N and reaches a characteristic maximum, x_m, in each of the 11×24 nuclide periods.

3. The alignment of spins on the excess protons obeys Hund's rule with respect to x_m. The normal multiplets $<x_m/2$ have the same magnetic spin quantum number, m_s. For $x > x_m/2$ inverted multiplets have paired spins.

4. In a closely spaced linear, or near linear, array of nuclei with $x < x_m/2$ a ferromagnetic transition at $T = T_c$ causes alignment of the surface layers and eliminates the periodic variation in nuclear potential, which is responsible for the electronic band structure of nearly-free-electron theory [129]. Superconductivity becomes possible.

Nuclear Spin

A crucial test of the proposed mechanism lies with the appearance of nuclear spin, which is summarized in Figure 8.6 and ascribed (Section 8.3.2) to the distribution of nucleons in an atomic nucleus. A remarkable correlation between those nuclides with an excess of $4n(\hbar/2)$ spins, due to nuclear distortion, and elemental superconductivity is immediately obvious. The relevant nuclides, (compare Table 8.5), with total spin in units of $\hbar/2$ are:

^{17}O(5), ^{25}Mg(5), ^{27}Al(5), ^{43}Ca(7), ^{45}Sc(7), ^{47}Ti(7), ^{49}Ti(7), ^{51}V(7), ^{59}**Co**(7), ^{67}Zn(5), ^{73}Ge(9), ^{75}As(5), ^{83}Kr(9), ^{87}Sr(9), ^{91}Zr(5), ^{93}Nb(9), ^{95}Mo(5), ^{97}Mo(5), ^{99}Ru(5), ^{101}Ru(5), ^{105}Pd(5), ^{113}In(9), ^{121}Sb(5), ^{123}Sb(7), ^{127}I(5), ^{133}Cs(7), ^{139}La(7), ^{141}**Pr**(5), ^{143}**Nd**(7), ^{145}**Nd**(7), ^{151}**Eu**(5), ^{153}**Eu**(5), ^{161}**Dy**(5), ^{163}**Dy**(5), ^{165}**Ho**(7), ^{167}**Er**(7), ^{173}Yb(5), ^{175}Lu(7), ^{177}Hf(7), ^{179}Hf(9), ^{181}Ta(7), ^{185}W(5), ^{209}Bi(9).

The symbols in bold print represent magnetic species which can be eliminated from the list for that reason. Others, such as ^{25}Mg are excluded by their excess values of x. The remaining nuclides include all of the single-isotope elements with high transition temperatures at ordinary pressure, i.e. Nb, V, Ta, La, Al, as well as the high-spin radioactive isotopes ^{115}In*(9/2) and ^{187}Re*(5/2).

In view of this result it appears all the more likely that superconduction is to be associated, not only with a proton surface excess, but also with high nuclear spin. It opens up the intriguing possibility that the two effects have a common origin, namely the packing of protons and neutrons in the nucleus and that superconduction is an almost exclusive property of nuclides with odd mass number. Transition temperatures for superconductivity, by this criterion, must then depend on the abundance and spin of odd-mass nuclei with $x < x_m/2$.

This proposition lies uneasily with a number of the classical superconductors such as Pb, Hg, Tl and Sn. Closer scrutiny reveals that three of these elements have at least one low-lying metastable isomer with high spin: ^{207m}Pb(13/2), ^{199m}Hg(13/2) and ^{119m}Sn(11/2). Several of the other exceptions also have high-spin isomers:

^{77m}Se(7/2), ^{89m}Y(9/2), ^{113m}Cd(11/2), ^{125m}Te(11/2), ^{137m}Ba(11/2), ^{183m}W(11/2), ^{189m}Os(9/2), ^{193m}Ir(11/2), ^{195m}Pt(13/2).

Although there is no known mechanism for promotion of these nuclides into those high-spin isomeric states on cooling, the regularity mocks coincidence and cannot be ignored. One possibility is that magnetic interaction may change only the spin state without activating the isomeric state. Three of the remaining high-mass exceptions are the twin-isotope elements: ^{69}Ga(60%), ^{71}Ga(40%); ^{79}Br(51%), ^{81}Br(49%); and ^{203}Tl(30%), ^{205}Tl(70%). The only element that lacks an odd-mass isotope and for which superconductivity has been reported, is Ce.

Most of the low-mass elements that turn superconducting on compression or structure modification have isotopes with spin 3/2, i.e. ^{7}Li, ^{9}Be, ^{11}B, ^{33}S, ^{53}Cr and ^{57}Fe. The only exceptions are ^{13}C, ^{29}Si and ^{31}P with spin 1/2.

In all cases superconduction is proposed to arise from the alignment of spins carried by the excess proton layer. Should the number of excess protons exceed $x_m/2$, cooperative alignment of the inverted multiplet spins is inhibited across the lattice and superconductivity is suppressed.

Reworking Table 10.1 in terms of odd-mass nuclides only leads to minor adjustments and confirms the previous identification of superconducting elements. Prediction of superconductivity, still unconfirmed by experiment, e.g. for Yb, is not of major concern. Observed superconductivity for nuclides, at variance with the rules of prediction, is more serious. The following nuclides in this class therefore need to be re-examined: ^{15}N, ^{17}O, ^{29}Si, ^{31}P, ^{57}Fe, ^{105}Pd, Ce.

The first clue to understanding of their anomalous behaviour is provided by the conditions under which Pd becomes superconducting, i.e. on low-temperature He$^+$ irradiation with a 400 kV ion accelerator [130]. Should this treatment lead to

$$^{105}_{46}\text{Pd} + ^{4}_{2}\text{He} \rightarrow ^{109}_{48}\text{Cd}$$

and cause the formation of trace amounts of high-spin ^{109}Cd(5/2) with half-life of 463 days, EC$\rightarrow^{109}$Ag, the intermediate nuclide would be the superconducting species. Enhancement of the effect, induced by α-particle bombardment, would provide experimental evidence to support the proposal.

Maybe a related effect can account for the observed superconduction of Ce. It is postulated that the pressure-induced effect, in this case, is not due to structural modification, but to reversible, pressure-induced electron capture:

$$^{140}_{58}\text{Ce} \rightleftharpoons ^{140}_{57}\text{La},\ 40\,\text{hrs}(\beta^-)$$

producing high-spin, odd–odd, ^{140}La(6/2), followed by bound-state β^- decay. Such an event would go undetected.

On a related theme, the record $T_c = 20\,$K for Li under pressure could relate to the 2/2 spin of odd–odd ^{6}Li, which means that both major isotopes carry spin. Likewise, both ^{10}B(6/2) and ^{11}B(3/2) would contribute to the superconductivity of MgB_2.

Stable odd-odd nuclides only occur for $A < 16$. Should the effective $x_m/2$ for this special subgroup be assigned at 1.7, half of the assumed 3.4 for the first period of nuclides, the failure of N to reach the superconducting state under pressure, needs no further explanation. However, if

$$^{16}_{8}\text{O} \rightleftharpoons ^{16}_{7}\text{N}(4/2)\ 7.13\,\text{s}(\beta^-)$$

occurs under pressure, oxygen is well within range, $x = 1.42$.

The low spin of the presumed active nuclides ^{13}C, ^{29}Si and ^{31}P does not adequately explain the effect of pressure in this instance. However, the reversible processes:

$$
\begin{aligned}
&{}^{12}_{6}\mathrm{C} \rightleftharpoons {}^{12}_{5}\mathrm{B}(2/2)\ 20.4\,\mathrm{ms}(\beta^-)\\
&{}^{28}_{14}\mathrm{Si}(0) \rightleftharpoons {}^{28}_{13}\mathrm{Al}(6/2)\ 2.24\,\mathrm{m}(\beta^-)\\
&{}^{31}_{15}\mathrm{P}(1/2) \rightleftharpoons {}^{31}_{14}\mathrm{Si}(3/2)\ 2.62\,\mathrm{h}(\beta^-)
\end{aligned}
$$

should they occur under high pressure, would promote superconduction.

The erratic $\mathrm{d}T_c/\mathrm{d}p$ response of many elements can be related to an interplay between the radioactive equilibrium proposed here, and pressure-induced structure modification. The critical temperature of sulphur increases with pressure, up to $T_c = 17\,\mathrm{K}$ at 160 GPa, in contrast to Se and Te which respond negatively at low pressure. This effect could be the result as the major isotope of S acquires spin in the process:

$$
{}^{32}_{16}\mathrm{S} \rightleftharpoons {}^{32}_{15}\mathrm{P}(3/2)\ 14.3\mathrm{d}(\beta^-)
$$

The supposedly active low-abundance isotope ^{57}Fe with $x(6.8) > x_m/2$ may also not be responsible for the observed superconductivity of non-magnetic Fe under pressure. The major isotope, if transformed reversibly,

$$
{}^{56}_{26}\mathrm{Fe} \rightleftharpoons {}^{56}_{25}\mathrm{Mn}(6/2)\ 2.58\,\mathrm{h}(\beta^-)
$$

produces a high-spin species with $x = 5.8 < x_m/2$.

For most metallic superconductors $\mathrm{d}T_c/\mathrm{d}p \leq 0$. The exceptions are Ca, Sr, Ba, Y, Sc, Lu, V and Zr. It probably means that electron capture could be a factor in these exceptional cases, as evidenced by the following processes, with respect to the most abundant isotope in each case:

$$
\begin{aligned}
&{}^{40}_{20}\mathrm{Ca} \rightleftharpoons {}^{40}_{19}\mathrm{K}(8/2)\ 1.28\,\mathrm{Gy}(\beta^-)\\
&{}^{88}_{38}\mathrm{Sr} \rightleftharpoons {}^{88}_{37}\mathrm{Rb}(4/2)\ 17.8\,\mathrm{m}(\beta^-)\\
&{}^{138}_{56}\mathrm{Ba} \rightleftharpoons {}^{138}_{55}\mathrm{Cs}(6/2)\ 32.2\,\mathrm{m}(\beta^-)\\
&{}^{89}_{39}\mathrm{Y}(1/2) \rightleftharpoons {}^{89}_{38}\mathrm{Sr}(5/2)\ 50.3\,\mathrm{d}(\beta^-)\\
&{}^{45}_{21}\mathrm{Sc}(7/2) \rightleftharpoons {}^{45}_{20}\mathrm{Ca}(7/2)\ 165\,\mathrm{d}(\beta^-)\\
&{}^{175}_{71}\mathrm{Lu}(7/2) \rightleftharpoons {}^{175}_{70}\mathrm{Yb}(7/2)\ 4.19\,\mathrm{d}(\beta^-)\\
&{}^{51}_{23}\mathrm{V}(7/2) \rightleftharpoons {}^{51}_{22}\mathrm{Ti}(3/2)\ 5.80\,\mathrm{m}(\beta^-)\\
&{}^{90}_{40}\mathrm{Zr} \rightleftharpoons {}^{90}_{39}\mathrm{Y}(4/2)\ 64.1\,\mathrm{h}(\beta^-)
\end{aligned}
$$

Even irrespective of spin, the x parameter is invariably reduced to more favourable values.

The magnetic field generated by aligned spins will be opposed to the applied field and its effects will be shown to account for the Meissner effect and the positive Hall coefficient of superconducting materials.

The prime objective of an alternative mechanism of superconductivity is that it should apply to all forms of superconductor. This is the major achievement of the golden excess alternative. In the second place it should be falsifiable and serve to identify and prescribe suitable procedures for the production of improved superconductors. In this case the method identifies, for each element, the isotope with maximum potential as a superconductor, by relating its proton excess to the optimum value of $x_m/2$ in its mass-number periodic group. By stipulating maximum homogeneity of the conduction chain, materials with the active conductor in the form of a single isotope is prescribed. Selection of the best matrix for support of the conduction chain depends on such interactions that minimize the interatomic distances within the chain.

Examples

Compounds of tin represent a promising class for further study. The element occurs in several allotropic forms and a mixture of ten isotopes, spread over two mass-number periods. The nuclide ^{119}Sn ($x = 7.2$) in isotopically pure form should have critical temperatures far in excess of the 3.72 K of natural tin. The nuclide ^{117}Sn has $x = 8.45$, fractionally higher than the first estimate of $x_m/2$, could acquire spin of 11/2, to become the prime target, despite its relatively low abundance. Another intriguing feature of tin is the low density of its low-temperature modification (5.75 gcm^{-3}), compared to 7.31 of the close-packed higher-temperature allotrope. Any dopant or matrix that promotes the stability of the high-density arrangement at low temperature, and the formation of a close-packed linear array, would yield an improved superconductor.

Another element whose performance as a superconductors should improve with isotope purification is Pb, with natural $T_c = 7.2$ K, and 22% abundance of the active ^{207}Pb isotope.

For immediate application it seems likely that the critical temperature of 1-2-3 Ba superconductors could be raised close to room temperature by isotope purification. The target isotope in this case is ^{137}Ba with 11% abundance and $x = 5.77$, which in isotopically pure form could improve 1-2-3 superconductivity dramatically.

The prognosis for 1-2-3 Sr is equally good. ^{87}Sr has $x = 7.64$ and abundance of 7%.

In the class of organic superconductors enrichment of ^{77}Se, with 8% abundance should be of obvious benefit.

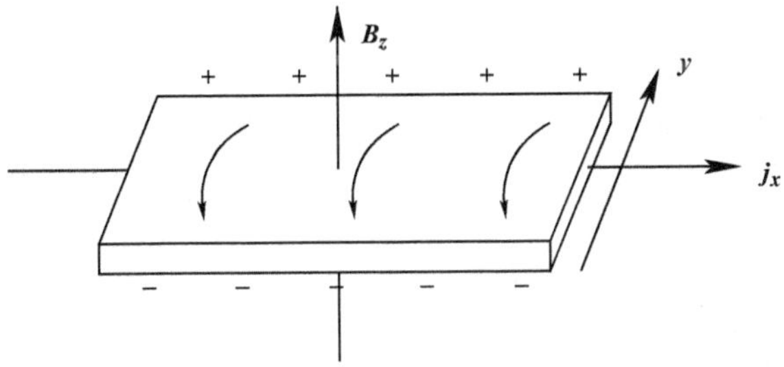

Figure 10.15: *Generation of an electric field in the direction* $\boldsymbol{j} \times \boldsymbol{B}$*, known as the Hall effect.*

Of the cheap industrially important metals, which could be considered for future isotopically pure superconducting use, the most obvious include Cr, Fe, Zn and Mo.

10.5.5 Hall Effect

Charged particles moving at right angles to a magnetic field experience a side-ways force and this gives rise to the Hall effect as shown in Figure 10.15. The magnetic force, $F_B = B_z e v_x = -j_x B_z/n$, where v_x is the velocity of the electrons in the x-direction. As no current flows in the y-direction the build-up of charge at the edges of the conductor amounts to an electric field $E_y = F_y/e$. In the steady state, the magnetic and electrical forces are balanced,

$$F_B = F_y \; : \; eE_y = -j_x B_z/n$$

$$\text{and the ratio} \quad R_H = \frac{E_y}{j_x B} = -\frac{1}{ne}$$

is known as the Hall coefficient of the material. A surprising aspect is that in some instances the Hall coefficient may be positive, $R_H > 0$. The standard explanation is that, in addition to electrons in the conduction band, positive holes in the valence band may also act as charge carriers.

The relative numbers of negative and positive charge carriers are assumed to depend on the electronic band structure of the conductor and can, in principle, be calculated by nearly-free electron methods. When positive carriers are in the majority, $R_H > 0$. However, there is no obvious explanation of the Hirsch correlation between positive Hall coefficient and superconductivity in terms of positive-hole charge carriers. It is not even sure that, irrespective of superconduction, energy band theory can actually predict meaningful correlation between hole concentration and positive R_H at all [131].

A totally different interpretation is possible in terms of golden excess. In this instance a magnetic field, applied perpendicular to a linear array of superconducting atoms, aligns the excess spin system

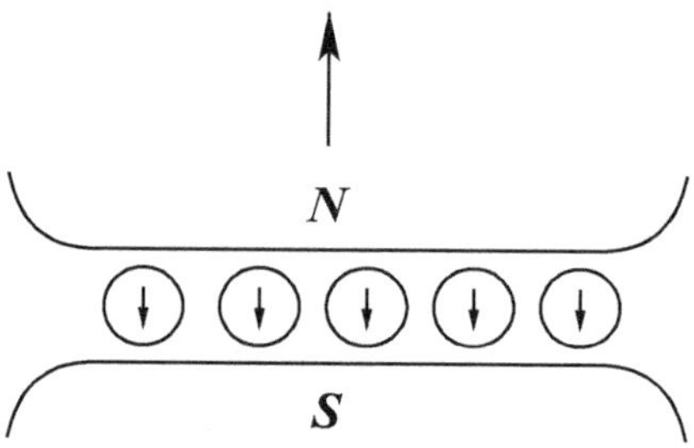

so as to generate an internal field in the opposite sense. The net effect is an increased R_H, which may even assume positive values for a sufficiently high excess.

The situation is more complicated in the case of 1-2-3 systems where both normal-state conductivity and Hall effect are anisotropic [132]. The Hall coefficient is positive when the magnetic field is applied parallel to the c-axis, and negative when the field is normal to c. In both cases the Hall density, defined as $n_H = 1/R_H$, is a linear function of temperature, precluding a direct identification of n_H with the carrier density [131].

The observed behaviour is in line with the golden-excess interpretation that defines the superconduction direction along the Ba/Sr chains, rather than along CuO_2 planes, as generally assumed. The anomalous conductivity ρ_c, compared to ρ_{ab} and to other, isotropic, superconductors also favours superconduction along Ba/Sr chains.

10.6 Nuclear Stability

This story has now come full circle. The original assumption of a fixed number of stable nuclides has often been disputed on the grounds that nuclear stability is a relative, rather than an absolute concept, as no atomic nucleus has an infinite half-life. This criterion is, admittedly, too strict to allow any useful definition of stability. In the present context it is therefore necessary to stipulate the thermodynamic parameters that may affect nuclear stability. In summary, it is the local curvature of space which determines, not only the arrangement of extranuclear matter, known as the electronic charge cloud, but also the configuration of each nucleus, and the overall atomic state of equilibrium. This curvature is fixed by the golden ratio. It is instructive to examine how this equilibrium is established.

Consider an atom consisting of a nucleus and a swarm of orbiting electrons. A given electron may reduce its potential energy by photon emission as it approaches the nucleus under coulombic attraction. The emission of each photon reduces the angular momentum by an amount $\hbar$. When the orbital angular momentum has been reduced to zero the radiation stops and the electron remains in a stationary state by spreading around the nucleus.

Because of the exclusion principle only two electrons, with paired spins, may share the space of minimum potential energy, which now constitutes an impenetrable shield between the nucleus and more energetic electrons. Should one of the base-level electrons be dislodged, its place is taken up by an outer electron, with emission of an x-ray photon. While the low-energy vacant site exists the atom is said to be in an excited state, which relaxes by the emission of an energy photon. The same thing happens in the nucleus. Because of previous perturbation (e.g. by cosmic-ray bombardment) a nucleus may find itself in an excited state. The occurrence of bound-state β^- decay (Section 8.3.1) reveals that an excited nuclear state can also result from disturbing the equilibrium between the nucleus and the extranuclear charge. In this instance, an otherwise stable nucleus is rendered radioactive by stripping off the electron shield that surrounds it. This event is quantified by a dramatic decrease of the observed nuclear half life.

Remarkably however, the nuclei in a given sample do not decay simultaneously, which means that they are not simultaneously in the excited state. This behaviour is not related to state of aggregation, as in chemical kinetics, and is also observed as the property of isolated nuclei in a storage ring. What emerges is a dynamic system that approaches equilibrium by random rearrangement steps. This rearrangement is entropy driven and the rate of decay, measured as the half-life, reflects how close or how far the system is from equilibrium. In near-equilibrium systems the rearrangement into the excited state, from where decay happens, takes a long time and the rate at which consecutive decay events occur, is slow. Systems, far from equilibrium may be considered highly excited and decay happens with minimum delay. The rate law is exponential, known as first order.

The stability of the electron shell, in turn, is subject to disturbance by nuclear imbalance, leading to events such as β capture. In this case the angular momentum that enables the emission of a γ-ray photon is carried by an antineutrino, released from the nucleus.

Nuclear stability, redefined in line with these principles, relies on the role of the golden ratio (τ) which dictates the optimum arrangement of both nucleons and extra-nuclear electrons, which together constitute a holistically balanced unit, conditioned by the geometry of space (π) and the rate of decay (e). A total of 264 nuclides satisfy the criteria and these are the nuclei, colloquially referred to as stable.

Chapter 11

Chemical Periodicity

11.1 Introduction

The major contribution of chemistry to theoretical science has been the periodic table of the elements. Although it enabled the classification of inorganic compounds, the anticipated impact of this discovery, on the understanding of chemical phenomena, has not been felt. Although some link between elemental periodicity and quantum theory became evident, the fundamental theory of chemistry, once thought secure, remained elusive. With the realization that the periodicity of atomic matter arises from number patterns, rather than quantum mechanics, another look at chemical concepts could well be useful. The unforeseen role of the golden ratio in the composition and function of atomic nuclei could conceivably recur in the laws that govern the behaviour of molecules and other chemical species. It would indeed be rather odd if the self-similarity that links atomic nuclei and botanical growth was absent from chemical systems, intermediate to these.

The quest for chemical theory remains focussed on chemical change and the nature of chemical interaction, involving electrons, atoms and molecules. Conventional wisdom has identified the key to solving this vexing problem as *chemical bonding*. It is exasperating to find that a body of self-perpetuating bonding jargon, supposedly based on quantum theory, has gained such respectability, by uncritical use among students of chemistry, that any re-examination of the situation is treated with deep suspicion. The hidden reality is that the little understanding there is, can be recognized as remnants of the nineteenth century theory of chemical affinity that developed in parallel with the periodic table. The modern concepts of valency, electronegativity and reactivity have their origins in this era and have never gained real quantum-mechanical respectability. It is proposed to revisit these concepts as periodic, rather than quantum, functions in an effort to rediscover their fundamental meaning.

11.2 Electronegativity

In historical context the electronegativity concept developed in a natural way from the early distinction between antagonistic elements, later loosely identified as metals and non-metals, and later still as electropositive and electronegative elements, classified as a function of their periodic atomic volumes [31] – (1953).

Lothar Meyer's atomic volume curve, in its modern form (Figure 6.5), still provides one of the most convincing demonstrations of elemental periodicity and an equally clear exposition of atomic electronegativity. The segmented regions occur in strict alternation as negatively and positively sloping curves. This is exactly the basis on which electropositive and electronegative elements used to be distinguished traditionally [31]. This theoretical notion, in one form or the other, has survived into the present, where, as will be shown, it provides a precise definition of electronegativity.

The first known attempt to quantify electronegativity was due to Pauling, who devised an empirical scale, on the basis of thermochemical data [133]. It relies on the simple idea that an *electrovalent* linkage between a pair of electropositive and electronegative atoms results from the transfer of an electron between the two:

$$\text{e.g.} \qquad \mathrm{Na + Cl \rightarrow Na^{+}Cl^{-}}$$

whereas a *covalent* linkage between a homopolar pair requires equal sharing of two electrons:

$$\text{e.g.} \qquad \mathrm{Cl\cdot + \cdot Cl \rightarrow Cl{:}Cl}$$

All possible diatomic combinations, AB, correspond to situations between the two extremes. The larger the difference in electronegativity, $|x_A - x_B|$, the larger is the electrovalent component that stabilizes the linkage, and the smaller the covalent contribution. For any diatomic interaction, such as A–B, in a molecule, the difference between the dissociation energy D(A–B) and the (arithmetic or geometric) mean of the energies D(A–A) and D(B–B) is assumed [134], to reflect the difference in electronegativity of the two atoms, interpreted as the power of an atom in a molecule to attract electrons to itself. To hold for both definitions of the covalent mean it is assumed that, ideally

$$\frac{1}{2}\left(D_{AA} + D_{BB}\right) = \sqrt{D_{AA} \cdot D_{BB}}$$

or squared:

$$D_{AA}^2 - 2D_{AA}D_{BB} + D_{BB}^2 = (D_{AA} - D_{BB})^2 = 0 \text{ or } \Delta_{AB}^2$$

such that $\Delta = 0$ for $A = B$, or $\Delta = |x_A - x_B|^2$ for $A \neq B$, defining electronegativities x_A and x_B through the relation

$$x_A - x_B = \sqrt{|D_{AA} - D_{BB}|}$$

The well-known Pauling scale of electronegativities results from this definition of x_A on specifying dissociation energies in units of electron volt.[1]

In a more fundamental approach Mulliken redefined electronegativity [135] as the average of the ionization potential and electron affinity of an atom. Although the Mulliken definition:

$$\chi_M(A) = \frac{1}{2}(I_A + E_A)$$

expresses electronegativity as an energy, rather than its square root as in the Pauling definition, it is widely interpreted that a linear relationship exists between the two scales, as in [136]:

$$\chi_M = 3.15\chi_P$$

This absurd assumption has never been formally challenged, with the result that the notion of a periodic electronegativity function has lost all credibility, despite sporadic efforts to revive it in terms of experimental variables such as hardness [137] and atomic polarizability [138], or the empirical formulae:-

$$\chi = 0.313\{(n + 2.60)/r^{2/3}\}$$
$$\chi = 0.31\{(n + 1)/r\} + 0.50$$
$$\chi \propto (n + 1)/\sqrt{2r}$$
$$\chi = 0.359\left(Z_f/r^2\right) + 0.744$$
$$\chi = 1.66(n/\alpha)^{1/3} + 0.37$$

of Liu [139], Gordy [140], Cottrell and Sutton [141], Allred and Rochow (AR) [142] and Nagle [138], respectively.

All of these formulations suffer from the same defect of being dimensionally inconsistent. They have no common basis and no sensible fudge factor

[1]In later work [134] Pauling used a value of 30 instead of the factor 23 that converts kilocalorie per mole into electron volt.

to project a theoretically sound relationship from these proposals, without addressing the problem of dimensional confusion, has been defined.

The scales of Pauling [134] and of AR [142] are routinely cited [143] and widely used on the assumption of being equivalent, despite huge numerical discrepancies. The Pauling scale measures the square root of an energy and the AR scale measures a force. Simple dimensional analysis shows that the two scales are related by a factor K, with dimensions such that:

$$K \cdot M^{1/2}LT^{-1} = MLT^{-2}$$

Charge, measured in coulomb, has dimensions $C = M^{1/2}L^{3/2}T^{-1}$, which means that K has dimensions $CL^{-3/2}$, or in terms of volume (L^3), $CV^{-1/2} = (C^2V^{-1})^{1/2}$. Neither charge nor volume, featuring in this complicated conversion, has been properly explained.

In reality, both scales are simply too erratic and unreliable to serve as a benchmark to guide theoretical modelling. On Pauling's scale there are too many criteria at play: bond energy or bond dissociation energy; arithmetic mean or geometric mean; different ligands around the central bond of interest; different oxidation states; standard state or valence state; molecular geometry – all of these variables have been implicated as environmental factors that could affect the estimate of atomic electronegativity in molecules. The AR scale, empirically based on variable single-bond covalent radii, is hardly more secure, but somewhat more regular as a periodic function, compared to the Pauling scale.

The simple truth is that the state of any atom within a molecule depends too critically on its special environment to support the idea of an atom with fixed electronegativity in all molecules. An alternative is to consider the valence state of a free atom as reference; more precisely, the first activated state of a neutral atom, before it enters into chemical interaction with another atom, electron or molecule. To demonstrate the validity of such theory it is not necessary to reproduce any of the empirical scales in exact detail. However, there should be strong qualitative agreement with the intuitive notion.

11.2.1 The Quantum-Potential Scale

A proper theoretical basis of electronegativity has in fact been identified before [89] as the ground-state energy E_g of a free valence-state electron, confined to its first ionization sphere. In order to avoid dimensional conflict, valence-state electronegativity may be defined equally well as the square root, $\sqrt{E_g}$, of the valence-state confinement energy, which, by definition, should relate to AR electronegativities by $\chi_v = \chi_{AR}/K$.

Confinement energies are calculated directly from atomic ionization radii, r_0, obtained by Hartree–Fock–Slater simulation of uniform compression of atoms [6] under the modified boundary condition $\lim_{r \to r_c} \psi = 0$, in which $r_c < \infty$. Increased pressure is simulated by decreasing the critical radius r_c of the impenetrable sphere. The simulation consists of multiplying all one-electron wave functions by the step function

$$S = \exp\left[-\left(\frac{r}{r_c}\right)^p\right], \text{ with } p >> 1$$

as part of the iterative procedure.

Simulated compression results in raising all electronic energy levels. Inter-electronic interaction leads to internal transfer of energy such that a single electron eventually reaches the ionization limit on sustained compression. At this point there is zero interaction between the ionized valence electron and the atomic core, although the valence electron remains confined to a sphere of radius r_0. The ground-state energy of such a confined particle amounts to

$$E_g = \frac{h^2}{8mr_0^2}$$

With energies in eV and radii in Å units the numerical value of the operational quantity $\sqrt{E_g} = 6.133/r_0$. The state of the confined particle is awkward to explain by conventional quantum theory but is readily understood in the Bohmian interpretation that describes the energy E_g, that derives from compressive work, as pure quantum potential energy [11]:

$$V_q = -\frac{h^2\nabla^2 R}{8\pi^2 mR}$$

Rearranged into the more familiar form of Schrödinger's amplitude equation:

$$\nabla^2 R + \frac{8\pi^2 m}{h^2} V_q R = 0$$

it correctly describes the confined particle, providing $V_q = E_g = h^2/8mr_0^2$ and R is the zero-order spherical Bessel function with first zero at r_0.

Since r_0 is characteristic of each atom, characteristic energies are predicted for atomic valence-state electrons. It is the atomic equivalent of the Fermi energy of an electron at the surface of the Fermi sea in condensed phases, and in that sense represents the chemical potential of the valence electron for each atom. Electronegativity has been defined independently [137] in almost identical terms before.

Relating E_g to electronegativity provides the theoretical basis of this concept, derived non-empirically and without assumptions from first principles. It is a function of the electronic configuration of atoms only and emerges naturally as the response of an atom to its environment. It is indeed the tendency of an atom to interact with electrons and the fundamental parameter that quantifies chemical affinity and bond polarity. It is precisely "the average one-electron valence shell energy of a ground-state free atom" that Allen proposed [144] as the third dimension of the periodic table.

The Valence State

Two independent sets of ionization radii for all atoms have been calculated [6, 11] using exponential parameters $p = 20$ and 100, respectively. In both cases a clear periodic relationship appears, but the values at $p = 100$ are consistently lower. This difference reflects the steepness of the barrier that confines the valence electron, shown schematically in Figure 11.1.

In a chemical environment the barrier is unlikely to be infinitely sharp and there should be some optimum value of p that describes the situation best. The choice of $p = 20$ was conditioned by comparison with independent chemical evidence that relates to ionization radii. A set of atomic radii, derived as an estimate to describe a single valence electron, uniformly spread over a characteristic sphere for each atom, has been known for a long time [145]. These radii were fixed by the assumption that electrostatic interaction between such uniform charge distributions, surrounding monopositive atomic cores, should correspond to the experimental values of binding energy for any pair of atoms.

The idea of relating E_g to electronegativity is supported by the interesting correspondence that it shows with atomic first ionization potentials [93]. On average

$$I_P/E_g = 0.42n$$

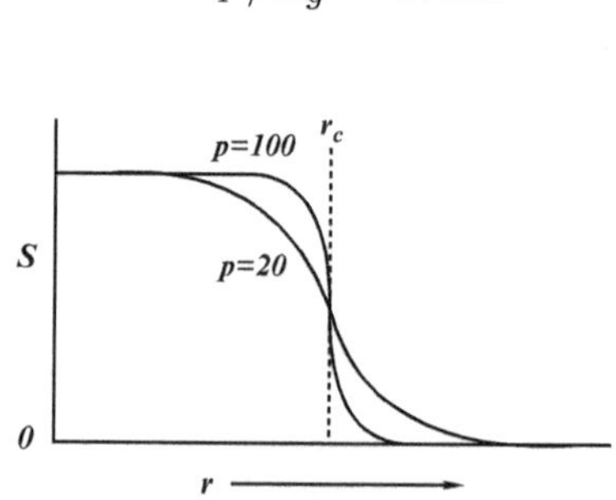

Figure 11.1: *The step function $S = \exp[-(r/r_c)^p]$ – schematic.*

where n is the number of the main periodic group in which the atom occurs. This relationship allows a fair estimate of radii in cases of uncertainty, e.g. for light elements like He.

It is instructive to examine the periodic variation of valence-state electronegativities $\chi_v = \sqrt{E_g}$, shown in Figure 11.2. The sequence, as a function of atomic number, fragments spontaneously into the same segments as the Lothar–Meyer curve. All segments, in this case, have positive curvature, sloping towards the origin on the left and towards closed-shell configurations on the right. The qualitative trends are immediately recognized as closely related to the known empirical trends of electronegativity scales. The slope of these curves, at an atomic position, represents the change in energy as a function of atomic number (electron count), and defines the chemical potential of the electrons, $\mathrm{d}E/\mathrm{d}N = -\mu$, at that point. When the AR scale is plotted on the same graph there is exact agreement with χ_v only at the positions of the closed-shell noble gases and a linear variation with atomic number within each period of elements.

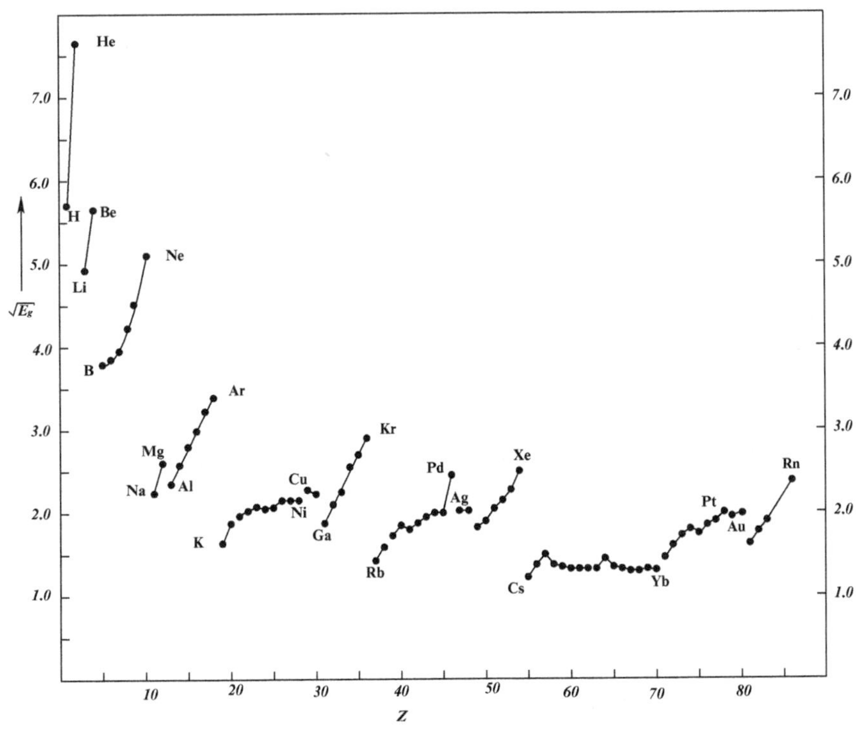

Figure 11.2: *Periodic variation of* $\chi_v = \sqrt{E_g}$ *as a function of atomic number.*

11.2.2 Derivation of a Common Scale

Convincing linear relationships between χ_{AR} and χ_v could be demonstrated immediately for all periodic families with p valence shells. A convenient physical interpretation of the relationship is to consider K as a valence-electron density, $\rho = 1$ for a closed shell and $\rho = x^{-\mu} < 1$ for non-closed shells with μ vacant sites. As an example, $x = 1.1$ for the $2p$ valence shell. Although it is tempting, and feasible, to scale all calculated values of χ_v to the AR scale, by assigning appropriate values to x, the chemical significance of electronegativity is obscured in the process. All evidence therefore supports the conclusion that $\sqrt{E_g}$ is a good measure of valence-state electronegativity. Not only does it parallel both the AR and Pauling scales convincingly, but it also relates directly to the Mulliken proposal.

For $n > 2$ the average formula, $E_g = 2.4I_P/n$ reduces to $E_g \simeq 0.5I_P$, and applies to the majority of elements. Because electron affinities are generally of smaller magnitude than I_P, the Mulliken energy-scale formula is explained directly. A sufficient number of high-precision atomic electron affinities, to test the Mulliken formula,

$$\chi_M^2 = E_g = 0.5(I_P + E_A)$$

are however, known [93]. By scaling χ_M against χ_v within periodic families, reliable values of unmeasurable electron affinities can be obtained by interpolation.

It is noted that electron affinity represents the energy difference between the ground state of a neutral atom and the lowest state of the corresponding negative ion. In many cases the negative-ion state is too unstable to allow experimental measurement of E_A. All noble gases with closed p sub-shells, and many elements with closed s sub-shells belong to this group. Interpolated values for such elements are parenthesized in Table 11.1.

Electronegativities on the Mulliken scale, as redefined here, agree remarkably well with the valence-state values calculated from ionization radii, and by inference, also with Pauling electronegativities, once the square-root relationship between the scales is properly recognized. The obvious general solution is to abanbon all efforts to scale towards traditional values of χ and to recognize $\chi_v = 6.133/r_0$ as the new electronegativity function.

11.3 Chemical Bonding

The ultimate test for any electronegativity scale, by definition, lies in its ability to rationalize the nature and energy of chemical interactions. As noted

Table 11.1: *Summary of ionization radii (r_0), characteristic radii (r_x), and Mulliken radii (r_M) in Å units; valence-state (χ_v) and Mulliken (χ_M) electronegativities in units of $\sqrt{eV}$; interatomic distance (r_e) or its estimate (d); measured (D_x) and calculated (D_c) dissociation energies, in kilojoule per mole, for diatomic molecules.*

Z	Sy	r_0	χ_v	χ_M	$r_e, (d)$	d/r_x	D_x	D_c	r_x	r_M
1	H	0.90	6.81	6.25	0.741	0.96	436	432	0.77	0.98
2	He	0.30	8.18	(0)						
3	Li	1.25	4.91	1.73	2.673	0.99	110	103	2.70	3.54
4	Be	1.09	5.63	(0.8)	(1.94)	1.14	59	68	1.70	
5	B	1.62	3.79	2.07	1.590	0.86	290	300	1.85	2.96
6	C	1.60	3.83	2.50	1.312	0.71	596	614	1.85	2.45
7	N	1.56	3.93	(1.2)	1.098	0.67	857	947	1.65	
8	O	1.45	4.23	2.75	1.208	0.80	498	496	1.51	2.23
9	F	1.36	4.51	3.23	1.412	1.03	159	162	1.37	1.90
10	Ne	1.20	5.11	(1.7)						
11	Na	2.73	2.25	1.69	3.079	1.03	75	74	3.00	3.63
12	Mg	2.35	2.60	(0)	(2.78)	1.32	9	15	2.10	
13	Al	2.61	2.35	1.79	2.466	0.95	133	139	2.60	3.43
14	Si	2.40	2.56	2.18	2.246	0.75	310	314	3.00	2.91
15	P	2.20	2.79	2.37	1.893	0.67	485	488	2.81	2.59
16	S	2.05	2.99	2.49	1.889	0.71	425	427	2.66	2.46
17	Cl	1.89	3.24	2.88	1.988	0.86	243	241	2.30	2.13
18	Ar	1.81	3.39	(3.0)						
19	K	3.74	1.64	1.56	3.905	1.04	57	56	3.74	3.93
20	Ca	3.26	1.88	1.75	(3.43)	1.18	15	30	2.90	3.50
21	Sc	3.13	1.96	1.83	(2.79)	0.89	163 ± 21	151	3.13	3.35
22	Ti	3.01	2.04	1.86	(2.52)	0.97	118	123	2.60	3.30
23	V	2.95	2.08	1.91	(2.28)	0.81	269	257	2.80	3.21
24	Cr	2.98	2.06	1.93	(2.17)	0.87	155	215	2.50	3.18
25	Mn	2.94	2.09	(0.2)	(2.38)	1.19	42	40	2.00	
26	Fe	2.87	2.15	2.01	(2.16)	1.03	118	109	2.10	3.05
27	Co	2.85	2.15	2.07	(2.18)	0.91	167	183	2.40	2.96
28	Ni	2.86	2.14	2.10	(2.17)	0.87	204	215	2.50	2.92
29	Cu	2.85	2.15	2.12	2.220	0.87	201	206	2.55	2.89
30	Zn	2.78	2.21	(0.2)	(2.32)	1.16	29	34	2.00	
31	Ga	3.29	1.86	1.79	(2.12)	1.01	138	129	2.10	3.43
32	Ge	2.94	2.09	2.14	(2.13)	0.73	274	361	2.90	2.87
33	As	2.62	2.26	2.30	2.103	0.72	382	375	2.92	2.67

Table 11.1: *(Continued)*

Z	Sy	r_0	χ_v	χ_M	$r_e, (d)$	d/r_x	D_x	D_c	r_x	r_M
34	Se	2.40	2.56	2.43	2.166	0.75	331	325	2.90	2.52
35	Br	2.28	2.69	2.75	2.281	0.88	290	300	2.59	2.23
36	Kr	2.12	2.89	(2.4)						
37	Rb	4.31	1.42	1.53	(4.31)	1.05	46	49	4.10	4.01
38	Sr	3.83	1.60	1.69	(3.74)	1.25	16	14	3.00	3.63
39	Y	3.55	1.73	1.81	(3.09)	0.87	159 ± 21	148	3.55	3.39
40	Zr	3.32	1.85	1.84	(2.77)	0.73	298	275	3.80	3.33
41	Nb	3.30	1.80	1.96	(2.49)	0.62	513	431	4.00	3.13
42	Mo	3.21	1.86	1.98	(2.37)	0.70	436	353	3.40	3.10
43	Tc	3.16	1.94	1.98	(2.35)	0.74	330	316	3.16	3.10
44	Ru	3.13	1.98	2.05	(2.31)					2.99
45	Rh	3.08	1.99	2.07	(2.34)	0.76	236	252	3.08	2.96
46	Pd	2.49	2.46	2.11	(2.39)	0.96	136	135	2.49	2.91
47	Ag	3.04	2.02	2.11	(2.51)	0.90	163	165	2.80	2.91
48	Cd	3.02	2.03	(0)	(2.59)	1.30	11	16	2.00	
49	In	3.55	1.73	1.74	(2.83)	0.98	100	104	2.90	3.52
50	Sn	3.26	1.88	2.06	(2.44)	0.87	195	192	2.80	2.98
51	Sb	3.01	2.04	2.20	(2.52)	0.74	299	293	3.40	2.79
52	Te	2.81	2.18	2.34	2.557	0.77	258	252	3.30	2.62
53	I	2.60	2.35	2.60	2.666	0.91	153	147	2.92	2.36
54	Xe	2.49	2.50	(1.5)						
55	Cs	4.96	1.24	1.48	4.47	1.04	44	48	4.30	4.14
56	Ba	4.48	1.37	1.64						3.74
57	La	4.13	1.48	1.74	(3.25)	0.72	247 ± 21	244	4.50	3.52
58	Ce	4.48	1.37	1,80	(3.18)	0.71	243 ± 21	254	4.48	3.41
59	Pr	4.53	1.35	1.79	(3.17)	0.91	130 ± 29	125	3.50	3.43
60	Nd	4.60	1.33	(1.0)	(3.16)	0.99	84 ± 29	89	3.20	
61	Pm	4.56	1.34	(1.1)						
62	Sm	4.56	1.34	(1.0)						
63	Eu	4.60	1.33	1.81	(3.47)	1.16	34 ± 17	33	3.00	3.39
64	Gd	4.22	1.45	(0.5)						
65	Tb	4.59	1.34	(0.7)	(3.07)	0.90	131 ± 25	136	3.40	
66	Dy	4.56	1.34	(0.7)						
67	Ho	4.63	1.32	(0.4)	(3.03)	1.01	84 ± 30	85	3.00	
68	Er	4.63	1.32	(0.3)						
69	Tm	4.62	1.33	1.90	(3.00)	1.11	54	52	2.70	3.23
70	Yb	4.66	1.32	1.76	(3.38)	1.13	21 ± 17	41	3.00	3.48

Table 11.1: *(Continued)*

Z	Sy	r_0	χ_v	χ_M	r_e, (d)	d/r_x	D_x	D_c	r_x	r_M
71	Lu	4.24	1.45	1.70	(2.99)	0.88	142 ± 33	150	3.40	3.61
72	Hf	3.83	1.60	(0.1)						
73	Ta	3.57	1.72	1.98	(2.49)	0.67	390 ± 96	372	3.70	3.10
74	W	3.42	1.79	2.08	(2.38)	0.68	486 ± 96	376	3.50	2.95
75	Re	3.38	1.81	2.00	(2.38)	0.68	386 ± 96	376	3.50	3.07
76	Os	3.37	1.82	2.18	(2.33)	0.67	415 ± 77	393	3.50	2.81
77	Ir	3.23	1.90	2.29	(2.36)	0.73	361 ± 68	324	3.23	2.68
78	Pt	3.16	1.94	2.35	(2.39)	0.75	308	295	3.20	2.61
79	Au	3.14	1.95	2.40	2.472	0.81	226	232	3.05	2.56
80	Hg	3.12	1.97	(0.9)	(2.61)	1.31	17	16	2.00	
81	Tl	3.82	1.61	1.75	(2.96)	1.06	63 ± 17	68	2.80	3.50
82	Pb	3.47	1.77	2.21	(3.05)	0.98	87	97	3.10	2.78
83	Bi	3.19	1.92	2.03	2.660	0.83	197	204	3.19	3.02
84	Po	3.14	2.06	2.27	(2.91)	0.83	186	187	3.50	2.70
85	At	3.12	2.16	(8.3)*						
86	Rn	3.82	32.31	(1.9)						

* Interpolated ionization potential.

Homonuclear bonds

Bond	d	d'	D_x	D_c	MM	d	d'	D_x	D_c
CH_3–CH_3	1.541	0.83	368	353	Cr	2.67	1.07	99	72
CH_2=CH_2	1.337	0.72	598	592	Mo	2.76	0.81	260	212
CH≡CH	1.202	0.65	836	815		2.50	0.74	411	300
NH_2–NH_2	1.47	0.89	296	294		2.21	0.65	593	443
HO–OH	1.48	0.98	213	200		2.12	0.62	759 ± 220	507
SiH_3–SiH_3	2.32	0.77	293	290	W	3.22	0.92	234	119
HS–SH	2.05	0.77	284	327		2.16	0.62	779 ± 220	493
GeH_3–GeH_3	2.41	0.83	259	225	Re	2.61	0.75	129	270
						2.21	0.63	562 ± 121	471

already, the thermochemical scheme devised for this purpose by Pauling, relies on too many variables to be representative of all elements. The quantum-potential valence-state scheme that only depends on atomic ionization radii, like the Mulliken scale, has a clear advantage in this respect and because of its more fundamental basis it is preferred over the empirical Mulliken scale.

The most representative class of experimentally well characterized relevant compounds that could be used as a benchmark are the diatomic molecules, which have been studied spectroscopically in terms of dissociation energy and equilibrium interatomic distance. A reliable, and frequently updated, list of these parameters is available from the Handbook of Chemistry and Physics [93]. Analysis of this data set by point-charge simulation of intramolecular interactions provides a direct procedure to assess the utility of ionization radius as a parameter that describes the strength of chemical bonds.

11.3.1 Point-Charge Model

In the same way that electronegativities determine the polarity of diatomic interactions, ionization radii should define the effective electronic charge clouds that interpenetrate to form diatomic molecules, as shown schematically in the following figure. The overlap of two such spheres defines a lens of focal lengths fixed by the ionization radii, r_1 and r_2, at an interatomic distance $d = x_1 + x_2$.

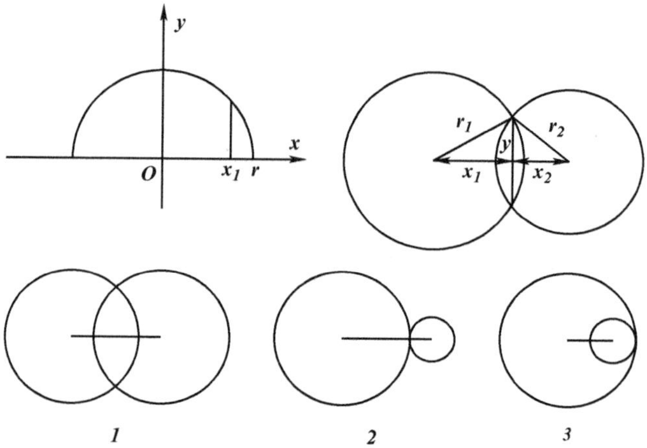

The lens consists of two parts, each of which is generated as a solid of revolution by rotating the semi-circle $x^2 + y^2 = r^2$ about Ox. The resulting solid body has the volume

$$\begin{aligned} v_1 &= \int_{x_1}^{r} \pi y^2 \mathrm{d}x \\ &= \int_{x_1}^{r} \pi \left(r^2 - x^2\right) \mathrm{d}x \\ &= \pi \left[r^2 x - \frac{1}{3} x^3 \right]_{x_1}^{r} \\ &= \pi \left[\frac{2}{3} r^3 - r^2 x_1 - \frac{1}{3} x_1^3 \right] \end{aligned}$$

In the problem of interest: $r_1^2 - x_1^2 = y^2$, $r_2^2 - x_2^2 = y^2$.

$$r_1^2 - r_2^2 = x_1^2 - x_2^2 \,,\; x_1 + x_2 = d \,,\; x_2 = d - x_1$$

Hence

$$\begin{aligned} r_1^2 - r_2^2 &= x_1^2 - d^2 + 2x_1 d - x_1^2 \\ &= 2x_1 d - d^2 = a\,(\text{say}) \\ x_1 &= \frac{d}{2} + \frac{a}{2d} \\ x_2 &= \frac{d}{2} - \frac{a}{2d} \end{aligned}$$

The overlap volume of the two spheres:

$$\begin{aligned} V_0 &= v_1 + v_2 \\ &= \pi\left[\frac{2}{3}\left(r_1^3 + r_2^3\right) - r_1^2\left(\frac{d}{2} + \frac{a}{2d}\right) + \frac{1}{3}\left(\frac{d}{2} + \frac{a}{2d}\right)^3 - r_2^2\left(\frac{d}{2} - \frac{a}{2d}\right) + \frac{1}{3}\left(\frac{d}{2} - \frac{a}{2d}\right)^3\right] \\ &= \pi\left[\frac{2}{3}(r_1^3 + r_2^3) - \frac{d}{2}(r_1^2 + r_2^2 - \frac{1}{4d}(r_1^2 - r_2^2)^2 + \frac{d^3}{12}\right] \end{aligned}$$

For the special cases:

1. $r_1 = r_2 \quad : \quad V_0 = \pi\left(\frac{4}{3}r^3 - r^2 d + \frac{d^3}{12}\right)$

2. $d = r_1 + r_2 \quad : \quad V_0 = 0$

3. $d \leq r_1 - r_2 \quad : \quad V_0 = \frac{4}{3}\pi r_2^3$

Each of the two overlapping spheres is considered to enclose a single valence electron that surrounds a monopositive atomic core. The electrostatic interaction is represented by point charges, related in magnitude to V_0, V_1 and V_2, by assuming that electron 1 is associated with nucleus 2 for a fraction $\epsilon = V_0/V_1$ of the time, such that a charge of ϵ^+ appears on nucleus 1 and a charge of $\delta^+ = V_0/V_1$ on nucleus 2 [145]. The foreign electrons which move in the field of nucleus 1 for a time fraction δ can be represented by a charge δ^- in the overlap region, and likewise by a charge ϵ^- with respect to nucleus 2.

The total electrostatic energy of interaction is made up of four components:

1. Nuclear–nuclear repulsion amounting to $\delta\epsilon/d$

2. Electron–electron repulsion, given by $\delta\epsilon/p$, where $p = r_1 + r_2 - d$, the thickness of the lens

3. Attraction between nucleus 1 and electron 2, given by $\delta\epsilon/b$, where $b = (r_1/r_2)(d/2)$, with $r_1 < r_2$

4. Interaction between nucleus 2 and electron 1, given by $\delta\epsilon/(d-b)$

From the expression for V_0 follows the formula for

$$\delta\epsilon = \left[\frac{d^3}{16} - \frac{3d}{8}\left(r_1^2 + r_2^2\right) - \frac{3}{16d}\left(r_2^2 - r_1^2\right)^2 + \frac{1}{2}\left(r_1^3 + r_2^3\right)\right]^2 / (r_1 r_2)^3$$

When the smaller cloud is completely enclosed by the bigger ($r_2 > r_1 + d$) the extra polarization amounts to an additional attraction between point charges of $\pm[1 - r_1/(r_2 - d)]$ at a distance d. The total energy of interaction reduces to:

$$E = K\left[\delta\epsilon\left\{\frac{1}{b} + \frac{1}{d-b} - \frac{1}{d} - \frac{1}{p}\right\} + X\right]$$

where K is a dimensional constant and $X = 0$, unless $r_2 > r_1 + d$, when $X = [1 - r_1/(r_2 - d)]^2/d$.

To demonstrate the validity of the point-charge model it is applied to homopolar diatomic molecules first. In this special case the overlap formulae reduce to:

$$V_0 = \pi\left[\frac{4}{3}\pi r^3 - r^2 d + \frac{d^3}{16}\right]$$

$$\epsilon = 1 - \frac{3d}{4r} + \frac{1}{16}\left(\frac{d}{r}\right)^3$$

$$E = K\epsilon^2\left[\frac{3}{d} - \frac{1}{2r - d}\right]$$

The interaction energy (E) between a homonuclear pair of atoms can be simulated by charges, defined in terms of $d' = d/r_0$, as

$$\varepsilon = 1 - 3d'/4 + (d')^3/16$$

$$E = K\varepsilon^2/r_0\left[\frac{3}{d'} - \frac{1}{2 - d'}\right] \tag{11.1}$$

Setting $K = r_0$, the binding energy (E'), obtained in diatomic units and shown as a negative quantity, varies as a function of d/r_0 as in Figure 11.3.

The energy function (11.1) remains attractive at all values of d' until ε reaches its natural limit, which by definition corresponds to exactly one pair of electrons, allowed by the exclusion principle, between the nuclei. An

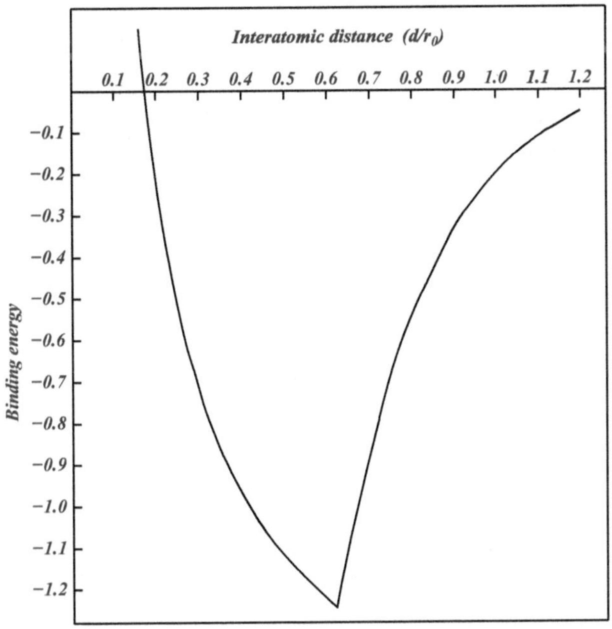

Figure 11.3: *Binding energy curve for homonuclear diatomic molecules, based on point-charge interaction; in diatomic units.*

obvious turning point, by this argument, is expected to occur when $\varepsilon = 1/2$, in which case each nucleus controls one half of the bonding pair. This occurs when $d' = 0.695$. However, trial calculations show several diatomic molecules having $d' < 0.695$. It simply means that the bonding pair at this point is spread over the total volume, defined by the overlap lens, which implies that the charge density in the internuclear region remains below the exclusion limit. Further calculations show that the minimum ratio for any diatomic molecule, $d/r_0 = d' > \tau$, approaches the golden mean.

The inference is clear: the exclusion principle kicks in at the point where the internuclear region is saturated with a spin-paired set of electrons. *Internuclear region* is interpreted as a volume that depends on both internuclear distance and atomic size, i.e. d/r_0 with a critical value of τ. In the same way that the golden ratio determines the phyllotaxis and packing density of nucleons and florets, it also regulates the formation of chemical bonds. In this case the effect is known as the Pauli exclusion principle, whose operation has remained a mystery until now. Being linked directly to τ, the riddle is solved: the exclusion principle is dictated by the curved structure of space.

11.3.2 The Diatomic Energy Function

According to (11.1) the point-charge simulation of homonuclear diatomic species reduces to a common basis in terms of a single characteristic potential-energy function. As additional factors need to be taken into account when dealing with heteronuclear diatomics, only homonuclear species will be considered here as a test of the valence-state scheme.

The point-charge density increases monotonically with decreasing d' until it reaches a maximum at the critical ratio of $d/r_0 = \tau$. At this point $\varepsilon = 0.5513$ and

$$E' = 0.3\left[\frac{3}{\tau} - \frac{1}{2-\tau}\right]$$

which rearranges into the expression

$$E' = 0.3\left[5.734 - \frac{1}{d'}\right]$$

that describes the energy for all $d' < \tau$. The total binding-energy curve therefore consists of attractive and repulsive parts that intersect at the point $d' = \tau$, $E' = 1.244$.

The dissociation energy of a homonuclear diatomic molecule (Figure 11.3), is shown in dimensionless units (D'), as a function of the interatomic distance, in units of the characteristic radius, $d/r_0 \equiv d'$. The energy minimum on the curve corresponds to the maximum possible dissociation energy, at the golden ratio $d' = \tau$. The dissociation energy in familiar units is obtained as

$$D_c = KD'/r_0$$

with r_0 in units of Å and the dimensional constant $K = 14.36$ or 1385.7 for units of electron volt or kilojoule per mole, respectively.

For a large number of, especially metallic, diatomic molecules equilibrium interatomic distances have not been measured spectroscopically. These elements could be included in the sample by noting that for those metals with measured r_e, it is related to δ, the distance of closest approach in the metal, by $r_e = 0.78\delta$. On assuming this to be generally valid, reference values of interatomic distance (d) became available for virtually all elements, as shown in Table 11.1.

Using ionization radii as characteristic radii, dissociation energies of the correct order were calculated for the majority of elements. Modest adjustment of r_0 to a set of empirical radii, r_x, resulted in the detailed match between observed (D_e) and calculated (D_x) dissociation energies, as shown in Table 11.1. A comparable set of Mulliken radii, calculated as $r_M = 6.133/\chi_M$, for each element, is also listed in the table. The variation of r_0, r_x and r_M across periodic groups is compared in Figure 11.4.

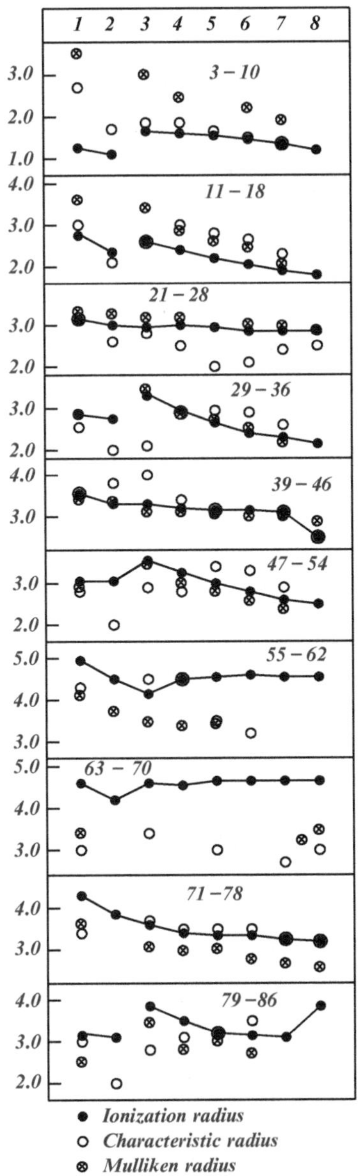

Figure 11.4: *Electronegativities as periodic functions.*

The most informative comparison is between ionization radii and the set of empirically adjusted radii. The Mulliken radii confirm that the observed trends reflect chemically meaningful effects. All of those elements that form diatomic molecules rather than metals (i.e. the traditional non-metals) have $r_x > r_0$. A number of traditional metals, which also have $r_x > r_0$, i.e. Li, Be, Na, Zr, Nb, Mo, La, Ta, W, Re and Os, are all known to form exceptionally stable diatomic molecules, compared to their congeners. In addition, the elements Mo, W and Re are known to form supershort, so-called quadruple dimetal bonds [146]. It has been stated [147] that niobium and tantalum "though metallic in many respects, have chemistries in the V oxidation state which are very similar to those of typical nonmetals."

The matching of characteristic radii to fit the curve of Figure 11.3 is more than one-dimensional parameterization. The radii, fixed by this procedure, can be used to derive dissociation energies for heteronuclear diatomic molecules and of all chemical bonds with known interatomic distances. As an example, dissociation energies of homonuclear bonds, as a function of interatomic distance, in compounds such as ethane, ethylene and acetylene, follow immediately from Figure 11.3, as tabulated in Table 11.1. The calculation for heteronuclear bonds is more complicated, but the results are equally convincing.

Despite considerable uncertainty in experimental dissociation energies of dimetal bonds of various order, the values calculated from observed interatomic distances agree reasonably well with measured values, summarized before [148].

11.3.3 Bond Order

The observation of a non-random discrepancy, between characteristic atomic radii r_x and ionization radii r_0, infers that the former also has real chemical significance, to be further explored. For electropositive atoms the relation $r_x < r_0$ indicates a tendency to form positive ions and metals rather than covalent bonds. For electronegative elements the discrepancy appears to be simply related to the properties commonly referred to as *bond order* and *multiple bonding.*

Although the picture of chemical bonding that emerges from the present analysis is radically different from the standard model, its elucidation of experimentally observed trends is no less general. The idea of bond order does not occur, as all chemical bonds are defined here in terms of a single pair of electrons, exchanged between chemically interacting atoms, as dictated by the exclusion principle. Cursory examination of Table 11.1 however, shows that the traditional classification in terms of bond order is paralleled by grouping chemical bonds in terms of interatomic distance, $d' = d/r_x$.

For bond orders

$$\begin{array}{rcl} <1 & : & 1.03 < d' \\ 1 & : & 0.8 < d' < 1.03 \\ 2 & : & 0.695 < d' < 0.8 \\ 3 & : & 0.65 < d' < 0.695 \\ 4 & : & 0.618 < d' < 0.65 \end{array}$$

This classification is neither more nor less rigorous than the definition of bond order as the number of electron pairs shared in the formation of multiple bonds, and has the distinct advantage of predicting a continuous range of non-integral bond orders. An alternative definition [149] that links the bond-order concept to molecular rearrangement explains the d' classification more logically than multiple bonding: "The elimination of a ligand pair from contiguous atoms in a molecule increases the bond order between them by unity ...". This happens because the distribution of valence electrons depends on the number of ligands. The valence density must obviously be different in ethane and ethene:

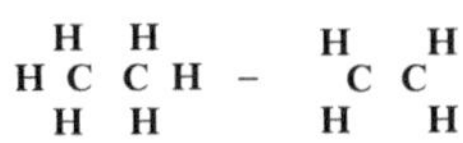

Since all covalent bonds are considered here as the exchange of a single pair of electrons between atomic cores, it is difficult to visualize a drastic modification of the interatomic C–C attraction caused by the transformation, but fairly obvious that the repulsion can be reduced considerably due to screening of the nuclei by the additional valence density. It amounts to modification of the effective nuclear charge [150]. The empirical screening function

$$k = -0.295b + 1.3 \tag{11.2}$$

has been demonstrated [149, 148] to relate binding energies as a function of bond order b, for all pairs of atoms, through the screening factor, k. In practice it was shown that two chemical bonds, which differ in order only, are both correctly described by the same Morse function, with allowance for screening according to (11.2).

In terms of the diatomic curve of Figure 11.3 a given bond, at two different orders is characterized by Morse curves, modified and centred at positions related by (11.2) [151]. This simulation produces correct dissociation energies as a function of observed interatomic distance, and without regard to

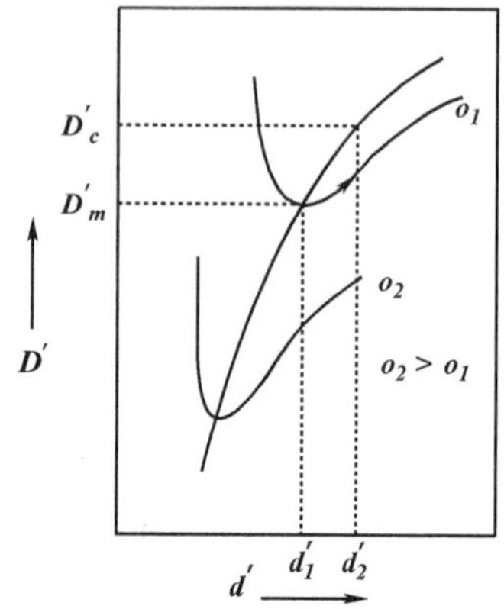

Figure 11.5: *Bonds of order* o_1 *and* o_2 *related by the curve* $d' = d/r_0$ *vs* D'.

steric effects related to ligand and valence-shell screening. When steric effects are taken into account, dissociation energies calculated from interatomic distance, expressed in units of ionization radius will be consistently too low.

Steric deformation of a chemical bond lowers the dissociation energy as indicated by the arrow along the local Morse curve in Figure 11.5, while stretching the interatomic distance from d'_1 to d'_2, the experimentally observed value. Calculation of D'_c, using the diatomic curve, gives a value which is too low compared to the measured D'_m. This discrepancy can be compensated for by assuming a characteristic radius $r_x > r_0$, bigger than the ionization radius, which measures interatomic distances free of strain.

Using the electronegative elements with open s–p valence shells, as an example, it is assumed that one electron from each atom mediates the covalent attraction, that paired electrons, which become part of the core, have only a steric effect, and the remaining unpaired electrons are responsible for screening.

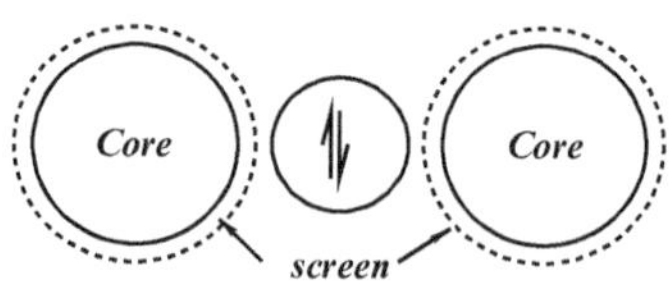

The bond orders predicted for the $2p$ diatomic molecules are B_2 – 1, C_2 – 2, N_2 – 3, O_2 – 2, F_2 – 1.

Screening reduces the interatomic distance along the diatomic curve and steric effects stretch the distance along the local Morse curve. Taking ethane as an example, assuming the interatomic distance free of strain as $d = 1.35$ Å, i.e. $d/r_0 = 0.84$, the dissociation energy is calculated as $D_c = 390$ kJmol^{-1}.

The difference of $\sim$40 kJmol^{-1} represents the energy required to stretch the C–C bond by $\sim$0.2 Å against a Hooke's law force constant of $\sim$5.5 mdyneÅ^{-1}.

11.3.4 Molecular Mechanics

The analysis of molecular properties by the methods of molecular mechanics is formulated in terms of three, essentially unobservable, Hooke's law parameters – force constant, bond length and dissociation energy of strain-free bonds [152]. These parameters are obtained as empirical estimates, only valid within closely related molecular families. The ideal parameter set, which is transferable across all elements, bond types and bond orders, is still unattainable.

Noting the use of screening constants, which relate different bond orders through a family of Morse curves, it follows that force constants can be derived for any bond. In principle, a fully transferable force field can hence be calculated directly from the diatomic energy function and atomic ionization radii.

11.4 Epilogue

Many of the claims made in this work will surely be considered extravagent at first sight. Some of the inferences may in fact be totally misconstrued. However, there is no doubt that a fundamental relationship between numbers and matter has been identified. This interaction is mediated by the golden ratio that shapes the world. It is no accident that self-similarity permeates the universe from the infinitesimally small to the immeasurably large. The totality is closed and appears periodic at all levels. This first observation is, inevitably, no more than the tip of an iceberg.

The numerical regularity in the structure of the solar system, known as the Bode–Titius law [78], represents an unexplored missing link in the self-similar symmetry of the cosmos. It occupies the gap between atomic nuclei, atoms, molecules and biological systems on the one hand, with galaxies, quasars and other mysterious objects, on the other. There is much to be explored before the interplay between π, e and τ can be appreciated.

Bibliography

1. M. Kaku, *Hyperspace*, 1994, University Press, Oxford.
2. M. Livio, *The Golden Ratio*, 2002, Review, London.
3. K.G. Ramanathan, *Acta Arith.*, 43 (1984) 209.
4. H. Rademacher, *Lectures on Elementary Number Theory*, 1977, Robert E. Krieger, Huntington, New York.
5. J.C.A. Boeyens, *J. Radioanal. Nucl. Chem.*, 257 (2003) 33.
6. J.C.A. Boeyens, *J. Chem Soc., Faraday Trans.*, 90 (1994) 3377.
7. A.W. Stewart, *Some Physico-chemical Themes*, 1922, Longmans Green and Co., London.
8. J.C.A. Boeyens, *Crystal Engin.*, 6 (2003) 167.
9. J.C.A. Boeyens, *Specul. Sci. and Technol.*, 15 (1992) 192.
10. E. Schrödinger, *Z. für Physik*, 12 (1922) 13.
11. J.C.A. Boeyens, *New Theories for Chemistry*, 2005, Elsevier, Amsterdam, The Netherlands.
12. M. Kline, *Mathematics in Western Culture*, 1953, Oxford University Press, Published in Pelican Books, London, 1972.
13. H. Pollard, *The Theory of Algebraic Numbers*, 1950, Mathematical Society of America/Wiley, New York.
14. J.V. Uspensky, *Theory of Equations*, 1948, McGraw-Hill, New York.
15. J.E. Maxfield and M.W. Maxfield, *Discovering Number Theory*, 1972, Saunders, Philadelphia.
16. P.J. Eccles, *An Introduction to Mathematical Reasoning*, 1997, University Press, Cambridge.
17. M.L. Stein, S.M. Ulam and M.B. Wells, *Amer. Math. Monthly*, 71 (1964) 516; 74 (1967) 43.
18. T. Goddard, *Ulam Spiral.* http://d4maths.lowtech.org/mirage/ulam.htm (accessed 17 July 2003).
19. M. Gardner, *Sci. Amer.*, 210 (1964) 120.
20. *Eric Weisstein's World of Mathematics.* Wolfram Research: mathworld.wolfram.com
21. H.M. Stark, *Michigan Math. J.*, 14 (1967) 1.

22. H.S.M. Coxeter, *Introduction to Geometry*, Classics edition, 1989, Wiley, New York.
23. R.B.J.T. Allenby and E.J. Redfern, *Introduction to Number Theory with Computing*, 1989, Routledge, New York.
24. P.M. Higgins, *Mathematics for the Imagination*, 2002, University Press, Oxford.
25. R. Penrose, *Math. Intelligencer*, 2 (1979) 32.
26. J.N. Ridley, *Math. Biosci.*, 79 (1986) 1.
27. I. Stewart, *Nature's Numbers: Unreal Reality of Mathematics*, 1995, Basic Books, New York.
28. D.W. Thomson, *On Growth and Form*, 1942, Cambridge University Press, Cambridge.
29. B.B. Mandelbrot, *Fractal Geometry of Nature*, 1988, Freeman, New York.
30. A. Deitmar, *A First Course in Harmonic Analysis*, 2nd ed., 2005, Springer, New York.
31. J.R. Partington, *Short History of Chemistry*, 1937; *A Textbook of Inorganic Chemistry*, 6th ed., 1953, Macmillan, London.
32. G.D. Parkes, *Mellor's Modern Inorganic Chemistry*, New revised edition, 1951, Longmans, London.
33. D. Mendeléeff, *The Principles of Chemistry*, Volume II, 1891, Translated from the Russian (5th ed.) by G. Kamensky, Longmans, London.
34. H.P.J. Thomsen, *Z. Anorg. Chem.*, 9 (1895) 283.
35. J.E. Reynolds, *J. Chem. Soc.*, 81 (1902) 612.
36. J.C.A. Boeyens, *Trans. Roy. Soc. S. Afr.*, 54 (1999) 323.
37. H. Haken and H.C. Wolf, *The Physics of Atoms and Quanta*, 4th ed., Translated by W.D. Brewer, 1994, Springer, Berlin.
38. A. van den Broek, *Physik. Z.*, 14 (1913) 32.
39. H.G.J. Moseley, *Phil. Mag.*, 26 (1913) 1024.
40. C.H. MacGillavry and G.D. Rieck (eds.), *International Tables for X-ray Crystallography*, Vol. III, 1968, Kynoch Press, Birmingham.
41. J.R. Rydberg, *Phil. Mag.*, 28 (1914) 144.
42. G. Holton, *Introduction to Concepts and Theories in Physical Science*, 1952, Addison-Wesley, Reading, MA.
43. G.N. Lewis, *J. Am. Chem. Soc.*, 38 (1916) 762.
44. I. Langmuir, *J. Am. Chem. Soc.*, 41 (1919) 868.
45. C.R. Bury, *J. Am. Chem. Soc.*, 43 (1921) 1602.
46. A. Sommerfeld, *Atombau und Spektrallinien*, 1924, Vieweg, Braunschweig.
47. G. Uhlenbeck and S. Goudsmit, *Nature*, 117 (1926) 264.

48. C.J.H. Schutte, *S.A. Chem. Process.*, (1974) CP2.
49. P.W. Atkins, *Physical Chemistry*, 6th ed., 1998, Oxford University Press, Oxford.
50. V.V. Varlamov, http://depni.sinp.msu.ru/cdfe/services/gsearch.html CDFE, Updated November 2002.
51. E.P. Wigner, *Phys. Rev.*, 51 (1937) 947.
52. D.W. Sciama, *Modern Cosmology*, 1971, Cambridge University Press, Cambridge.
53. D. ter Haar, *Revs. Mod. Phys.*, 22 (1950) 119.
54. R.A. Alpher and R.C. Herman, *Revs. Mod. Phys.*, 22 (1950) 153.
55. R.A. Alpher, H. Bethe and G. Gamow, *Phys. Rev.*, 74 (1948) 1737.
56. W.D. Harkins, *J. Am. Chem. Soc.*, 39 (1917) 856.
57. W.D. Harkins, *J. Am. Chem. Soc.*, 42 (1920) 1965.
58. W.D. Harkins, *Phil. Mag.*, 42 (1921) 305.
59. W.D. Harkins, *Phys. Rev.*, 38 (1931) 1270.
60. W.D. Harkins and M. Popelka, *Phys. Rev.*, 76 (1949) 989.
61. E. Anders and M. Grevesse, *Geochem. Cosmochim. Acta*, 53 (1989) 197.
62. D. Arnett, *Supernovae and Nucleosynthesis*, 1996, Princeton University Press, New Jersey.
63. R.A. Alpher and R.C. Herman, *Ann. Rev. Nuclear Sci.*, 2 (1953) 1.
64. R.D. Evans, *The Atomic Nucleus*, 1955, McGraw-Hill, New York.
65. H.E. Suess and H.C. Urey, *Revs. Modern Phys.*, 28 (1956) 53.
66. G. Gamow, *Phys. Rev.*, 70 (1946) 572L.
67. E.M. Burbidge, G.R. Burbidge, W.A. Fowler and F. Hoyle, *Revs. Mod. Phys.*, 29 (1957) 547.
68. J. A. Simpson, *Ann. Rev. Nucl. and Part. Sci.*, 33 (1983) 323.
69. W.M. Elsasser, *J. Phys. Radium*, 5 (1933) 389.
70. W.N. Cottingham and D.A. Greenwood, *An Introduction to Nuclear Physics*, 1986, Cambridge University Press, Cambridge.
71. M. Goeppert Mayer, *Phys. Rev.*, 78 (1950) 16; 22.
72. L. Pauling, *Science*, 150 (1965) 297.
73. J.C.A. Boeyens, *The Theories of Chemistry*, 2003, Elsevier, Amsterdam, The Netherlands.
74. H. Yukawa, *Proc. Phys. Math. Soc. Japan*, 17 (1935) 48.
75. K. Gottfried and V.F. Weisskopf, *Concepts of Particle Physics*, 1986, Clarendon Press, Oxford.
76. G.D. Coughlan and J.E. Dodd, *The Ideas of Particle Physics*, 2nd ed., 1993, Cambridge University Press, Cambridge.
77. H. Fritzsch, *Quarks*, 1984, Penguin Books, London.

78. G.E. Satterthwaite, *Encyclopedia of Astronomy,* 1970, Hamlyn, London.
79. J.V. Narlikar, *An Introduction to Cosmology,* 3rd. ed., 2002, University Press, Cambridge.
80. *Ensiklopedie Afrikana, Boek van die Jaar,* 1993, Wêreldspektrum, Johannesburg.
81. R.J. Stephenson, *Mechanics and Properties of Matter,* 2nd ed., 1960, Wiley, New York.
82. F. Hoyle, G. Burbidge and J.V. Narlikar, *A Different Approach to Cosmology,* 2000, University Press, Cambridge.
83. V.I. Sokolov, *Comp. Maths. with Appls.*, 12B (1986) 547.
84. R.A. Hegstrom and D.K. Kondepundi, *Sci. Am.*, Jan. 1990, 108.
85. R.J.T. Bell, *An Elementary Treatise on Coordinate Geometry of Three Dimensions,* 1931, Macmillan, London.
86. G. Thierrin, *J. Theoretics,* 3–4 (2001).
87. D. Bohm, *Wholeness and the Implicate Order,* 1980, Routledge, New York.
88. T. Stonier, *Information and the Internal Structure of the Universe,* 1990, Springer, Berlin.
89. J.C.A. Boeyens and J. du Toit, *Electronic J. Theor. Chem.*, 2 (1997) 296.
90. S. Goldman and C. Joslin, *J. Phys Chem.* 96 (1992) 6021.
91. M.P. do Carmo, *Differential Geometry of Curves and Surfaces,* 1976, Prentice-Hall, New Jersey, p. 135.
92. A-E.B. de Chancourtois, (in translation), *Nature,* 41 (1889) 186.
93. D.R. Lile (ed.) *Handbook of Chemistry and Physics,* 86th ed., 2005, CRC Press, Baco Raton, FL.
94. A. Beck, M.N. Bleicher and D.W. Crowe, *Excursions into Mathematics,* 2000, AK Peters.
95. M. Jung *et al.*, *Phys. Rev. Lett.*, 69 (1992) 2164.
96. F. Bosch *et al.* (20 authors) *Phys. Rev. Lett.*, 77 (1996) 5190.
97. K.S. Krane, *Introductory Nuclear Physics,* 1988, Wiley, New York.
98. S. Sugano, Y. Tanabe and H. Kamimura, *Multiplets of Transition-Metal Ions in Crystals,* 1970, Academic Press, New York.
99. H.C. Arp, G. Burbidge, F. Hoyle, J.V. Narlikar and N.C. Wickramasinghe, *Nature,* 346 (1990) 807.
100. *New Scientist,* 11 May 2002, p. 29.
101. J.C.A. Boeyens, *Cryst. Eng.*, 4 (2001) 61.
102. P. Plichta, *God's Secret Formula,* 1998, Translated from German, Element Books, Boston.

103. B. Matthias, in C.J. Gorter (ed.) *Progress in Low-Temperature Physics*, Vol. 2, 1957, North-Holland, Amsterdam, The Netherlands.
104. C.J. Gorter (ed.) *Progress in Low-Temperature Physics*, Vol. 2, 1957, North-Holland, Amsterdam, The Netherlands.
105. D. Pines, *Phys. Rev.*, 109 (1958) 280.
106. C. Buzea and K. Robbie, *Supercond. Sci. Technol.*, 18 (2005) R1.
107. J. Bardeen, L.N. Cooper and J.R. Schrieffer, *Phys. Rev.*, 108 (1957) 1175.
108. J.E. Hirsch, *Phys. Rev. B*, 55 (1997) 9007.
109. D. Bohm and B.J. Hiley, *The Undivided Universe. An Ontological Interpretation of Quantum Theory*, 1993, Routledge, London.
110. A.H. Morrish, *The Physical Principles of Magnetism*, 1965, Wiley, New York.
111. M. Kociak, A. Yu. Kasumov, S. Guéron, B. Reulet, I.I. Khodos, Yu. B. Gorbatov, V.T. Volkov, L. Vaccarini and H. Bouchiat, *Phys. Rev. Lett.*, 86 (2001) 2416.
112. M. Schlüter, M. Lannoo, M. Needels, G.A. Baraff and D. Tománek, in H.W. Kroto, J.E. Fischer and D.E. Cox (eds.), *The Fullerenes*, 1993, Pergamon, Oxford.
113. H.W. Kroto, J.E. Fischer and D.E. Cox, *The Fullerenes*, 1993, Pergamon, Oxford.
114. P.W. Stephens, L. Mihaly, P.L. Lee, R.L. Whetten, S-M. Huang, R. Kaner, F. Deiderich and K. Holczer, *Nature*, 351 (1991) 632.
115. J.D. Jorgensen, D.G. Hinks, S. Short, *Phys. Rev., B - Condens. Matt.*, 63 (2001) 2245221.
116. J. Nagamatsu, T. Muranaka, Y. Zenitani and J. Akimitsu, *Nature*, 410 (2001) 63.
117. A.W. Sleight, J.L. Gillson and P.E. Bierstedt, *Solid State Comm.*, 17 (1975) 299.
118. J.G. Bednorz and K.A Müller, *Z. Phys. B*, 64 (1987) 189.
119. M.K. Wu, J.R. Ashburn, C.J. Torng, P.H. Hor, R.L. Meng, L. Gao, Z.J. Huang, Y.Q. Wang and C.W. Chu, *Phys. Rev. Lett.*, 58 (1987) 908.
120. E.M. Engler, R.B. Beyers, V.Y. Lee, A.I. Nazzal, G. Lim, S.S.P. Parkin, P.M. Grant, J.E. Vazquez, M.L. Ramirez and R.D. Jacowitz, in D.L. Nelson, M.S. Whittingham and T.F. George (eds.), *Chemistry of High-Temperature Superconductors, ACS Symposium Series 351*, 1987, Washington, DC.
121. D.L. Nelson, M.S. Whittingham and T.F. George, *Chemistry of High-Temperature Superconductors, ACS Symposium Series 351*, 1987, Washington, DC.

122. A.W. Sleight, in D.L. Nelson, M.S. Whittington and T.F. George (eds.) *Chemistry of High-Temperature Superconductors*, ACS Symposium Series 351, 1987, Washington, Dc.
123. H. Takagi, S. Uchida, K. Kitazawa and S. Tanaka, *Jpn. J. Appl. Phys.*, 26 (1987) L123.
124. H.H. Wang, U. Geiser, R.J. Thorn, K.D. Carlson, M.A. Beno, M.R. Monaghan, T.J. Allen, R.B. Proksch, D.L. Stupka, W.K. Kwok, G.W. Crabtree and J.M. Williams, *Inorg. Chem.*, 26 (1987) 1192.
125. A. Santoro, in J.F. Lynn (ed.), *High Temperature Superconductivity*, 1990, Springer, New York.
126. J.F. Lynn (ed.), *High Temperature Superconductivity*, 1990, Springer, New York.
127. S.C. Moss, K. Forster, J.D. Axe, H. You, D. Holwein, D.E. Cox, P.H. Hor, R.L. Meng and C.W. Chu, *Phys. Rev.*, B35 (1987) 7195.
128. T. Ishiguro and K. Yamaji, *Organic Superconductors*, 1990, Springer, Berlin.
129. C. Kittel, *Introduction to Solid State Physics*, 5th ed., 1976, Wiley, New York.
130. B. Stritzker, *Phys. Rev. Lett.*, 41 (1979) 1769.
131. J.E. Crow and N.-P. Org, in J.F. Lynn (ed.), *High Temperature Superconductivity*, 1990, Springer, New York.
132. S.W. Tozer, A.W. Kleinsasser, T. Penney, D. Kaiser and F. Holtzberg, *Phys. Rev. Lett.*, 59 (1987) 1768.
133. L. Pauling, *J. Am Chem. Soc.*, 54 (1932) 3570.
134. L. Pauling, *Nature of the Chemical Bond*, 3rd ed., 1960, Cornell University Press, Ithaca.
135. R.S. Mulliken, *J. Chem. Phys.*, 2 (1934) 782.
136. H.O. Pritchard and H.A. Skinner, *Chem. Revs.*, 55 (1955) 745.
137. R.G. Parr and R.G. Pearson, *J. Am. Chem. Soc.*, 105 (1983) 7512.
138. J.K. Nagle, *J. Am. Chem. Soc.*, 112 (1990) 4741.
139. T.-S. Liu, *J. Chinese Chem. Soc.*, 9 (1942) 119.
140. W. Gordy, *Phys. Rev.*, 69 (1946) 604.
141. T.L. Cottrell and L.E. Sutton, *Proc. Roy. Soc.* A207 (1951) 49.
142. A.L. Allred and E.G. Rochow, *J. Inorg. Nucl. Chem.*, 5 (1958) 264.
143. *Periodic Table of the Elements*, VCH, 1985.
144. L.C. Allen, *J. Am. Chem. Soc.*, 111 (1989) 9003; 114 (1992) 1510.
145. J.C.A. Boeyens, *J. S. Afr. Chem. Inst.*, 26 (1973) 94.
146. F.A. Cotton and R.A. Walton, *Multiple Bonds Between Metal Atoms*, 1982, Wiley, New York.

147. F.A. Cotton and G. Wilkinson, *Advanced Inorganic Chemistry*, 2nd ed., 1967, Interscience, New York.
148. J.C.A. Boeyens and D.J. Ledwidge, *Inorg. Chem.*, 22 (1983) 3587.
149. J.C.A. Boeyens, *J. Crystallogr. Spectrosc. Res.*, 12 (1982) 245.
150. J.C. Slater, *Phys. Rev.*, 36 (1930) 57.
151. J.C.A. Boeyens, F.A. Cotton and S. Han, *Inorg. Chem.*, 24 (1985) 1750.
152. P. Comba and T.W. Hambley, *Molecular Modeling of Inorganic Compounds*, 2nd ed., 2001, Wiley-VCH, Weinheim.

Index

K

L

M

N

Q

R

S

W

X

Y

Z

SPRINGER NATURE

GPSR Compliance

The European Union's (EU) General Product Safety Regulation (GPSR) is a set of rules that requires consumer products to be safe and our obligations to ensure this.

If you have any concerns about our products, you can contact us on ProductSafety@springernature.com

In case Publisher is established outside the EU, the EU authorized representative is:

Springer Nature Customer Service Center GmbH
Europaplatz 3
69115 Heidelberg, Germany

Papier | Fördert gute Waldnutzung
FSC® C083411

Kolibri 360 GmbH
Ferdinand-Jühlke-Straße 7
99095 Erfurt, Deutschland
produktsicherheit@kolibri360.de